Advances in Intelligent Systems and Computing

Volume 1033

AF895526

The series "Advances in Intelligent Systems and Computing" contains publications on theory, applications, and design methods of Intelligent Systems and Intelligent Computing. Virtually all disciplines such as engineering, natural sciences, computer and information science, ICT, economics, business, e-commerce, environment, healthcare, life science are covered. The list of topics spans all the areas of modern intelligent systems and computing such as: computational intelligence, soft computing including neural networks, fuzzy systems, evolutionary computing and the fusion of these paradigms, social intelligence, ambient intelligence, computational neuroscience, artificial life, virtual worlds and society, cognitive science and systems, Perception and Vision, DNA and immune based systems, self-organizing and adaptive systems, e-Learning and teaching, human-centered and human-centric computing, recommender systems, intelligent control, robotics and mechatronics including human-machine teaming, knowledge-based paradigms, learning paradigms, machine ethics, intelligent data analysis, knowledge management, intelligent agents, intelligent decision making and support, intelligent network security, trust management, interactive entertainment, Web intelligence and multimedia.

The publications within "Advances in Intelligent Systems and Computing" are primarily proceedings of important conferences, symposia and congresses. They cover significant recent developments in the field, both of a foundational and applicable character. An important characteristic feature of the series is the short publication time and world-wide distribution. This permits a rapid and broad dissemination of research results.

**** Indexing: The books of this series are submitted to ISI Proceedings, EI-Compendex, DBLP, SCOPUS, Google Scholar and Springerlink ****

More information about this series at http://www.springer.com/series/11156

Józef Korbicz · Roman Maniewski ·
Krzysztof Patan · Marek Kowal
Editors

Current Trends in Biomedical Engineering and Bioimages Analysis

Proceedings of the 21st Polish Conference on Biocybernetics and Biomedical Engineering

Springer

Editors
Józef Korbicz
Institute of Control and Computation Engineering
University of Zielona Gora
Zielona Góra, Poland

Roman Maniewski
Committee of Biocybernetics and Biomedical Engineering
Polish Academy of Sciences
Warsaw, Poland

Krzysztof Patan
Institute of Control and Computation Engineering
University of Zielona Góra
Zielona Góra, Poland

Marek Kowal
Institute of Control and Computation Engineering
University of Zielona Góra
Zielona Góra, Poland

ISSN 2194-5357 ISSN 2194-5365 (electronic)
Advances in Intelligent Systems and Computing
ISBN 978-3-030-29884-5 ISBN 978-3-030-29885-2 (eBook)
https://doi.org/10.1007/978-3-030-29885-2

This Springer imprint is published by the registered company Springer Nature Switzerland AG
The registered company address is: Gewerbestrasse 11, 6330 Cham, Switzerland

Preface

This book constitutes the proceedings of the 21st *Polish Conference on Biocybernetics and Biomedical Engineering (PCBBE)* held in Zielona Góra, Poland, from September 25 to 27, 2019. The conference was organized by the Committee on Biocybernetics and Biomedical Engineering of the Polish Academy of Sciences, the Polish Society of Biomedical Engineering, and the University of Zielona Góra (the Faculty of Computer, Electrical and Control Engineering and the Faculty of Mechanical Engineering), in cooperation with the International Federation for Medical and Biological Engineering (IFMBE), Lubuskie Scientific Society in Zielona Góra, and the Polish Society of Theoretical and Applied Electrical Engineering. The biannual ***Polish Conference on Biocybernetics and Biomedical Engineering*** has been held for nearly four decades and has continually attracted scientists and professionals in the fields of engineering, medicine, physics, and computer science. This year the conference was hosted for the first time by the University of Zielona Góra, Poland.

The 21st *PCBBE* was a great opportunity for the exchange of ideas and presentation of the latest developments in all areas within the field of biomedical engineering. Biocybernetics and biomedical engineering are currently considered a promising approach to improve healthcare and consequently the quality of life. Innovative technical solutions not only respond to the needs of caregivers but also stimulate the development of medical sciences by supporting medical practitioners, and we are currently witnessing a profound change in the role of medicine, which has become ubiquitous in everyday life thanks to the recent technological advances. The development of civilization manifests itself in a growing focus on investigating all aspect of human life and continuous efforts to support it, and in mimicking biological systems in engineering.

The volume is divided into six parts:

- signal processing,
- image analysis,
- computer-aided medical diagnosis,
- e-health and telemedicine,

- modeling of physiological processes,
- other bioengineering studies,

and consists of 30 carefully selected full papers. Every submission was subjected to a thorough peer review process (2 to 3 independent reviews per paper), and only those with a consistent and strong recommendation from reviewers were accepted and included in the proceedings. In this regard, we would like to express our gratitude to the authors for their valuable submissions and to the reviewers from the Program Committee for their inestimable effort in this process.

The *PCBBE 2019* conference also witnessed outstanding keynote presentations given by distinguished guest speakers:

- Prof. **Jos Vander Sloten**, Catholic University Leuven, Leuven-Heverlee, Belgium
  Personalised medical treatments enabled by image-based biomechanical models
- Prof. **Jorg Vienken**, Technical University Mittelhessen, Giessen, Germany
  Polymer properties to comply with requirements for medical devices
- Dr **Dirk Grosenick**, Physikalisch-Technische Bundesanstalt, Berlin, Germany
  Near-infrared spectroscopy and fluorescence imaging of tissue in vivo
- Prof. **Piotr Augustyniak**, AGH University of Science and Technology, Kraków, Poland
  Multimodal measurement systems for health and behavior analysis in living environment.

We believe that this collective work will become a great reference tool for scientists and researchers working in the area of biocybernetics and biomedical engineering. Readers are kindly encouraged to contact the corresponding authors for further details concerning their research and presented results.

June 2019

Józef Korbicz
Roman Maniewski
Krzysztof Patan
Marek Kowal
Zielona Góra

Acknowledgment

We would like to acknowledge the technical and administrative support of Kinga Włoch (financial manager), Ewa Lehmann (organizational manager), and Paweł Kasza (webmaster) for their commitment and effort put into making the conference a very successful event.

Contents

E-health and Telemedicine

Modelling of Physiological Processes

Other Bioengineering Studies

Signal Processing

Analysis of Amped Heart Rate Sensor Performance Under the Influence of Various External Factors and User Characteristics

Aleksandra Królak(✉) and Julia Biskupiak

Lodz University of Technology, Institute of Electronics, Lodz, Poland
aleksandra.krolak@p.lodz.pl

Abstract. Nowadays optical sensors in the form of wearable devices are omnipresent. People use them mainly to measure heart rate during everyday activities and often base their training schedules on these measurements. Therefore the performance of these devices should be reliable. Amped optical sensor is Arduino-dedicated device, widely used in DYI implementations of pulse meters. In this research the performance of this sensor was investigated under the influence of the following factors: contact force between the sensor and measurement site, low hand temperature, skin thickness, dark skin color, and various lighting conditions, namely without light, in artificial light, in daylight and with flashlight directed to the sensor. Obtained results show that there is no significant influence of the lighting conditions on the performance of investigated sensors, apart from the severe degradation of performance in case of strong light application to the measurement site. The significant decrease in the accuracy of the device was observed in case of users with dark skin color and with low hand temperature. The influence of skin thickness and contact force on the sensor performance was also detected.

Keywords: Heart rate · Photoplethysmography (PPG) · Amped sensor · Factor influencing PPG

1 Introduction

The heart rate is a name for temporary local stretches of the artery appearing rhythmically which are consistent with heart contractions. The radiant expansion of the aorta by the ejection of blood from the left ventricle causes a pressure wave (heart rate) that travels at a certain speed along the aorta and its branches. Most common methods for constant heart rate monitoring that are used in hospitals are electrocardiography (ECG) and photoplethysmography (PPG). PPG is also applied in smart watches that are nowadays used as one of the most popular method of heart rate monitoring in everyday life.

J. Korbicz et al. (Eds.): PCBEE 2019, AISC 1033, pp. 3–14, 2020.
https://doi.org/10.1007/978-3-030-29885-2_1

Photoplethysmography is a method of optical measurement that is used to investigate volumetric changes in peripheral circulation of blood. It is an inexpensive and noninvasive technique that is based on measurements performed at the skin surface. The basic device based on PPG technology is built only with a few optoelectrical elements: a source of light for skin illumination and a photodetector to detect small fluctuations in the intensity of light that result from changes in the vascular volume due to the blood flow. Peripheral pulse is considered as the most detectable feature of PPG waveform as it is adjusted to each heartbeat. Although, a lot of studies are devoted to the origin of PPG signal, there are still some components that are not entirely understood. However, this method provides a lot of essential information that can be used by health care and users themselves in everyday life, e.g. data concerning cardiovascular system or the value of heart rate during daily activities [1].

This technique allows investigating even the smallest variations in blood volume; however it cannot provide information about the quantity of the blood. There are several elements that concur to the PPG signal, namely: a DC component that displays the optical property of tissues and small changes in the body energy, changes in the arterial blood volume which are related to heart activity, volumetric variations in venous blood which are PPG signal modulators [2].

There are many factors that may influence the accuracy of the measurement with PPG method, such as lighting conditions, contact force between sensor and measurement site, hand temperature, skin thickness or skin color. The goal of this research is to investigate the influence of the abovementioned factors on the performance of Amped HR sensor. In the second section of the paper related works concerning the factors influencing PPG accuracy are presented. Section 3 describes the experimental setup. Section 4 presents and discusses the results and Sect. 5 concludes the paper.

2 Factors Affecting Photoplethysmographic Measurements

PPG sensors are built in different ways with changes in their designs, however, all of them are made to estimate variations in blood volume and produce similar data. A basic type of PPG sensor illuminates the measurement site using one or more LEDs. Although there is a wide selection of LED colors that can be used in PPG sensors, green and red seem to be the most popular ones [3].

PPG sensors using green light can be found in optical heart rate monitors (OHRM). This choice is due to the following: there is a lot of knowledge and information about the technology, the value of signal-to-noise ratio (SNR) is considerably large, vulnerability to motion artifacts is low. However, usage of green light carries some disadvantages. The green light is highly absorbed by human skin, so limited amount of it penetrates the tissue what reduces signal amplitudes. Light emitted by green diode is greatly absorbed by hemoglobin preventing the penetration into deeper tissue [4].

PPG sensor using red light employs light in near-infrared spectroscopy (NIRS). It is very often used for medical purposes in hospitals and health care centers where the accuracy is highly desirable. Due to the fact that red light is not absorbed by human body it has the ability to pass through tissue layers ten times deeper than green light, what allows determining various biometric signals. Moreover, local changes in skin pigment like freckles, moles or tattoos, as well as melanin, do not interrupt output of red light from PPG sensors. What is more, the operation wavelengths in case of these sensors vary between 650 nm and 1100 nm. This broad range allows to measure different physiological parameters from one sensor at the same time [4,5].

In case of wearable PPG sensors two operation modes can be distinguished: transmission and reflectance. In devices with transmission mode, the photodiode which is located on the opposite site to the light source (LED) measures the transmitted light that passed through the tissue. This method allows for acquiring good signal but the measurement site must assure immediate detection of the transmitted light, e.g. fingertip. In the reflectance mode, photodiode and light source are located on the same site and light is measured after being reflected from blood vessel, tissue or bone. In this mode the wide selection on measurement site can be chosen but the sensor readings can be more often influenced by motion artifacts and pressure disturbances [6]. There are 3 most common locations of PPG sensors placement: ear, finger and wrist.

Temperature of skin surface also has influence of the PPG sensor readings. The cooling process causes the walls of blood vessel to decrease their contractive capacity while heating process increases this capacity. It was investigated that in the skin temperatures between 22 °C and 40 °C, the value of DC components of PPG signal were remarkably smaller while AC components were significantly larger. Changes among PPG components were connected with the variations in the body temperature and the phenomenon in which under changing temperature of the skin surface body temperature is maintained due to blood flow control [7].

Other factor that can significantly disturb the PPG signal is contact force between the sensor and the measurement site. External force in the range of 0.2∼1.8N acting on the PPG sensor can influence the geometry of blood vessels causing a variation in PPG amplitude up to 70%. Some additional parts are incorporated into devices measuring heart rate with PPG sensors to reduce contact force, such as: springchips, stretchable bands or miniaturized force regulators [8].

PPG sensors are considered to be very vulnerable to motion artifacts which influence the sensor readings. In case of wearable devices, which are supposed to provide continuous information about heart rate during everyday activities, disturbances caused by motion artifacts are extremely problematic. They can be produced in various situations e.g. by hemodynamic effects, by variations in tissue compression or due to sensor movement and deformation caused by changes in user's posture. In order to reduce motion artifacts it is important to choose optimal measurement site for the PPG sensor location to obtain the most accurate PPG readings [9,10].

3 Materials and Methods

3.1 Signal Acquisition with Amped Sensor

Amped is a sensor dedicated for Arduino boards used to detect heart rate from fingertip or earlobe. It measures heart rate basing photoplethysmography (PPG) technique. Green light emitted by the LED mounted in this pulse sensor is absorbed by the blood in blood vessels. Then it is reflected back to the sensor which detects the relative variations in the light intensity. The output signal from Amped pulse sensor is an analog fluctuation in voltage and from its wave the heart rate value can be estimated. The more amount of light is detected by the sensor the higher is the signal amplitude. Amped sensor is connected to the Arduino board through 3 male header connectors for power supply (5 V or 3.3 V), grounding (GND) and analog signal pin. The measurements were performed using Arduino Uno board.

In order to automatically calculate the heart rate values a simple application was written with Arduinno IDE programming environment. The heart rate was calculated at the end of the measurement process based on the detection of local maxima fulfilling the following conditions: the distance between the subsequent peaks had to be at least 230 ms and the value of the local maximum had to exceed the value of 80% of the global maximum in the signal analyzed.

3.2 Experimental Design

The measurement setup consisted of pulse oximeter Contec CMS50DL1 and Amped sensor connected through the Arduino Uno board to the laptop ASUS UX410UAK. The data acquired from pulse oximeter was considered as the reference while the heart rate information from Amped sensors was taken into investigation.

Additionally, the Sensor Puck from Silicon Labs was used during every heart rate measurement to measure the value of light intensity. This sensor measures the intensity of ambient light in lux unit. It works wirelessly with the smartphone application via Bluetooth connection where all the necessary data is sent and displayed, such as humidity, temperature and illumination.

Lighting Conditions
First, accuracy of the optical sensor for heart rate estimation under various lighting condition was investigated. For this purpose heart rate measurements from 10 subjects were performed in 4 different lighting conditions: in the dark, in the artificial light, in the day light and with the strong, artificial light affecting the sensors.

Intensity of light varied significantly among all measurements: the dark indicated 0 lux, artificial light was achieved in the range between 60 lux and 450 lux and day light in the range between 900 lux and 1850 lux. The measurement with strong artificial light was performed using the flashlight working on Amped sensor. Source of light was placed about 5 cm from the working sensors giving the light intensity equal to 10000 lux.

Contact Force

Next investigated condition was connected with the contact force between the operating sensor and a fingertip. The heart rate measurements of 10 subjects were performed during which the slight force was applied to the Amped sensor. There were no reference measurements performed and the force applied was subjective.

Hand Temperature

The impact of the low temperature of hands on the accuracy of optical sensors for heart rate measurements was investigated. For this purpose, 10 people suffering from the circulatory problems that contribute to permanently cold hand were subjected to the heart rate measurements. Described condition was visualized with the infrared camera FLIR P660 as presented in Fig. 1.

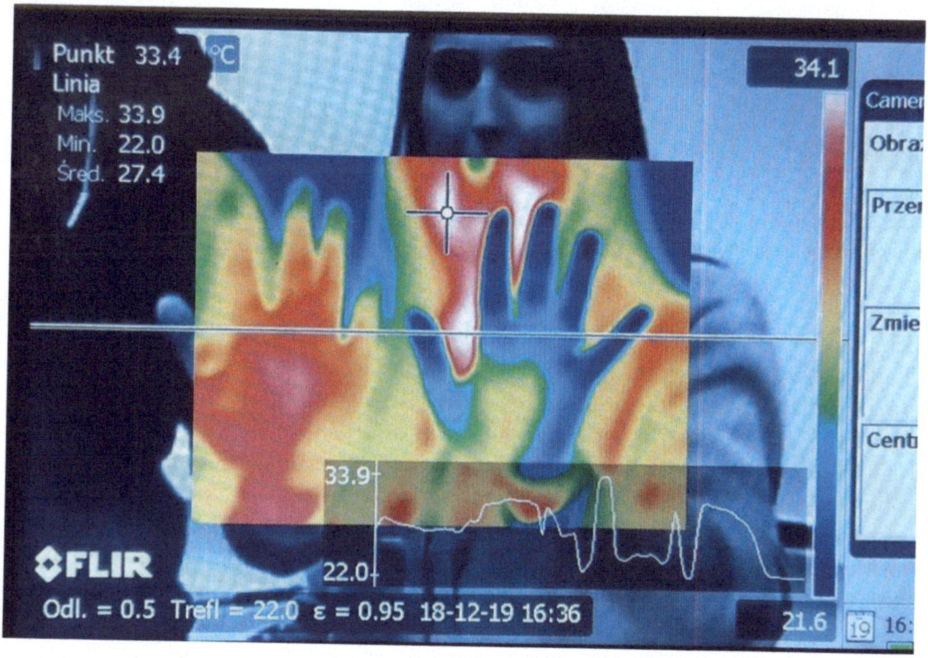

Fig. 1. The view from the infrared camera FLIR P660 showing warm hand versus cold hand of subjects participating in experiment.

Skin Thickness

In this part of the experiment the influence of the thick skin on the accuracy of optical sensors for heart rate measurements was investigated. The group of 10 people with the thick skin on their hands that result from manual work (Fig. 2) was subjected to the heart rate measurements.

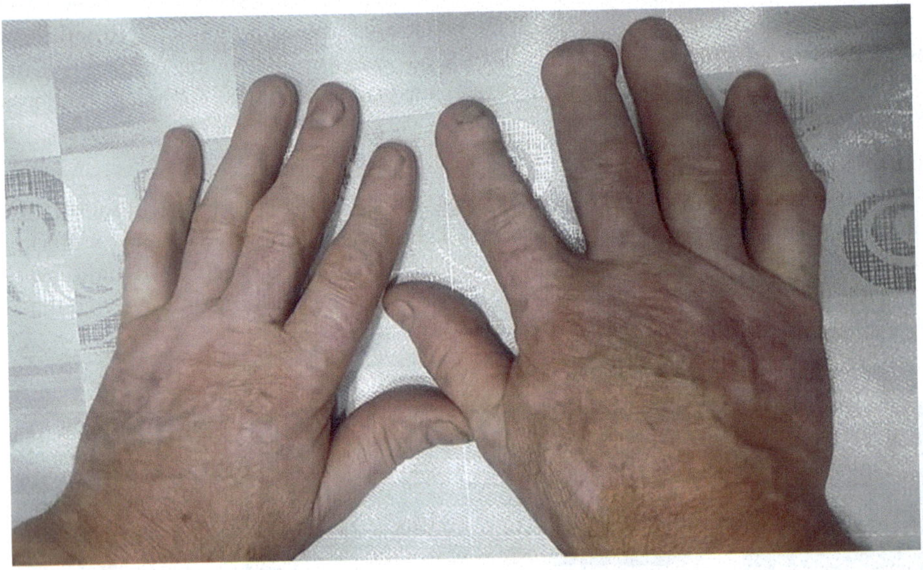

Fig. 2. The hand condition of subject taking part in experiment part related to thick skin influence on the performance of the Amped sensor.

Skin Color

The last investigated condition was connected with the darker skin color and its impact on the optical sensors accuracy for pulse estimation. Here, the heart rate measurements were performed on the group of 10 subjects with higher amount of the skin pigment. After every measurement concerning the darker skin color a photo of reference subject's hand with light skin color next to subject's hand was taken. For each photo the mean RGB values of a Region of Interest (ROI) containing reference hand and subject's hands were calculated and the RGB ratio of the two means was calculated.

4 Results and Discussion

In this section the obtained results for analysis of PPG Amped sensor under the influence of above described factors are presented. For every measurement done with investigated device a value of relative error was calculated from Eq. 1.

$$Relative\ error = \frac{Measured\ HR - Reference\ HR}{Reference\ HR} \cdot 100\%. \quad (1)$$

Where:

Measured HR – heart rate value obtained either from Amped sensor

Reference HR – correlating heart rate value obtained from pulse oximeter.

The values of calculated relative errors for all investigated conditions are equal to 1.77% in case of measurements in the dark, 1.5% for artificial light (60–450 lux), 0.92% in the daylight (900–1850 lux) and 27.21% when the sensor was illuminated with strong flash light (~10 000 lux). For persons with thick skin the relative error was equal to 8.3%, when slightly larger than typical force was applied to the sensor the error value reached 39.78%, in case of dark skin color subjects the measurements showed error of 51.26%. The largest relative error was calculated in case of persons with cold hands, equal to 91.81%.

The results of the measurements performed under various lighting condition are presented in Figs. 3 and 4. The dependence between the error in mean heart rate difference measured and values of light intensity is shown in Fig. 5.

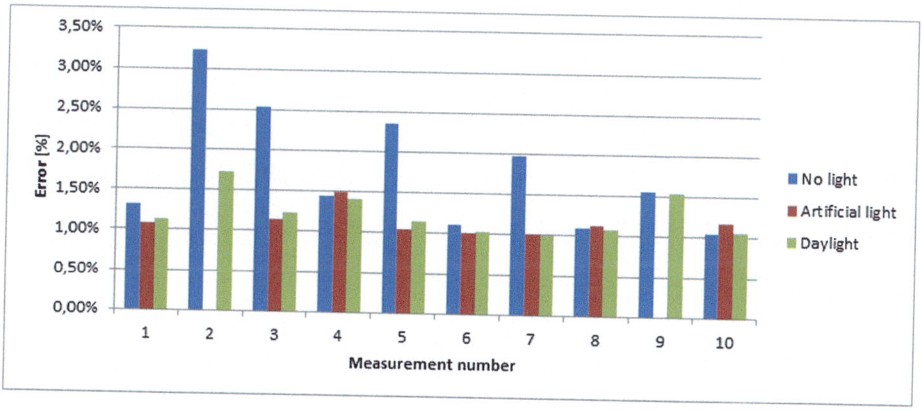

Fig. 3. Mean relative error for examined lighting conditions.

Analyzing charts in Figs. 3 and 4 it is easily noticeable that in case of measurements done without light, in daylight and in artificial light there are no significant differences between the heart rate values. Amped sensor produced results almost the same as the pulse oximeter. However, graph displaying measurements done with flashlight working on investigated sensors presents significant differences between heart rate obtained from Amped sensor and pulse oximeter. This difference is also clearly visible in plot presented in Fig. 5.

The comparison of reference heart rate measurements done with pulse oximeter and measurements obtained with Amped sensor influenced by other investigated factors, namely slight pressure applied to the sensor, low hand temperature, thick skin and dark skin color, are presented in Fig. 6. Additionally the graph showing the heart rate measurement errors versus the RGB ratio is presented in Fig. 7.

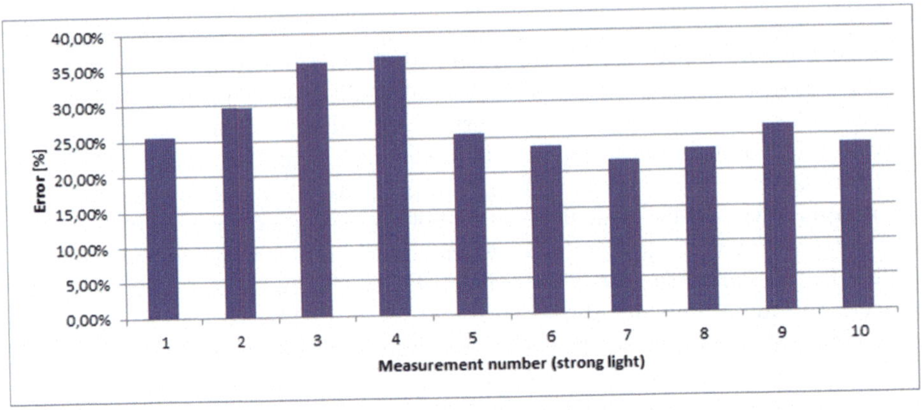

Fig. 4. Mean relative error in case of strong light (10000 lux) directed to the sensor.

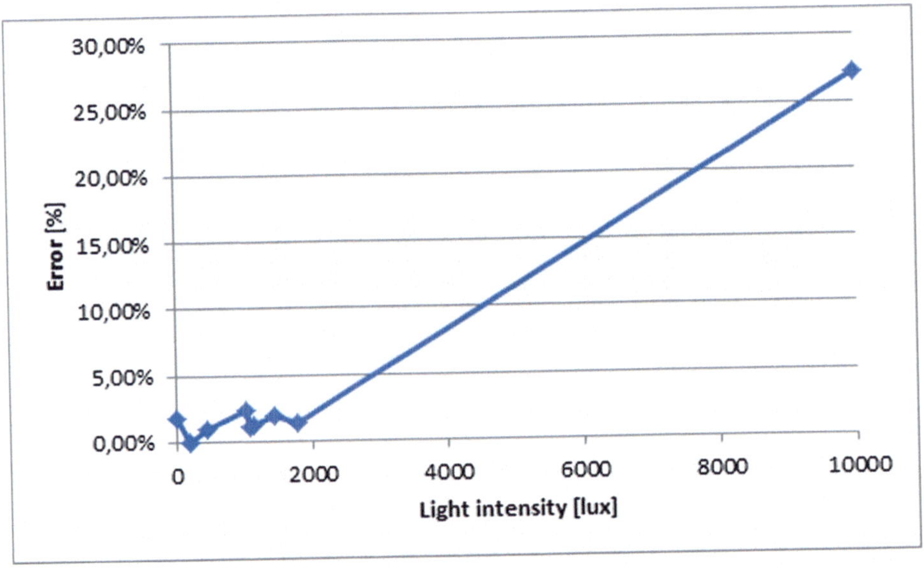

Fig. 5. Mean relative error vs. light intensity of the surrounding.

For each investigated lighting condition the accuracy of Amped sensor was calculated as the number of measurements for which the heart rate difference (between the Amped sensor value and reference value) was less than 1bpm to the total number of measurements collected. The accuracy values obtained are equal to 70% in case of no light, 100% in daylight, 80% in artificial light and 0% with strong illumination of the flashlight. In case of measurements conducted without light, in artificial light (60–450 lux) and in daylight (900–1850 lux) the heart rate difference between the pulse oximeter and Amped sensor varied

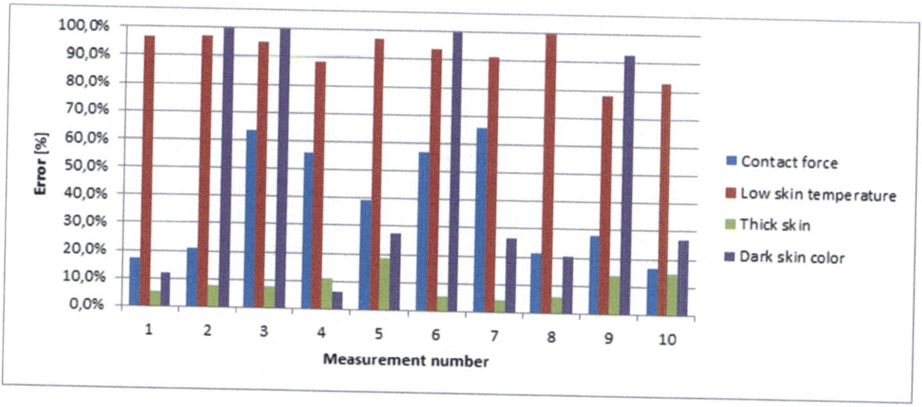

Fig. 6. Mean relative error for 4 conditions: contact force, cold hands, thick skin, dark skin.

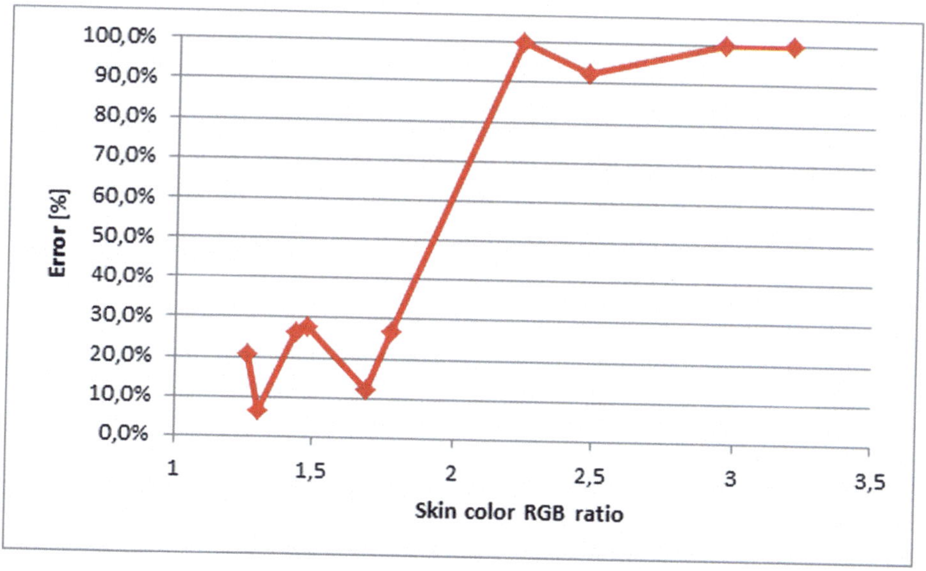

Fig. 7. Mean relative error vs. skin color RGB ratio.

between 0–2 heart beats giving the value of mean relative error in a range from 0.81% to 1.77% and the accuracy between 70% and 100%.

These results show that there is no noticeable influence of mentioned lighting conditions on the performance of Amped sensor. On the other hand, more significant differences were presented in case of measurements performed with the flashlight illuminating the investigated sensor with the light intensity equal to 10000 lux. Heart rate difference referring to Amped sensor increased rapidly in case of flashlight measurements and varied between 15 and 27 heart beats with the mean relative error equal to 27.21% and accuracy equal to 0%.

This significant degradation of the performance of the sensor can be caused by the fact that intense light emitted by the flashlight might have influenced the operation of electronics part in Amped sensor that is not shielded in any way. In such situation flashlight had direct access to the source of light and photodetector used in the sensor, what increased the possibility of disrupting its performance. What is more, the illumination of the sensor with such strong, artificial light might have limited the amount of light emitted from sensor that actually reached the skin. This led to the decreased amount of the light absorbed by the blood in vessels and as a result less amount of light reflected back to sensor.

In case of applied slight contact force the mean relative error was equal to ~40%. Such differences between the investigated and reference device were caused by the fact that even the slight contact force applied between the sensor and the measurement site can influence the geometry of blood vessels and deform them. This phenomenon introduces variations in PPG amplitude and leads to the unreliable output values.

Heart rate measurements done on people with cold hands presented the largest differences among all factors investigated in this experiment. Heart rate differences referring to Amped sensor varied in range 50–93 heart beats with the value of mean relative error equal to 91.81%. Presented significant changes in the output of investigated sensors might have been caused by the narrowing of the blood vessels in fingertips that contribute to cold hands. Shrinking of the blood vessels causes a decrease in the amount of LED light absorbed by the blood in the vessels and as a result less light is reflected and detected by the sensor. This leads to disturbances in the signal provided by sensor that contributes to unreliable results, in some cases even close to 0.

Heart rate measurements done with people with thick skin illustrated noticeable differences between investigated and reference devices. However, these differences were smaller than in case of other conditions presented in this paper. Heart rate difference varied between 3 and 17 heart beats with the mean relative error value equal to 8.3%. This may have been caused by the fact that thick skin absorbed too much light illuminated by LED reducing the amount of light coming to blood vessels that led to the disturbances in the output signal and unreliable result provided by sensor.

Heart rate measurements done on people with dark skin also presented significant differences between the performances of investigated sensors. Here, heart rate differences referring to Amped sensor varied between 5 and 95 heart beats with the value of mean relative error equal to 51.26%. What is more, it was illustrated that with the rise in RGB ratio the heart rate difference between investigated and reference devices also increases. However, after reaching RGB ratio value equal to 1.7 the error increased significantly, corresponding to measurements in which almost no heart beats were detected. This might have been caused by higher amount of melanin present in the subject's skin which absorbed the majority of the light emitted by the sensor. Melanin is a natural pigment, created by melanocytes, that gives color to human skin. It absorbs green light in very high degree. It means that the darker is the skin, the more of green light it absorbs. This can contribute to inaccurate reading since light cannot get to dipper tissue layers. The same problem will occur during measurement in place of local changes in skin pigment like tattoos, moles or freckles [5].

5 Conclusions

In this paper the performance of Amped optical sensor for heart rate estimation under the influence of several factors was investigated. Data acquired from pulse oximeter were considered as the reference. The performance of the device was investigated in the following conditions: applied slight contact force between sensor and measurement site, cold hands, thick and dark skin color, as well as 4 different lighting conditions: without light, artificial light, daylight and with flashlight operating on investigated sensors. For each investigated condition the heart rate measurements of 10 subjects were performed. Obtained results showed that there is no significant influence of the lighting conditions on the performance of the investigated sensor. For measurement without light, in artificial light and in daylight the value of mean relative error varied between 0.81% and 1.77% while the accuracy ranged from 70% to 100%. The only exception was presented in case of measurements with flashlight where the mean relative error for Amped sensor was equal to 27.21% with the accuracy of 0%. The mean relative error in the rest of the investigated conditions was equal to: 8.30% for thick skin, 39.78% for contact force, 51.26% for dark skin and 91.81% for cold hands. Obtained results indicate that the unshielded PPG-based HR sensor is characterized by rather low performance in conditions even slightly different from standard ones. Moreover application of green light in the PPG sensor may provide very inaccurate measurements for users with dark skin color.

References

1. Allen, J.: Photoplethysmography and its application in clinical physiological measurement. Physiol. Meas. **28**(3), R1–R39 (2007)
2. Elgendi, M.: On the analysis of fingertip photoplethysmogram signals. Curr. Cardiol. Rev. **8**(1), 14–25 (2012)
3. Castaneda, D., Esparza, A., Ghamari, M.: A review on wearable photoplethysmography sensors and their potential future applications in health care. Int. J. Biosens. Bioelectron. **4**(4), 195–202 (2018)
4. Lee J., Matsumura K., Yamakoshi K., Rolfe P., Tanaka S., Yamakoshi T.: Comparison between red, green and blue light reflection photoplethysmography for heart rate monitoring during motion. In: Conference Proceedings of IEEE Engineering in Medicine and Biology Society 2013, pp. 1724–1727 (2013)
5. Weiler, D. T., Villajuan, S. O., Edkins, L., Cleary, S., Saleem, J. J.: Wearable heart rate monitor technology accuracy in research: a comparative study between PPG and ECG technology. In: Proceedings of the Human Factors and Ergonomics Society 2017 Annual Meeting, pp. 1292–1296 (2017)

6. Tamura, T., Maeda, Y., Sekine, M., Yoshida, M.: Wearable photoplethysmographic sensors - past and present. Electronics **3**, 282–302 (2014)
7. Jeong, I.C., Yoon, H., Kang, H., Yeom, H.: Effects of skin surface temperature on photoplethysmograph. J. Healthcare Eng. **5**(4), 429–438 (2014)
8. Sim, J.K., Ahn, B., Doh, I.: A contact-force regulated photoplethysmography (PPG) platform. Am. Inst. Phys. Adv. **8**, 045210.1-7 (2018)
9. Lee, Y., Han, H., Kim, J.: Influence of motion artifacts on photoplethysmographic signals for pulse rate. In: IEEE International Conference on Control, Automation and Systems, vol. 962–965 (2008)
10. Couceiro, R., Carvalho, P., Paiva, R.P., Henriques, J., Quintal, I., Muehlsteff, J.: Detection of motion artifacts in photoplethysmographic signals: algorithms comparison. In: The International Conference on Health Informatics. IFMBE Proceedings, vol. 42, pp. 327–330 (2014)

Comparison of Indices Derived from Poincaré Maps on Electrocardiograms and Seismocardiograms

Szymon Sieciński(✉), Paweł Kostka, Natalia Piaseczna, and Marta Wadas

Faculty of Biomedical Engineering, Department of Biosensors and Processing of Biomedical Signals, Silesian University of Technology, Roosevelta 40, 41-800 Zabrze, Poland
{szymon.siecinski,pawel.kostka,natalia.piaseczna,marta.wadas}@polsl.pl

Abstract. Heart rate variability (HRV) is the physiological variation of time between heart beats caused by the activity of autonomic nervous system. Heart rate variability analysis has found various applications in the diagnosis and treatment of different clinical and functional conditions. One of the prominent approaches in HRV analysis are Poincaré map. HRV analysis is traditionally performed on electrocardiograms (ECG) although seismocardiograms can also be used. In this study we compare indices derived from Poincaré maps on electrocardiograms and seismocardiograms found in CEBS database available on PhysioNet.org. Poincaré map is a non-linear method of HRV analysis which uses diagrams in which inter-beat intervals are plotted as a function of previous inter-beat intervals. We found that there are no significant differences of indices of Poincaré maps calculated on electrocardiograms and seismocardiograms, which indicates the reliability of seismocardiogram as a source signal in non-linear HRV analysis using Poincaré maps.

Keywords: Poincaré maps · HRV analysis · Seismocardiography · Electrocardiography

1 Introduction

Heart rate variability (HRV) is the physiological phenomenon of variation of time between heartbeats [23] caused by the activity of autonomic nervous system [27]. HRV has been frequently used in the analysis of physiological signals in different clinical and functional conditions [17,20], including atrial fibrillation [1], sleep bruxism [13] diagnosis, assessing the progress of rehabilitation of patients after ischemic brain stroke [28] and identification of various cardiac diseases [4]. One of available methods of HRV analysis are Poincaré maps [3,11,12,33].

HRV analysis is traditionally performed on electrocardiogram (ECG) [15], however, technological improvements of accelerometers have provided the reasons for using less obtrusive devices to monitor heart rate, such as seismocardiography [6,21,32]. Seismocardiography (SCG) is a technique of recording and analyzing

J. Korbicz et al. (Eds.): PCBEE 2019, AISC 1033, pp. 15–24, 2020.
https://doi.org/10.1007/978-3-030-29885-2_2

cardiac activity by measuring precordial acceleration. Recordings are taken using accelerometer on subjects in supine position [31]. In the past, SCG was mainly a tool for physiologists, due to the need of complex recording devices [5].

The feasibility of HRV analysis on seismocardiograms has been described in papers [14,15,21,24,26]. In 2012 Ramos-Castro et al. [21] performed the first HRV analysis on seismocardiograms. Laurin et al. [15] proved the validity of HRV indices obtained from SCG signal and Tadi et al. [26] and Siecinski et al. [24] studies showed high correlation between HRV indices obtained from ECG and SCG. To the best of authors' knowledge HRV analysis on SCG signals using Poincaré maps has never been performed.

The aim of this study is to compare indices derived from Poincaré maps obtained on twenty electrocardiograms and seismocardiograms from CEBS (Combined measurement of ECG, breathing and seismocardiogram) database publicly available at PhysioNet.org [7–9].

2 Materials and Methods

2.1 Data Set

CEBS Database used in this study contains 60 multi-channel signals acquired on 20 healthy volunteers. Each recording consists of four channels sampled with a frequency of 5 kHz: ECG (lead I and II), respiratory signal and SCG. Electrocardiogram (ECG) and respiratory signal were registered using Biopac MP36 data acquisition system. ECG was recorded with a bandwidth between 0.05 Hz and 150 Hz and channel 4 (SCG) was recorded using the tri-axial accelerometer LIS334ALH by ST Microelectronics and the bandwidth of 0.5–100 Hz [7–9].

Volunteers were asked to be awake and stay still in supine position on a bed during the measurement. After attaching the sensors, the basal state (before playing the music) was acquired for five minutes (recordings b001-b020). Then, the subjects started listening to music for 50 min (recordings m001-m020). Finally, the subjects were monitored for 5 min after the music ended (recordings p001-p020) [8,9]. Recordings b001-b020 were used to compare the indices derived from Poincaré maps obtained from ECG and SCG.

2.2 Signal Processing

Heart Rate Estimation on Electrocardiograms. To detect R waves in ECG lead I in analyzed signals, we applied Pan-Tompkins algorithm [18] implemented by Wedekind [29]. This algorithm consists of the following steps: band-pass filtering (to reduce noise, baseline wandering, muscle noise etc, power grid interference), differentiation, squaring of samples, applying of moving average filter, and correlation analysis [10,18]. Amplitude thresholding is applied after signal preprocessing in order to identify R waves in the ECG signal. The inter-beat intervals are calculated as differences between time of occurrence of successive R waves as in Eq. (1):

$$t_{RR,i} = t_n - t_{n-1} \tag{1}$$

where $t_{RR,i}$ is the i-th cardiac interval in ECG and t_n denotes the occurrence of n-th R wave.

Heart Rate Estimation on Seismocardiograms. Detection of heart beats in seismocardiograms is based on nearly periodic appearance of fiducial points [30]. We determine the heart beat as the occurrence of aortic valve opening (AO) wave which indicates the start of ventricular contraction and is usually visible as a single sharp wave [31].

To detect heart beats we use the algorithm proposed by Tadi *et al.* in paper [26], which uses R waves in ECG signal as reference points for heart beats in seismocardiograms and is based on the windowing method proposed in papers [19,25]. The first step of the algorithm is applying a band-pass filter with cut-off frequencies of 4 Hz and 50 Hz.

Then, the SCG signal is smoothed using a moving average filter with the window width of 10–20 ms. The R waves in the ECG signal are localized using Pan-Tompkins algorithm and serve as the reference points. The location of AO wave of a cardiac cycle is determined as a local maximum value of the SCG signal within a 90 ms window. The inter-beat intervals in SCG are calculated using the Eq. (1) where t_n denotes the occurrence of n-th AO wave in SCG signal instead of R wave in ECG signal.

2.3 Poincaré Map Analysis

The Poincaré map is a diagram in which inter-beat intervals are plotted as a function of previous inter-beat intervals [11]. It is a geometrical technique taken from nonlinear dynamics, which portrays the nature of HRV [3,12,33]. The example of Poincaré map (signal b007) is shown in Fig. 1 (ECG lead I) and Fig. 2 (SCG). Blue circles of Poincaré maps indicate scatter points of inter-beat intervals and orange dashed line shows the $y = x$ axis.

Several indices can be used to describe the Poincaré map, such as SD_1, SD_2, SD_1/SD_2, vector angular index and vector length index. A popular technique of analysis is fitting an ellipse to the shape of Poincaré plot and measuring the dispersion of points [22,33]. SD_1 is defined as the width of the ellipse and is calculated as standard deviation of the distances from $y = x$ axis (see Eq. (2)), indicating the short-term variability.

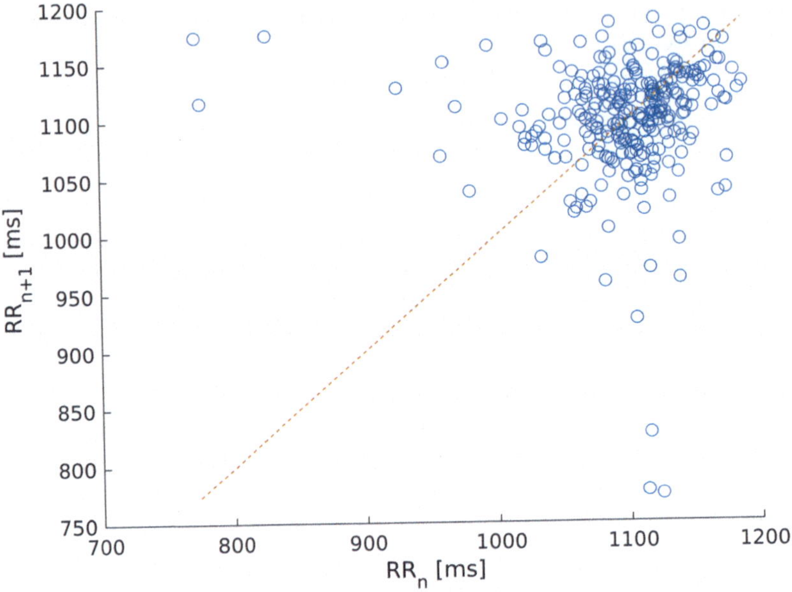

Fig. 1. Poincaré map on ECG lead I for recording b007.

SD_2 which indicates the long term variability, is defined as the length of the ellipse and is calculated as the standard deviation of the distances of points from $y = -x + 2\overline{NN}$ axis (see Eq. (3)), where $\overline{NN}$ is the mean of all inter-beat intervals [3,22].

$$SD_1 = stddev\left(\frac{|NN_{i+1} - NN_i|}{\sqrt{2}}\right) \tag{2}$$

$$SD_2 = stddev\left(\left|\frac{NN_{i+1} - NN_i}{\sqrt{2}} - 2\overline{NN}\right|\right) \tag{3}$$

where NN_i is an inter-beat interval series with $i = 1, 2, \ldots N - 1$, NN_{i+1} is the next inter-beat interval to NN_i, $\overline{NN}$ is the mean of all inter-beat intervals and $stddev()$ is the standard deviation.

SD_1/SD_2 is defined as the ratio of SD_1 and SD_2 and represents the randomness in HRV time series [2]. Vector Angular Index (VAI) and Vector Length Index (VLI) are two indices derived from Poincaré plot proposed by Ruan *et al.* in paper [22]. VAI measures the angular dispersion of scatter points and is defined as the mean of all the absolute value of angular differences between the lines plotted from every scatter point to the original point and the diagonal line as in Eq. (4):

$$VAI = \frac{1}{N}\sum_{i=1}^{N}|\theta_i - 45°| \tag{4}$$

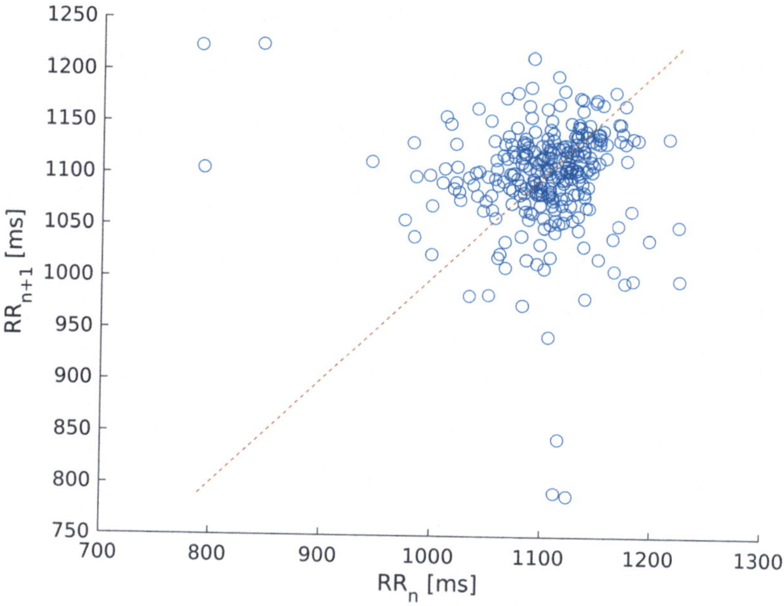

Fig. 2. Poincaré map on SCG for recording b007.

where θ_i is the angle between the line plotted from i-th point to the original point and the x-axis and N is the number of points.

VLI measures the distance dispersion of scatter points and is defined as the standard deviation of all distances of scatter points from the original point, as shown in Eq. (5):

$$VLI = \sqrt{\frac{1}{N}\sum_{i=1}^{N}(l_i - L)^2} \tag{5}$$

where l_i is the length between i-th scatter point and the original point, L is the mean of all l_i and N is the number of points.

3 Results

Due to the lack of annotations of recordings from CEBS Database [16], the heart beats in ECG signal and SCG signal were annotated using the algorithms described in Subsect. 2.2. SD_1, SD_2, SD_1/SD_2, VAI and VLI were calculated for entire signals b001-b020.

Differences between the indices of Poincaré maps derived from ECG lead I and SCG signal shown in Tables 1 and 2 are insignificant for each analyzed signal.

Table 1. Indices describing Poincaré plot calculated from ECG lead I.

Signal	SD_1	SD_2	SD_1/SD_2	VAI	VLI
b001	25.6380	37.9359	0.6758	1.2810	62.0643
b002	17.4687	31.3234	0.5577	1.1054	53.1401
b003	59.7934	73.5490	0.8130	2.1066	129.6599
b004	13.9699	24.9292	0.5604	1.0094	52.8160
b005	32.4302	50.9893	0.6360	1.8557	95.6363
b006	11.2452	18.2798	0.6152	0.6423	66.0045
b007	17.1902	30.0707	0.5717	1.1581	67.2752
b008	13.5598	20.6610	0.6563	0.7644	77.5070
b009	7.8209	13.9218	0.5618	0.6306	45.8446
b010	20.1847	37.7379	0.5349	1.4568	73.5263
b011	12.8855	24.4927	0.5261	0.8720	51.9759
b012	24.3128	35.4235	0.6863	1.0741	104.7814
b013	15.5289	20.1058	0.7724	0.7975	57.9535
b014	40.6278	51.9255	0.7824	1.2228	50.6318
b015	24.7270	41.4962	0.5959	1.3961	109.5244
b016	13.5586	27.0900	0.5005	1.1463	48.1259
b017	53.0670	57.1329	0.9288	0.8739	82.6737
b018	15.8030	33.9682	0.4652	1.4037	57.4209
b019	18.6717	36.2089	0.5157	1.2924	66.7963
b020	9.9936	19.5339	0.5116	0.7474	44.5606

We also expressed each analyzed index derived from Poincaré maps as mean±standard deviation and tested using Student's t-test. Differences between values of analyzed HRV indices shown in Table 3 are insignificant by $p = 1$ and significance level $\alpha = 0.05$, however, the most prominent differences are in SD_1 and SD_2.

Table 2. Indices describing Poincaré plot calculated from SCG.

Signal	SD_1	SD_2	SD_1/SD_2	VAI	VLI
b001	26.4598	39.4945	0.6700	1.3449	62.8023
b002	33.7794	61.9751	0.5450	2.2312	59.8728
b003	58.9173	72.7176	0.8102	2.0964	129.2176
b004	18.2259	30.8532	0.5907	1.2242	51.1496
b005	32.7154	51.8632	0.6308	1.8997	97.0998
b006	20.6264	37.3750	0.5519	1.3973	66.4817
b007	18.6835	32.6117	0.5729	1.2555	68.5059
b008	13.8818	21.3339	0.6507	0.7939	77.7043
b009	25.7795	38.3134	0.6729	1.5633	49.9283
b010	20.9848	39.6427	0.5293	1.5367	75.4846
b011	15.9421	28.1708	0.5659	0.9727	50.8871
b012	24.2529	35.4980	0.6832	1.0819	105.6278
b013	15.6656	20.3452	0.7700	0.8106	58.1170
b014	42.1407	54.7904	0.7691	1.3151	50.6887
b015	25.1164	42.1986	0.5952	1.4220	110.0463
b016	14.4678	29.2033	0.4954	1.2412	49.9772
b017	50.3223	58.3549	0.8624	1.2434	80.9677
b018	28.7467	47.8483	0.6008	1.7620	61.6313
b019	40.3629	62.5558	0.6452	1.9834	70.7600
b020	17.4935	30.8478	0.5671	1.1293	46.5272

Table 3. Statistical comparison of indices calculated from Poincaré maps on ECG and SCG signals.

	ECG		SCG	
	Mean	Standard deviation	Mean	Standard deviation
SD_1	22.4238	14.0693	27.2282	12.4756
SD_2	34.3388	14.9965	41.7997	14.5164
SD_1/SD_2	0.6234	0.1218	0.6389	0.0993
VAI	1.1418	0.3851	1.4152	0.4053
VLI	69.8959	62.0643	71.1739	23.0850

4 Conclusion and Discussion

We found no significant differences between values of indices derived from Poincaré maps (SD_1, SD_2, SD_1/SD_2, VAI and VLI) on electrocardiograms and seismocardiograms. High correlation between HRV indices derived from ECG and SCG occurs when the differences between inter-beat intervals computed on ECG and SCG are small (under 10 ms) [21,26]. This means that seismocardiogram can be used as a reliable source signal in non-linear analysis of HRV using Poincaré plots.

Statistical comparison of SD_1, SD_2, VAI and VLI indicates that in this study we analyzed the signals acquired on healthy volunteers. Differences between indices in our study and in paper [22] are caused by the fact that we considered entire 5-min signals instead of 1-min windows. Another limitation is using only one heart beat detector on seismocardiograms to determine heart beats used further in HRV analysis.

In future studies more signals from CEBS database and another data sets containing ECG and SCG signals should be analyzed and compared to our results to investigate further the similarity of HRV indices obtained from Poincaré plots.

References

1. Barauskiene, V., Rumbinaite, E., Karuzas, A., Martinkute, E., Puodziukynas, A.: Importance of heart rate variability in patients with atrial fibrillation. J. Cardiol. Clin. Res. **4**(6), 1080 (2016)
2. Biala, T., Dodge, M., Schlindwein, F.S., Wailoo, M.: Heart rate variability using poincaré plots in 10 year old healthy and intrauterine growth restricted children with reference to maternal smoking habits during pregnancy. In: 2010 Computing in Cardiology, pp. 971–974 (2010)
3. Brennan, M., Palaniswami, M., Kamen, P.: Do existing measures of poincare plot geometry reflect nonlinear features of heart rate variability? IEEE Trans. Biomed. Eng. **48**(11), 1342–1347 (2001). https://doi.org/10.1109/10.959330
4. Budzianowski, Z., Tkacz, E., Oleksy, W., Garbacik, M.: The higher-order spectra as a tool for the identification of patients diagnosed with various cardiac diseases. In: Gzik, M., Tkacz, E., Paszenda, Z., Piętka, E. (eds.) Innovations in Biomedical Engineering, pp. 193–203. Springer, Cham (2018). https://doi.org/10.1007/978-3-319-70063-2_21
5. Castiglioni, P., Faini, A., Parati, G., Rienzo, M.D.: Wearable seismocardiography. In: 2007 29th Annual International Conference of the IEEE Engineering in Medicine and Biology Society, pp. 3954–3957 (2007). https://doi.org/10.1109/IEMBS.2007.4353199
6. Castiglioni, P., Meriggi, P., Rizzo, F., Vaini, E., Faini, A., Parati, G., Merati, G., Rienzo, M.D.: Cardiac sounds from a wearable device for sternal seismocardiography. In: 2011 Annual International Conference of the IEEE Engineering in Medicine and Biology Society, pp. 4283–4286 (2011). https://doi.org/10.1109/IEMBS.2011.6091063
7. García-González, M.A., Argelagós, A., Fernández-Chimeno, M., Ramos-Castro, J.: Differences in QRS locations due to ECG lead: relationship with breathing. In: Roa Romero, L.M. (ed.) XIII Mediterranean Conference on Medical and Biological Engineering and Computing 2013, pp. 962–964. Springer, Cham (2014). https://doi.org/10.1007/978-3-319-00846-2_238

8. García-González, M.A., Argelagós-Palau, A., Fernández-Chimeno, M., Ramos-Castro, J.: A comparison of heartbeat detectors for the seismocardiogram. In: Computing in Cardiology 2013, pp. 461–464 (2013)
9. Goldberger, A.L., Amaral, L.A.N., Glass, L., Hausdorff, J.M., Ivanov, P.C., Mark, R.G., Mietus, J.E., Moody, G.B., Peng, C.K., Stanley, H.E.: PhysioBank, PhysioToolkit, and PhysioNet: components of a new research resource for complex physiologic signals. Circulation **101**(23), e215–e220 (2000). https://doi.org/10.1161/01.CIR.101.23.e215. Circulation Electronic Pages: http://circ.ahajournals.org/content/101/23/e215.fullPMID:1085218
10. Hamilton, P.S., Tompkins, W.J.: Quantitative investigation of QRS detection rules using the MIT/BIH arrhythmia database. IEEE Trans. Biomed. Eng. **BME–33**(12), 1157–1165 (1986). https://doi.org/10.1109/TBME.1986.325695
11. Karmakar, C.K., Gubbi, J., Khandoker, A.H., Palaniswami, M.: Analyzing temporal variability of standard descriptors of poincaré plots. J. Electrocardiol. **43**(6), 719–724 (2010). https://doi.org/10.1016/j.jelectrocard.2010.09.001. http://www.sciencedirect.com/science/article/pii/S0022073610003821
12. Kitlas Golińska, A.: Poincaré plots in analysis of selected biomedical signals. Stud. Logic Grammar Rhetoric **35**(1), 117–127 (2013). https://doi.org/10.2478/slgr-2013-0031
13. Kostka, P.S., Tkacz, E.J.: Multi-sources data analysis with sympatho-vagal balance estimation toward early bruxism episodes detection. In: 2015 37th Annual International Conference of the IEEE Engineering in Medicine and Biology Society (EMBC), pp. 6010–6013 (2015). https://doi.org/10.1109/EMBC.2015.7319761
14. Landreani, F., Morri, M., Martin-Yebra, A., Casellato, C., Pavan, E., Frigo, C., Caiani, E.G.: Ultra-short-term heart rate variability analysis on accelerometric signals from mobile phone. In: 2017 E-Health and Bioengineering Conference (EHB), pp. 241–244 (2017). https://doi.org/10.1109/EHB.2017.7995406
15. Laurin, A., Blaber, A., Tavakolian, K.: Seismocardiograms return valid heart rate variability indices. Comput. Cardiol. **2013**, 413–416 (2013)
16. Li, Y., Tang, X., Xu, Z.: An approach of heartbeat segmentation in seismocardiogram by matched-filtering. In: 2015 7th International Conference on Intelligent Human-Machine Systems and Cybernetics, vol. 2, pp. 47–51 (2015). https://doi.org/10.1109/IHMSC.2015.157
17. Montano, N., Porta, A., Cogliati, C., Costantino, G., Tobaldini, E., Casali, K.R., Iellamo, F.: Heart rate variability explored in the frequency domain: a tool to investigate the link between heart and behavior. Neurosci. Biobehav. Rev. **33**(2), 71–80 (2009). The Inevitable Link between Heart and Behavior: New Insights from Biomedical Research and Implications for Clinical Practice. https://doi.org/10.1016/j.neubiorev.2008.07.006, http://www.sciencedirect.com/science/article/pii/S0149763408001176
18. Pan, J., Tompkins, W.J.: A real-time QRS detection algorithm. IEEE Trans. Biomed. Eng. **BME–32**(3), 230–236 (1985)
19. Pandia, K., Inan, O.T., Kovacs, G.T.A., Giovangrandi, L.: Extracting respiratory information from seismocardiogram signals acquired on the chest using a miniature accelerometer. Physiol. Meas. **33**(10), 1643–1660 (2012). https://doi.org/10.1088/0967-3334/33/10/1643
20. Pumprla, J., Howorka, K., Groves, D., Chester, M., Nolan, J.: Functional assessment of heart rate variability: physiological basis and practical applications. Int. J. Cardiol. **84**(1), 1–14 (2002). https://doi.org/10.1016/S0167-5273(02)00057-8. http://www.sciencedirect.com/science/article/pii/S0167527302000578

21. Ramos-Castro, J., Moreno, J., Miranda-Vidal, H., García-González, M.A., Fernández-Chimeno, M., Rodas, G., Capdevila, L.: Heart rate variability analysis using a seismocardiogram signal. In: 2012 Annual International Conference of the IEEE Engineering in Medicine and Biology Society, pp. 5642–5645 (2012). https://doi.org/10.1109/EMBC.2012.6347274
22. Ruan, X., Liu, C., Liu, C., Wang, X., Li, P.: Automatic detection of atrial fibrillation using r-r interval signal. In: 2011 4th International Conference on Biomedical Engineering and Informatics (BMEI), vol. 2, pp. 644–647 (2011). https://doi.org/10.1109/BMEI.2011.6098492
23. Saykrs, B.: Analysis of heart rate variability. Ergonomics **16**(1), 17–32 (1973). https://doi.org/10.1080/00140137308924479. PMID: 4702060
24. Siecinski, S., Kostka, P.S., Tkacz, E.J.: Heart rate variability analysis on CEBS database signals. In: 2018 40th Annual International Conference of the IEEE Engineering in Medicine and Biology Society, pp. 5697–5700 (2018). https://doi.org/10.1109/EMBC.2018.8513551
25. Tadi, M.J., Koivisto, T., Pänkäälä, M., Paasio, A.: Accelerometer-based method for extracting respiratory and cardiac gating information for dual gating during nuclear medicine imaging. Int. J. Biomed. Imaging **2014**(690124), 1–11 (2014). https://doi.org/10.1155/2014/690124
26. Tadi, M.J., Lehtonen, E., Koivisto, T., Pänkäälä, M., Paasio, A., Teräs, M.: Seismocardiography: toward heart rate variability (HRV) estimation. In: 2015 IEEE International Symposium on Medical Measurements and Applications (MeMeA) Proceedings, pp. 261–266 (2015). https://doi.org/10.1109/MeMeA.2015.7145210
27. Task Force of the European Society of Cardiology the North American Society of Pacing Electrophysiology: Heart rate variability. standards of measurement, physiological interpretation, and clinical use. Circulation **93**, 1043–1065 (1996). https://doi.org/10.1161/01.CIR.93.5.1043
28. Tkacz, E., Budzianowski, Z., Oleksy, W.: The higher-order spectra as a tool for assessing the progress in rehabilitation of patients after ischemic brain stroke. In: Rocha, Á., Guarda, T. (eds.) Proceedings of the International Conference on Information Technology & Systems (ICITS 2018), pp. 874–882. Springer, Cham (2018). https://doi.org/10.1007/978-3-319-73450-7_83
29. Wedekind, D.: qrsdetector (2014). https://github.com/danielwedekind/qrsdetector
30. Zanetti, J.M., Poliac, M.O., Crow, R.S.: Seismocardiography: waveform identification and noise analysis. In: [1991] Proceedings Computers in Cardiology, pp. 49–52 (1991). https://doi.org/10.1109/CIC.1991.169042
31. Zanetti, J.M., Salerno, D.M.: Seismocardiography: a technique for recording precordial acceleration. In: [1991] Computer-Based Medical Systems, Proceedings of the Fourth Annual IEEE Symposium, pp. 4–9 (1991). https://doi.org/10.1109/CBMS.1991.128936
32. Zanetti, J.M., Tavakolian, K.: Seismocardiography: past, present and future. In: 2013 35th Annual International Conference of the IEEE Engineering in Medicine and Biology Society (EMBC), pp. 7004–7007 (2013). https://doi.org/10.1109/EMBC.2013.6611170
33. İşler, Y., Kuntalp, M.: Combining classical HRV indices with wavelet entropy measures improves to performance in diagnosing congestive heart failure. Comput. Biol. Med. **37**(10), 1502–1510 (2007). QT Variability & Heart Rate Variability. https://doi.org/10.1016/j.compbiomed.2007.01.012. http://www.sciencedirect.com/science/article/pii/S0010482507000285

Research of Ocular Hemodynamics by Optical Coherence Tomography and Transpalpebral Rheoophthalmography Methods

Elena N. Iomdina[1], Olga A. Kiseleva[1], Daniel D. Khoziev[1], Petr V. Luzhnov[2(✉)], Anna A. Kiseleva[2], and Dmitry M. Shamaev[2]

[1] Moscow Helmholtz Research Institute of Eye Diseases, Sadovaya-Chernogryazskaya 14/19, 105062 Moscow, Russia
iomdina@mail.ru
[2] Bauman Moscow State Technical University, 2-nd Baumanskaya 5, 105005 Moscow, Russia
peterl@hotmail.ru

Abstract. The main task of this work was to find the connection between ocular hemodynamic parameters and the lamina cribrosa parameters of the eyes with primary open angle glaucoma (POAG). The study was conducted on 51 eyes of 34 subjects, including 11 eyes without ophthalmopathology (control group, average age 49.4 ± 17.5 years), 15 eyes of patients with the 1st (initial) stage of POAG (group 1, average age 67.0 ± 7.6 years), 15 eyes with II (developed) stage of glaucoma (group 2, average age 75.4 ± 6.1 years) and 10 eyes with III (advanced) disease stage (group 3, average age 70.3 ± 5.4 years). The aim of this work was the study of the parameters analyzed in the diagnosis of POAG, which were obtained in two complementary ways: optical coherence tomography (OCT) with angiography function and transpalpebral rheoophthalmography (TP ROG). A pronounced positive correlation established between the sclera lamina cribrosa in OCT and rheographic parameters indicates that the TP ROG reflects, among other things, the hemodynamics in medium and large choroidal vessels passing through the retinal nerve fiber layer. The results indicate the promise of further research in this area to develop new effective individual criteria for the early detection and monitoring of POAG.

Keywords: Primary open angle glaucoma · Optical coherent tomography angiography · Transpalpebral rheoophthalmography · Ocular hemodynamics · Eye

1 Introduction

Primary open-angle glaucoma (POAG) is a chronic disease characterized by optical neuropathy, progressive degeneration of ganglion cells and retinal nerve fibers

J. Korbicz et al. (Eds.): PCBEE 2019, AISC 1033, pp. 25–34, 2020.
https://doi.org/10.1007/978-3-030-29885-2_3

layer. The irreversible loss of visual function is the consequence of this process. The modern approach provides two research directions for further understanding the mechanism of the emergence and progression of POAG [1]. Firstly, the study of pathogenetic role of the corneoscleral (fibrous) shell changes. Secondly, the study of circulatory eye tissues disorders. Circulatory disorders can be an independent factor in the progression of glaucomatous lesions [2,3].

As it is known, an increase in intraocular pressure (IOP) is associated with impaired microcirculation of the eye. On the other hand, the pathological changes of the sclera lamina cribrosa (LC), as part of the corneoscleral shell, are directly related to the pathogenesis of glaucoma. Axons of retinal ganglion cells pass through its perforated structure, that is why the deformed LC is considered to be the main site of their damage [2]. This causes the thinning of the nerve fiber layer and, as a result, the narrowing of the peripheral visual fields. So, it is the most important factor in the progression of glaucoma. In this regard, it is fundamentally important a thorough study of ocular hemodynamics and the LC for a more detailed understanding of the POAG pathogenesis.

Rheoophthalmography (ROG) is electroimpedance diagnostic method that can be used to measure ocular hemodynamics in myopia, diabetic retinopathy, glaucoma, and other eye diseases [4]. With ROG, the state of hemodynamics in the ciliary body, and, integrally, in the eye can be evaluated. In [5] electrodes are placed on the closed eye, i.e. on the eyelid, which drastically simplifies the ROG technique. This method of diagnostics, called transpalpebral rheoophthalmography (TP ROG), is used in ophthalmologic practice now [6].

Using modern technology of eye structures visualization, in particular, optical coherence tomography (OCT) with increased penetration depth beams allows obtaining images of the front surface of the LC, examine its changes during glaucomatous process [7,8]. This method provides not only quality assessment, but also provide a quantitative assessment of eye microcirculation. Color Doppler technique was originally used for the visualization of the eye blood flow parameters. Possessing a number of indisputable advantages, this method is sensitive only to the blood flow, which is parallel to the direction of the probe beam. The information content of this method is reduced due to the fact that the movement of blood flow in the retina and choroid is mainly perpendicular to the probe beam direction. OCT with an angiography module is used for increasing sensitivity to transverse blood flow [9]. The disadvantages of this method are primarily the likelihood of artifacts caused by the imposition of capillary plexuses on each other or shadows of large vessels in the superficial plexus deep capillary network. OCT with angiography function can give only an estimate of fluid flow through the vessels, without further imaging of the blood flow character. The last limitation of the OCT with angiography function can be eliminated by using the technique of TP ROG [10]. It gives the possibility not only to quantify the hemodynamics of the eye, but also to determine the type of the blood flow by analyzing the recorded pulse oscillations [11,12].

At the moment there is no definite answer informative use of a particular method for glaucoma, especially for early diagnosis. As a result, is relevant and

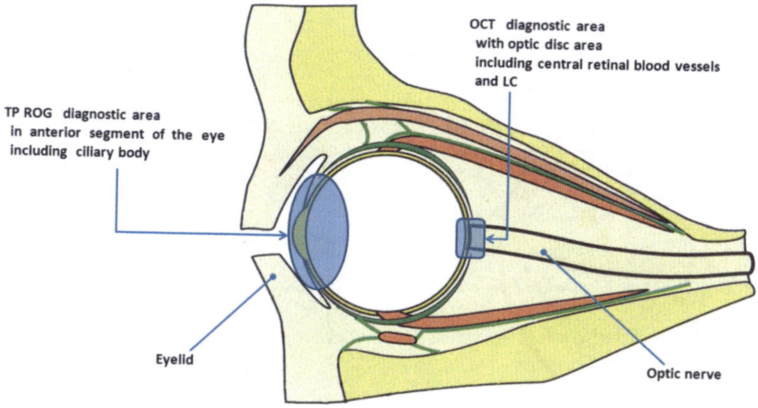

Fig. 1. The research areas for simultaneous analysis of OCT and TP ROG diagnostic data

promising to conduct a joint analysis of data obtained by different diagnostic methods (see the Fig. 1).

The aim of this work was to study diagnostic parameters of POAG, which were obtained in two complementary ways: OCT with angiography function and TP ROG.

2 Materials and Methods

In studies of the ocular hemodynamics, up to 16 of the largest arteries can be distinguished, as shown in the Fig. 2 and in the Table 1. The vascular supply to the eye proceeds to the internal carotid artery and then to the ophthalmic artery [13,14]. The ophthalmic artery branches to the short posterior ciliary vessels, the long posterior ciliary vessels and the central retinal artery [15].

During the studies of TP ROG, the quantitative indicators are obtained for the anterior segment of the eye. Also in the diagnosis process, an integrated assessment of ocular blood flow is carried out. This assessment represents the blood flow in all parts of the eye. However, for the posterior segment and LC, the blood flow is determined indirectly. For this purpose, it is used mathematical models, methods of nonlinear dynamics.

In our work the rheographic index (RI) and the stroke volume of blood (SV) were estimated by the TP ROG method using a tetrapolar electrodes system. It means that four electrodes are positioned on an eyelid surface, the pair of measuring electrodes is between pair of current electrodes. The scheme of the electrodes location is presented in papers [6,10]. Our TP ROG device utilize 100 kHz probing current, with amplitude 3 mA. A measurement of biological tissues full impedance (base impedance, BI) and its change at pulse blood filling

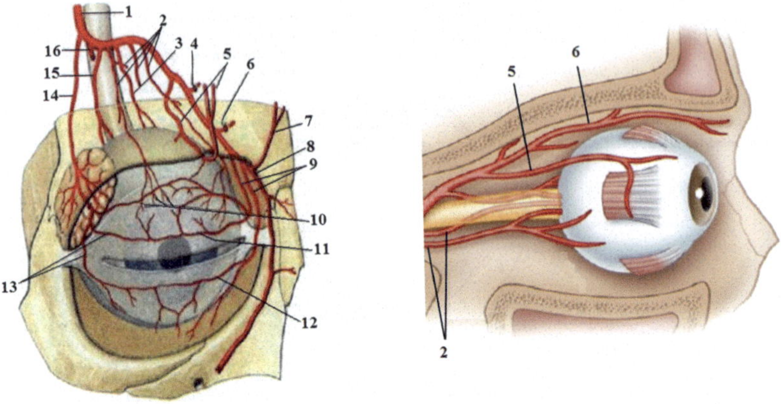

Fig. 2. The spatial location of the major eye arteries on an example of the left eye

(rheographic index, RI) lies in basis of rheographic methods (see Fig. 3). BI was obtained as following: electroimpedance signals were low-pass filtered with

Table 1. The structure of the arterial vascular eye system

Position	Artery	Included in TP ROG diagnostic area	Included in OCT diagnostic area
1	Ophthalmic artery	No	Yes
2	Muscular arteries	No	Yes
3	Long posterior ciliary artery	Yes	No
4	Posterior ethmoidal artery	No	No
5	Supraorbital arteries	No	No
6	Anterior ethmoidal artery	No	No
7	Supratrochlear artery	No	No
8	Dorsal nasal artery	No	No
9	Medial arteries of the eyelids	Partly	No
10	Episcleral arteries	Partly	No
11	Upper arterial arch	Partly	No
12	Lower arterial arch	Partly	No
13	Lateral eyelid arteries	Partly	No
14	Lacrimal artery	Yes	No
15	Short posterior ciliary artery	Yes	Yes
16	Central retinal artery	Partly	Yes

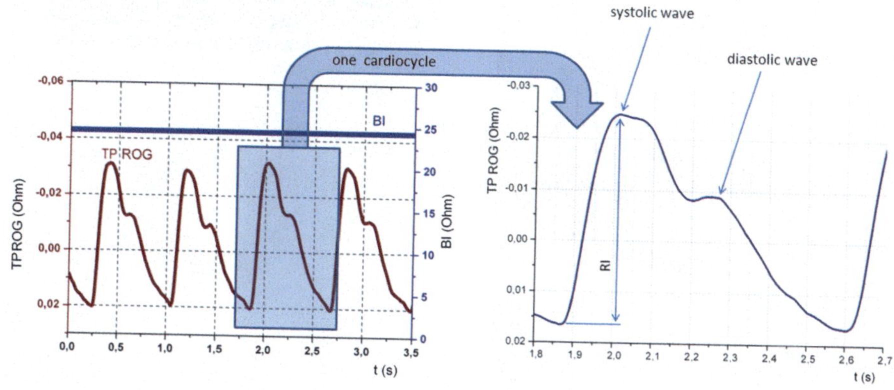

Fig. 3. The examples of TP ROG diagnostic signals

cut-off frequency 0.1 Hz. For RI electroimpedance signals were band-pass filtered from 0.1 Hz to 100 Hz.

The most informative parameter of ROG is the rheographic index. It is proportional to the size of systolic pulse blood filling (see the Fig. 3). The analysis of the signals obtained by the TP ROG method is performed in the time domain by the amplitude indices of the RI and the SV [5,6], as well as by the method of nonlinear dynamics [16].

The blood supply to the optic nerve is complex and varies by optic nerve segment. The LC study was conducted on a Spectralis OCT2 device (Heidelberg Engineering, Germany) in the mode of increased image depth in our work. The depth of the LC position was determined as a perpendicular, lowered from the center of the reference line (the line connecting the end points of the Bruch membrane) to the front surface of the LC. Estimation of surface (SVL) and deep (DVL) vascular layer density in the investigated peripapillary area of 3×3 mm (see the Fig. 4) was performed on a Spectralis OCT2 device with an angiography module (Heidelberg Engineering, Germany) using Angiotool software.

In order to quantify the blood flow of the eye in OCT with angiography function, a split-spectrum amplitude-decorrelation angiography algorithm is used (SSADA). The principle of operation of this algorithm is based on decomposing OCT - images into spectral bands, which further allows increasing the number of processed images [17]. Each new frame has a lower axial resolution than the previous one and is less sensitive to eye movements. Each spectral component has its own pattern and carries independent information about the blood flow. When the image is restored, a multitude of spectral bands are summed, the amplitude is decorrelated and the blood flow signal is amplified. Because the four-division of the spectrum, this algorithm can improve the signal/noise ratio twice that reduces the scanning time fourfold. The main characteristics of the blood flow obtained by this method include: blood flow index (average value of amplitude decorrelation), area of perfusion zones, index of microcapillary density per unit area (see the Fig. 5).

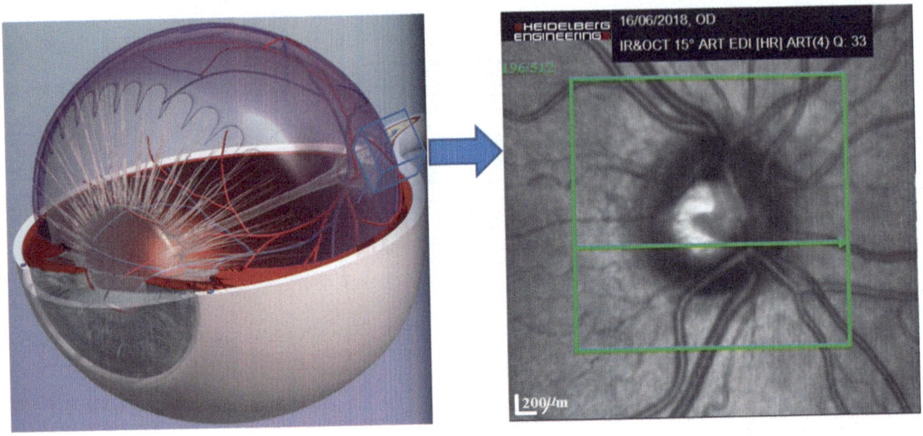

Fig. 4. The investigated optic disc area in OCT diagnostics

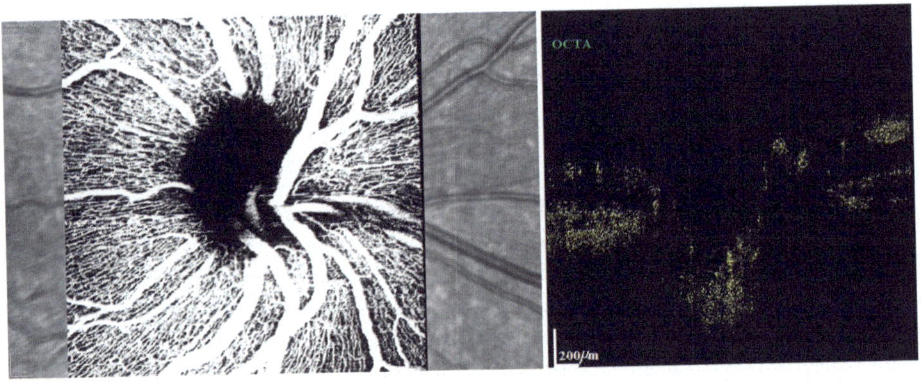

Fig. 5. The examples of OCT diagnostics with angiography function

An example of the processing OCT images from the patient with first stage of POAG is shown in the Fig. 6. The optic disc area and the OCT angiographic images of the macula are presented in the Fig. 6.

The study was conducted on 51 eyes of 34 subjects, including 11 eyes without ophthalmopathology (control group, average age 49.4 ± 17.5 years), 15 eyes of patients with the 1st (initial) stage of POAG (group 1, average age 67.0 ± 7.6 years), 15 eyes with II (developed) stage of glaucoma (group 2, average age 75.4 ± 6.1 years) and 10 eyes with III (advanced) disease stage (group 3, average age 70.3 ± 5.4 years). True IOP (Po) in patients with POAG was normalized and was, respectively, in the control group - 12.3 ± 2.4 mm Hg, in the first group - 13.4 ± 4.2 mm Hg, in the second group - 14.4 ± 2.5 mm Hg, in the third group - 16.4 ± 3.4 mm Hg.

Fig. 6. The examples of OCT images of the patient with first stage of POAG

Patients with POAG have not previously undergone surgical treatment and received antihypertensive therapy with beta-blockers, alpha-adrenomimetics, carbonic anhydrase inhibitors, prostaglandin analogues.

An example of automated software processing by the vascular layers of a patient from the second group is presented in the Fig. 7. The calculation of LC parameters is shown in the Fig. 7.

Statistical processing was performed quantitative results in the program IBM SPSS Statistics 20 calculating the mean value and mean deviation parameter (M ± SD), when comparing the groups using U - Mann-Whitney test. The differences were considered significant at $p < 0.05$.

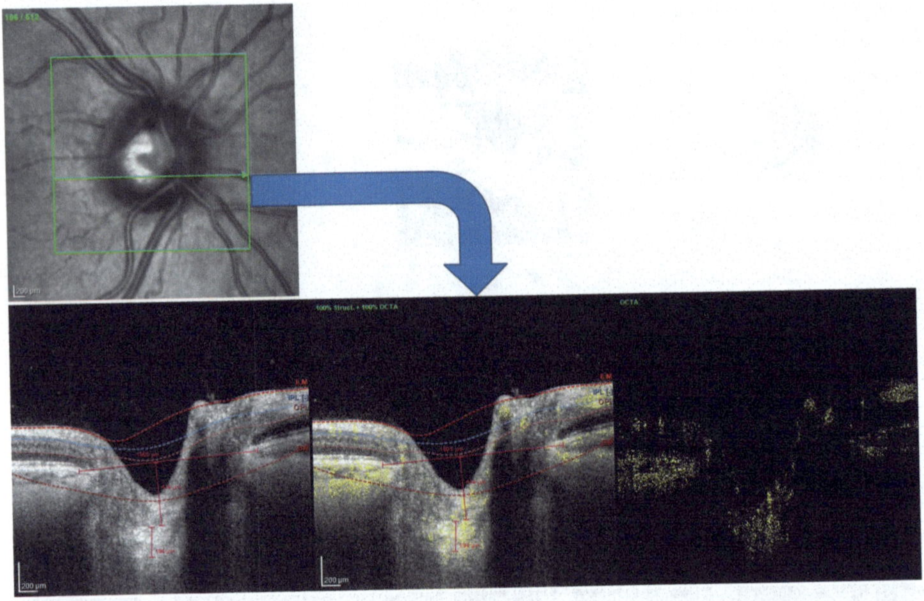

Fig. 7. The examples of software processing by the vascular layers

3 Compliance with Ethical Requirements

This study was performed in accordance with the Declaration of Helsinki and was approved by the Local Committee of Biomedical Ethics of the Moscow Helmholtz Research Institute of Eye Diseases. A written informed consent was obtained from all participants.

4 Results

As the analysis of data obtained in the control group showed, a change in some indicators is observed with age: a decrease in the thickness of the LC ($p = 0.006$), density of SVL ($p = 0.000$), DVL ($p = 0.019$) and RI ($p = 0.006$) are observed.

It is found that increasing stage of glaucoma is accompanied by an increase in the true IOP. At the same time, changes in the indices of the corneoscleral shell and hemodynamic parameters of the eye vascular system in patients with stages I, II, and III of POAG were more pronounced than age-related changes in the control group (see Table 2).

The data in the Table 2 indicate that as the POAG develops, the morphometric parameters of the LC, which are closely related to its biomechanical properties, change: as the glaucoma stage increases, the LC thickness decreases significantly and its depth increases.

When evaluating hemodynamic parameters, a significant decrease in RI was found ($p = 0.030$), as well as a pronounced tendency to decrease in SV with the

Table 2. Parameters of LC and TP ROG in control and POAG groups (M ± SD)

Indicators	Control group	I stage POAG	II stage POAG	III stage POAG
Po [mm Hg]	16.7 ± 5.1	18.4 ± 6.0	16.2 ± 4.5	13.8 ± 5.2
LC thickness [μm]	229.9 ± 19.7	205.4 ± 9.6	201.8 ± 12.2	176.1 ± 28.6
LC depth [μm]	405.9 ± 59.5	432.5 ± 45.9	449.2 ± 36.6	530.2 ± 75.3
SVL [%]	52.4 ± 3.0	49.7 ± 3.3	44.6 ± 7.8	43.1 ± 8.5
DVL [%[	45.8 ± 1.6	41.6 ± 5.2	37.3 ± 6.3	38.4 ± 6.7
RI [mOhm]	69.1 ± 51.7	53.1 ± 17.7	40.6 ± 20.0	26.8 ± 13.1
SV [μl]	1.5 ± 0.8	1.2 ± 0.3	1.2 ± 1.0	0.7 ± 0.3

development of POAG ($p = 0.070$). At stage III of POAG, a decrease in the density of the surface SVL ($p = 0.047$) and deep vascular layer density DVL ($p = 0.019$) was detected.

5 Conclusions

A pronounced positive correlation established between SVL and RI ($p = 0.02$) indicates that the RI reflects, among other things, the hemodynamics in medium and large choroidal vessels passing through the retinal nerve fiber layer.

The results indicate the promise of further research in this area to assess the relative contribution of each of the studied factors to the pathogenesis of POAG, as well as to develop new effective individual criteria for the early detection and monitoring of POAG.

Conflict of Interest. The authors declare that they have no conflict of interest. The paper was supported by a grant from RFBR (No.18-08-01192).

References

1. Flammer, J., Orgul, S., Costa, V.P., et al.: The impact of ocular blood flow in glaucoma. Prog. Retin. Eye Res. **21**, 359–393 (2002)
2. Kurysheva, N.I., Parshunina, O.A., Shatalova, E.O., et al.: Value of structural and hemodynamic parameters for the early detection of primary open-angle glaucoma. Curr. Eye. Res. **42**(3), 411–417 (2017)
3. Schmetterer, L.: Ocular perfusion abnormalities in glaucoma. Russ. Ophthalmol. J. **4**, 100–109 (2015)
4. Lazarenko, V.I., Komarovskikh, E.N.: Results of the examination of hemodynamics of the eye and brain in patients with primary open-angle glaucoma. Vestnik oftalmologii **120**(1), 32–36 (2004). https://doi.org/10.21687/0233-528X-2017-51-3-22-30
5. Luzhnov, P.V., Shamaev, D.M., Iomdina, E.N., et al.: Transpalpebral tetrapolar reoophtalmography in the assessment of parameters of the eye blood circulatory system. Vestn Ross Akad Med Nauk **70**(3), 372–377 (2015). https://doi.org/10.15690/vramn.v70i3.1336

6. Luzhnov, P.V., Shamaev, D.M., Iomdina, E.N., et al.: Using quantitative parameters of ocular blood filling with transpalpebral rheoophthalmography. In: IFMBE Proceedings, vol. 65, pp. 37–40 (2017). https://doi.org/10.1007/978-981-10-5122-7_10
7. Kurysheva, N.I., Maslova, E.V.: Optical coherence tomography angiography in glaucoma diagnosis. Vestnik oftalmologii **132**(5), 98–102 (2016)
8. Liu, L., Jia, Y., Takusagawa, H.L., et al.: Optical coherence tomography angiography of the peripapillary retina in glaucoma. JAMA Ophthalmol. **133**(9), 1045–1052 (2015). https://doi.org/10.1001/jamaophthalmol.2015.2225
9. Suh, M.H., Zangwill, L.M., Manalastas, P.I.C., et al.: Optical coherence tomography angiography vessel density in glaucomatous eyes with focal lamina cribrosa defects. Ophthalmology **123**(11), 2309–2317 (2016). https://doi.org/10.1016/j.ophtha.2016.07.023
10. Shamaev, D.M., Luzhnov, P.V., Iomdina, E.N.: Modeling of ocular and eyelid pulse blood filling in diagnosing using transpalpebral rheoophthalmography. In: IFMBE Proceedings, vol. 65, pp. 1000–1003 (2017). https://doi.org/10.1007/978-981-10-5122-7_250
11. Luzhnov, P.V., Shamaev, D.M., Kiseleva, A.A., et al.: Analyzing rheoophthalmic signals in glaucoma by nonlinear dynamics methods. In: World Congress on Medical Physics and Biomedical Engineering. In: IFMBE Proceedings, vol. 68(2), pp. 827–831 (2018). https://doi.org/10.1007/978-981-10-9038-7_152
12. Shamaev, D.M., Luzhnov, P.V., Iomdina, E.N.: Mathematical modeling of ocular pulse blood filling in rheoophthalmography. In: IFMBE Proceedings, vol. 68(1), pp. 495–498 (2018). https://doi.org/10.1007/978-981-10-9035-6_91
13. Krstic, R.V.: Human Microscopic Anatomy. Springer, Heidelberg (1991). https://doi.org/10.1007/978-3-662-02676-2
14. Roebuck, J.: Anatomy 360: The Ultimate Visual Guide to the Human Body. Thunder Bay Press, San Diego (2015)
15. Siesky, B., Harris, A., Ehrlich, R., Kheradiya, N., Lopez, C.R.: Glaucoma risk factors: ocular blood flow. In: Schacknow, P., Samples, J. (eds.) The Glaucoma Book. Springer, New York (2010). https://doi.org/10.1007/978-0-387-76700-0_11
16. Luzhnov, P.V., Shamaev, D.M., Kiseleva, A.A., et al.: Using nonlinear dynamics for signal analysis in transpalpebral rheoophthalmography. Sovremennye tehnologii v medicine **10**(3), 160–167 (2018). https://doi.org/10.17691/stm2018.10.3.20
17. Jia, Y., Tan, O., Tokayer, J., Potsaid, B., et al.: Split-spectrum amplitude-decorrelation angiography with optical coherence tomography. Opt. Express **20**(4), 4710–4725 (2012). https://doi.org/10.1364/OE.20.004710

A New Method and Device for Differentiating Elastic and Resistive Properties of the Respiratory System

Krzysztof Jakub Pałko(✉), Tomasz Gólczewski, Maciej Kozarski, Barbara Stankiewicz, and Marek Darowski

Nalecz Institute of Biocybernetics and Biomedical Engineering, Polish Academy of Sciences, Warsaw, Poland
kpalko@ibib.waw.pl

Abstract. Obstructive lung diseases can be caused by bronchi narrowing or loss of lung tissue elasticity or both. The aim of the work was to develop a new measurement method and an equipment which enables to evaluate the obstruction causes. This Added Compliance and Resistance Method (ACRM) determines the fundamental parameters of the respiratory system mechanics that are the total respiratory system compliance (C_{rs}) and resistance (R_{rs}). Each case of obstruction can be characterized by the corresponding point location on the R_{rs}-C_{rs} plane. ACRM was verified by means of experiments performed with the use of our artificial patient developed previously, a commercial spirometer and the developed equipment. ACRM was compared with forced spirometry being the fundamental screening method for diagnosis of obstructive lung diseases and basing on the forced expiratory volume in 1st second (FEV1). Spirometry could not distinguish between pulmogenic and bronchogenic causes of an obstruction. For example, the value of FEV1 did not enable to differentiate moderate, mild and no bronchial obstruction from severe, moderate and mild obstruction, respectively, accompanied with twice increase of C_{rs}. On the other hand, a C_{rs} fall increased FEV1, and thus FEV1 underestimated the bronchial obstruction severity if the severity was determined on the base of its value. In contrast, ACRM could determine the cause of an obstruction case and the higher the severity of the obstruction, the better the differentiation. Concluding, the proposed method should supplement the forced spirometry to determine the patient state more precisely.

Keywords: Added Compliance and Resistance Method · Screening · Total respiratory system compliance · Total respiratory system resistance

1 Introduction

In functional diagnostics of the respiratory system, the emergence of new research techniques has contributed to significant progress in medicine, particularly, in pulmonology [1–3]. These new techniques under development are useful not only for pulmonologists, but also for cardiologists, surgeons and even general practitioners, as well as for education.

J. Korbicz et al. (Eds.): PCBEE 2019, AISC 1033, pp. 35–44, 2020.
https://doi.org/10.1007/978-3-030-29885-2_4

Among respiratory system diseases, a significant increase in the number of obstructive and restrictive diseases has been observed. The chronic inflammatory process in bronchi during the chronic bronchitis disease, associated with bronchial asthma, leads to airway remodeling over many years, resulting in a permanent (poorly reversible) disturbance of airflow through the airways. This leads to the formation of air traps, i.e., exclusion of parts of the lungs from the gas exchange process, development of emphysema, irreversible damage to the lung structure. They involve increased air spaces located peripherally from the final bronchiole, and are associated with alveolar wall destruction. Diseases of the airways and other structures of the lungs are known as chronic respiratory diseases. Some of the most common are chronic obstructive pulmonary disease (COPD), asthma, occupational lung diseases and pulmonary hypertension. According to WHO estimates, 65 million people in the world have moderate or severe COPD. More than 3 million people died of COPD in 2015, corresponding to 5% of all deaths. COPD is now the third leading cause of death in the world (Fig. 1), after ischemic heart disease and stroke [4]. In Poland, 2 million people suffer from it, but as much as 80% do not know about this and go untreated [5]. The socioeconomic status of the patients has a significant impact on coping with this disease [6].

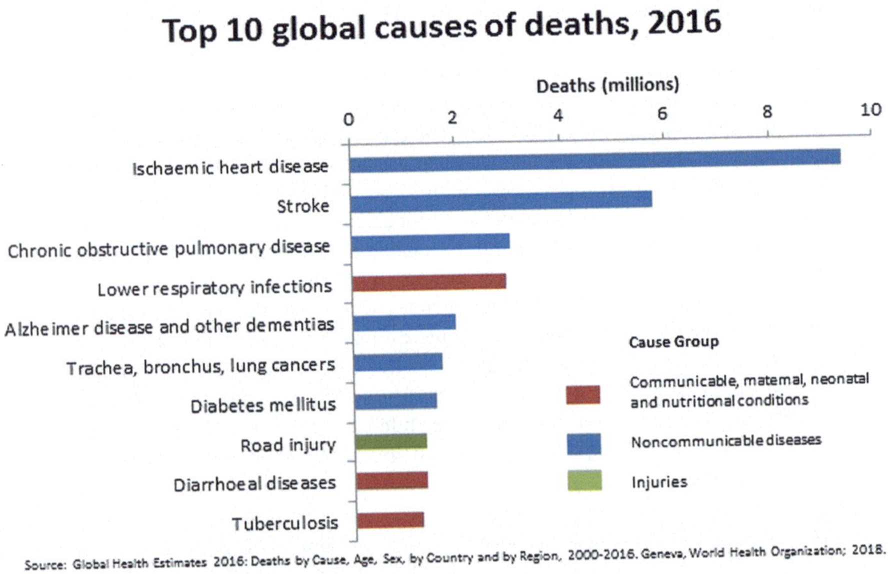

Fig. 1. WHO data about the causes of deaths where COPD ranks third [4].

Forced spirometry is a fundamental, relatively easy diagnostic method of COPD. It enables to determine the air flow-volume dependence during forced expiration. The average airflow in the first second of expiration (FEV1) is the main parameter used in diagnosis of airway obstruction. Obstructive diseases are associated with a FEV1 value decrease. For the purpose of diagnostic assessment, the ratio of FEV1 and the forced vital capacity (FVC) is used. If FEV1/FVC is below the lower limit of normal, an obstructive disease is diagnosed. In such a case, the disease severity is determined by the ratio of the measured FEV1 value to the FEV1 value predicted for patient's age, height, sex and ethnicity (Table 1).

COPD is a complex disease caused by bronchial obstruction, leading to an increase in airways resistance (Raw), or by emphysemic changes of the lung parenchyma, leading to a decrease of lung tissue elasticity, or usually by both pathologies. The processes are irreversible in the chronic stage and lead to permanent lung degeneration. Therefore, early detection and proper treatment of this disease are very important.

A serious disadvantage of spirometry is its inability to objectively assess COPD severity resulting from lung elasticity decrease (lung compliance increase), so important for diagnosis and monitoring of treatment effectiveness.

This paper presents a new method of determining the basic parameters of respiratory mechanics, i.e., total respiratory system (lung and chest) compliance (C_{rs}) and total respiratory system resistance (R_{rs}). The method is based on electrical circuit modelling of the respiratory system, with added compliance (pneumatic condenser) and/or resistance (pneumatic resistor). The presented method and device enable assessment of resistive and elastic properties of the respiratory system, and are intended as a practical tool for screening and diagnostics of lung and bronchial diseases. They could be considered as a useful supplement to commonly used spirometry.

2 Materials and Methods

A method of measuring the mechanical parameters of the respiratory system was developed at the Nalecz Institute of Biocybernetics and Biomedical Engineering, Polish Academy of Sciences (IBBE PAS). It consists of a pressure method of added compliance using computer data analysis. Research into this method followed from earlier works by Darowski et al. on the flow method of added compliance [7, 8].

The preliminary study of this method was carried out on a simple linear RC model, describing the mechanics of the respiratory system during artificial ventilation [8, 9]. It reflects the fundamental features of the real respiratory system (R_{rs} – the resistance of the respiratory system characterizing energy dissipation, C_{rs} – the total compliance of the lungs and thorax characterizing potential and kinetic energy accumulation).

The added compliance method described in [9–13] was a subject of two research projects. It consists of determining the C_{rs} and R_{rs} values.

In this study, it is assumed that the measurement is performed at the maximal lung volume, i.e. after maximum inspiration; on the one hand, the C_{rs} and R_{rs} values depend on the lung volume due to nonlinearity. On the other hand, FEV1 is the volume exhaled after maximal inspiration, and the proposed method is intended to help to interpret FEV1.

2.1 The Idea of Added Compliance and Resistance Method (ACRM)

By modifying the previously tested method of the added compliance using the added resistance maneuver (presented in Fig. 2), based on the interrupter technique proposed over 80 years ago [14] but developed in recent years, simulation tests were carried out as described in detail in [15].

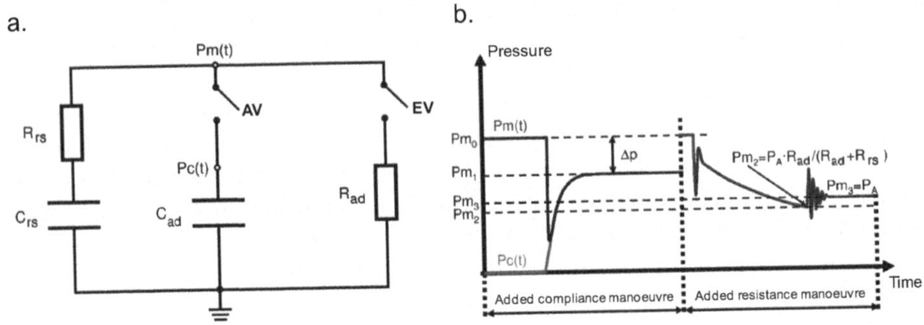

Fig. 2. The ACRM idea; (a) Electrical analogue of the ACRM, where C_{rs} - compliance of the respiratory system, C_{ad} - added compliance, R_{rs} - total resistance of the respiratory system, R_{ad} - added resistance, AV and EV - added compliance and exhalation valves, respectively; (b) Time courses of the pressure during test using the developed method. Pm(t) - pressure measured in the mouthpiece, Pc(t) - pressure measured in C_{ad}; Pm_0, Pm_1, Pm_2, Pm_3 - pressure values used to determine the parameters C_{rs} and R_{rs}, P_A - alveolar pressure. Both maneuvers are performed in the early phase of quiet exhalations.

The following Eqs. (1) and (2) are used to calculate C_{rs} and R_{rs} parameters respectively.

$$C_{rs} = C_{ad} \cdot Pm_1 / \Delta P \tag{1}$$

where: C_{ad} - the added compliance, Pm_0 and Pm_1 - the pressure values at the steady state, at the end of inhalation and after opening the added compliance valve (AV) when the expiratory valve (EV) is closed (Fig. 2), respectively, ΔP - the pressure drop between Pm_0 and Pm_1.

$$R_{rs} = R_{ad} \cdot (P_A - Pm_2)/Pm_2 \tag{2}$$

where: R_{ad} - added resistance, P_A - alveolar pressure, P_{m2} - pressure at the steady state after opening the added compliance valve (AV) (Fig. 2), and after opening EV valve for approx. 250–300 ms, and then the valve is closed again [15]. In the absence of air flow (occlusion before and after closing the valve) the pressure measured in the mouthpiece stabilizes and is equal to alveolar pressure P_A (Fig. 2), while the change of pressure (decrease after opening and increase after closing the valve) depends on airway resistance Raw. We treat R_{ad} and R_{rs} connected in series as a pressure divider.

2.2 The Method of ACRM Validation

A previously elaborated artificial patient (AP) was used in ACRM validation (Fig. 3). AP is a hybrid, i.e., physical-numerical, model of the respiratory system. Movement of the piston that is the main part of the physical part of AP is controlled by data supplied by the virtual respiratory system (VRS) that can be used, among others, in spirometry analysis and interpretation [15, 17]. Such a way of validation seems to be much better than experiments on real patients due to both ethical and technical reasons. In particular, we know the numerical values of all VRS parameters, and thus we can compare results of measurements with this values. Such full knowledge on patients' physiological properties is impossible, and thus reliability of measurements is difficult to evaluate.

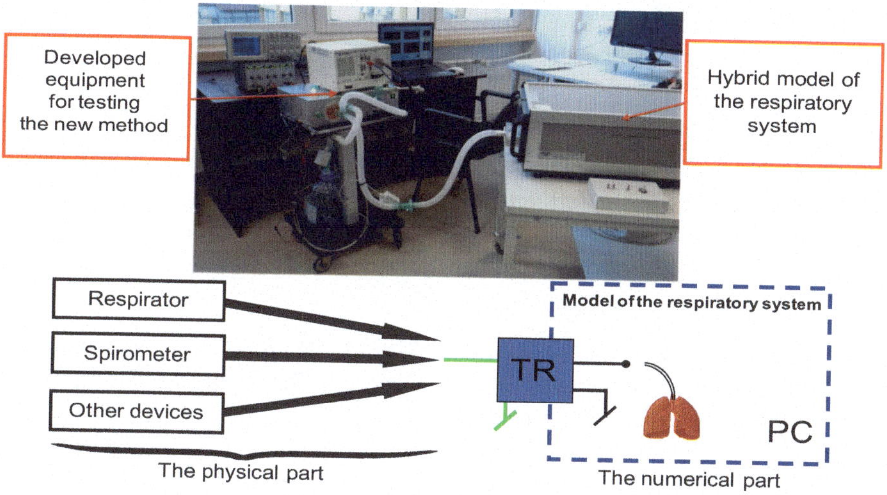

Fig. 3. The device connected to the hybrid model of the respiratory system with a built-in piston; view of the measuring stand. Hybrid model of the respiratory system, in which the TR - impedance transformer connects the physical and numerical parts.

The validation tests were carried out on an artificial population prepared to be implemented in an artificial patient according to the following steps:

- 16 combinations of age and height were randomly selected (8 for women and 8 for men);
- with the help of a computer program using equations elaborated for the Polish population [17], the predicted values of spirometric indices for these combinations of sex, age and height were found;
- using the tgol.e-spirometry computer simulation system basing on VRS [17], for each set of the predicted values, the set of respiratory system parameters was found, for which the virtual spirometry gave results equal to predicted ones;
- these sixteen sets of parameters created a population of 16 artificial healthy individuals,
- by changing parameters for healthy individuals, pathologies were introduced (Table 1);
- AP with those sets of parameters introduced into the AP numerical part created an artificial population of patients with obstructive and restrictive diseases.

Table 1. Chronic obstructive pulmonary disease severity [19].

Severity	FEV1% range
Mild obstruction	FEV1% $\geq$ 80%
Moderate obstruction	50% $\leq$ FEV1% < 80%
Severe obstruction	30% $\leq$ FEV1% < 50%
Very severe obstruction	FEV1% < 30%

FEV1% - forced expiratory volume in one second expressed as the percentage of the value predicted for a particular patient

Both initial tests [20] and the final validation of the device that is reported here were made with the use of such an artificial population.

3 Results

The results of the validation are shown in Figs. 4, 5 and 6. Tests of measurement repeatability were the first step the validation (Fig. 4). The obtained relative error was less than 10% and results were comparable.

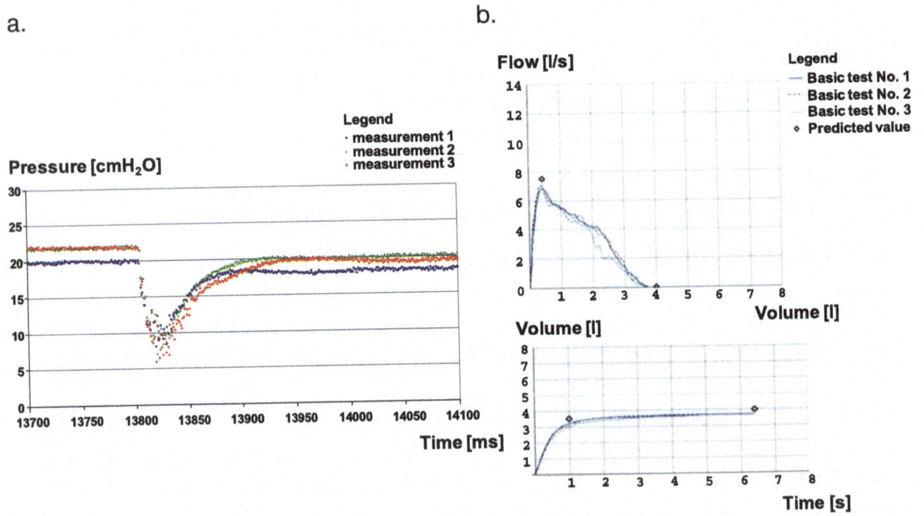

Fig. 4. Tests of the repeatability for a subject; (a) sample measurement result, added compliance maneuver; C_{rs} = 0.02 ± 0.0015 ml/cmH_2O, (b) an example of spirometry test results.

Figure 5 presents values of FEV1 for an individual artificial subject, for various elastic and resistive properties (they are quantified by means of percentages of the normal properties expected for this subject). FEV1 has the smallest value for severe obstruction and increased compliance (%C200%R800). As shown, the FEV1 value made it impossible to differentiate severe bronchial obstruction with normal compliant properties (%C100%R800) from mild bronchial obstruction with increased compliance (%C200%R400). The same problem appeared for moderate obstruction and normal compliance (%C100%R400) and those with mild obstruction and increased compliance (%C200%R200), as well as for patients with mild obstruction (%C100%R200) and people with healthy bronchi but with increased compliance (%C200%R100). As shown in (Fig. 6), ACRM can differentiate the above cases.

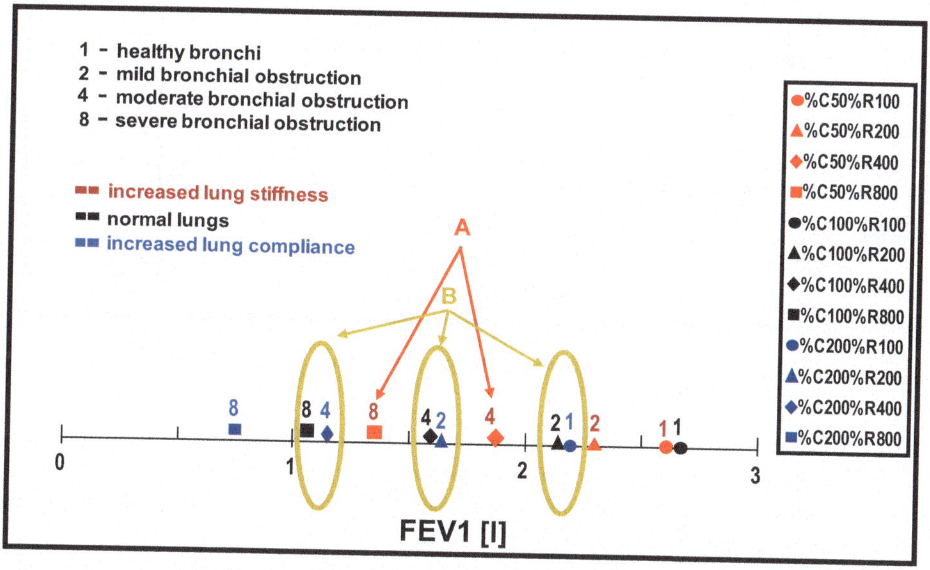

Fig. 5. Spirometry of an artificial subject for the combination of different severities of bronchial obstruction (increased R) and compliances (C) of the pulmonary tissue; (A) The decrease in lung compliance (increase in stiffness/elasticity) facilitates exhaustion in obstruction. (B) FEV1 poorly differentiates increases in airway resistance and compliance.

4 Discussion and Conclusions

In modern medicine and health policy there is a clear shift of focus from the exclusive treatment of advanced pathology to activities aimed at disease prevention [4, 21]. Prophylactic treatment has been recognized as equally and sometimes even more important than the treatment of already diagnosed diseases and health disorders. Effective long-term preventive actions require the preparation of an integrated strategy aimed at simultaneous improvement of many aspects of population health status at a population scale. Spirometry belongs to the group of listed services guaranteed by the

Ministry of Health in the field of primary health care. COPD is a serious public health problem in Poland [19]. This is one of a few chronic non-infectious diseases of the respiratory system, whose incidence still increases in Poland. As many as 2.5 million Poles suffer from COPD. Meanwhile, this disease is poorly recognized in Polish society. Only three out of a hundred Poles have even encountered the term COPD.

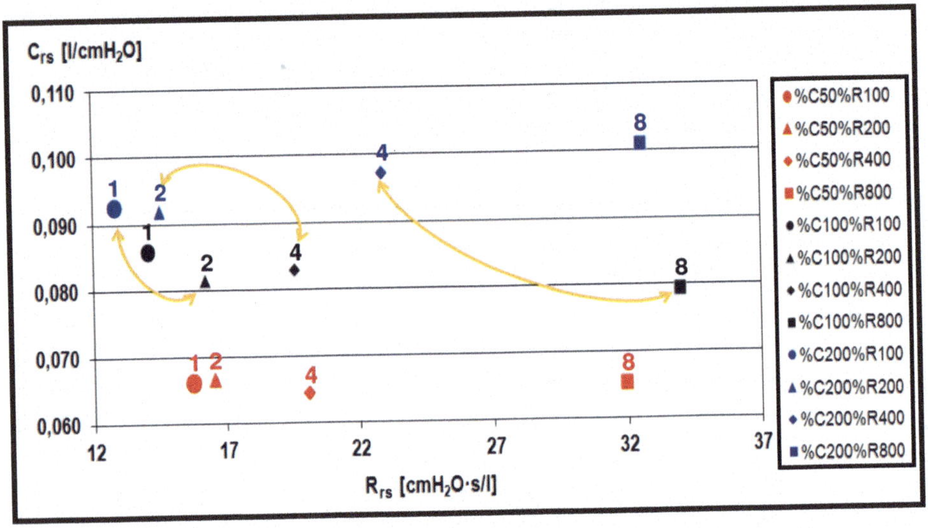

Fig. 6. Presentation of the results from Fig. 5 on the C_{rs}-R_{rs} plane.

The GOLD guidelines increasingly draw attention to the multidisciplinary approach to COPD disease and worsening the course of co-morbid disease that may affect treatment and require a comprehensive approach [21]. Correct diagnosis and treatment also positively affect the course of COPD itself.

COPD is one of the main causes of chronic morbidity and mortality in the world [4, 21]. It ranks third among the causes of deaths, and its significance is likely to grow as the population ages and continues to be exposed to risk factors. Currently, it is forecasted that COPD will be the third most common cause of death in the world. Due to the rapidly increasing morbidity and mortality because of this pathology, it is necessary to take all and appropriate actions to prevent, detect and implement a proper, multidisciplinary approach. It is important to increase public awareness and initiate discussion on solutions aimed at improving the quality of life of patients suffering from this deadly disease.

The results of the study using AP show some artifacts in the form of oscillations. They are probably related to the dynamics of the piston in the physical part of AP and the regulation error of this piston. Spirometry measurements were performed in a single maneuver procedure, because the results of previous tests showed good measurement repeatability on the simulator with the connected spirometer (Fig. 4).

By changing compliant and/or resistive properties, it was possible to simulate different lung pathologies and assess how the developed device realizing ACRM operates (Fig. 6). As shown, forced spirometry, in which only the values of FEV1/FVC and FEV1 are taken into account, slightly differentiate increased resistance and compliance. Additionally, restriction caused by decreased compliant properties increases FEV1 (Fig. 5). Moreover, comorbidity of obstructive and resistive pathologies makes forced spirometry useless because a restrictive disease decreases FVC, bronchial obstruction decreases FEV1, and thus their ratio, i.e., FEV1/FVC, may have a correct value, i.e. the value above the lower limit of normal. For those reasons, supplementation of forced spirometry with the proposed ACRM makes it possible to interpret patient's state better. The validation of results confirmed in a documented manner and in accordance with the assumptions that the patented ACRM method and the device [22] can be used to differentiate elastic and resistance properties as a supplement to spirometric parameters in the assessment of lung diseases.

We conclude that the Added Compliance and Resistance Method enables differentiation of lung pathologies related to disturbed resistive or elastic lung properties. It may also be complementary to spirometry tests. The artificial patient seems to be a good tool for preliminary validation of measurement methods due to the full knowledge on its parameters.

Acknowledgements. Authors of this paper would like to express their thanks to Dr. Krzysztof Zieliński and Zofia Sawecka-Gólczewska, M.Sc. for help in the implementation of measurements, and special thanks to Prof. Tadeusz Pałko for providing equipment and a laboratory.

References

1. Kowalski, J., Radwan, L.: Development of respiratory function investigation in Poland - historical review. Adv. Respir. Med. **77**(5), 487–493 (2009)
2. Alejos-Palomares, R., Ramírez Cortes, J.M., Domínguez-Martinez, N.: Digital spirometer with LabView interface. In: 18th International Conference on Electronics, Communications and Computers, CONIELECOMP, Cholula, pp. 105–110. IEEE (2008)
3. Shaikh, J.G., Daimiwal, N.M.: Respiratory parameter measurement and analysis using differential pressure sensor. In: International Conference on Communication and Signal Processing (ICCSP), pp. 0845–0848 (2017)
4. WHO Homepage. https://www.who.int/en/news-room/fact-sheets/detail/chronic-obstructive-pulmonary-disease-(copd). Accessed 24 Mar 2019
5. Zejda, J., Brozek, G.: Chronic obstructive pulmonary disease in Poland - a need for population-based epidemiological studies. Adv. Respir. Med. **84**(4), 203–204 (2016)
6. Jankowska-Polańska, B., Kasprzyk, M., Chudiak, A., Uchmanowicz, I.: Relation between illness acceptance and quality of life in patients with chronic obstructive pulmonary disease (COPD). Adv. Respir. Med. **84**(1), 3–10 (2016)
7. Darowski, M., Rudowski, R., Rawicz, M., Hedenstierna, G.: A new method for calculating total respiratory system compliance theory and model experiments. Acta Anaesthesiol. Scand. **29**(4), 395–399 (1985)
8. Darowski, M., Gottlieb-Inacio, I., Ludwigs, U., Hedenstierna, G.: Assessment of respiratory system compliance by a flow recording method. Acta Anaesthesiol. Scand. **39**(4), 462–466 (1995)

9. Darowski, M., Pałko, K.J., Jucha, A., Kozarski, M.: Expiratory pressure curve analysis for estimation of lungs mechanics. In: Darowski, M., Ferrari, G. (eds.) Lecture Notes of the ICB Seminars Assessment of Mechanical Support of Heart and Lungs, pp. 88–97. International Center of Biocybernetics, Warsaw (2001)
10. Darowski, M., Pałko, K.J., Jucha, A., Kozarski, M.: Expiratory pressure curve analysis for estimation of lungs mechanics. Biocybern. Biomed. Eng. **23**(1), 91–102 (2003)
11. Pałko, K.J., Kozarski, M., Darowski, M.: Identification of mechanical parameters of respiratory system during ventilatory support of lungs. Biocybern. Biomed. Eng. **25**(1), 73–81 (2005)
12. Pałko, K.J.: A comparison of two methods for assessment of lung mechanical parameters. Biocybern. Biomed. Eng. **29**(3), 19–30 (2009)
13. Pałko, K.J., Rogalski, A., Zieliński, K., Glapiński, J., Kozarski, M., Pałko, T., Darowski, M.: RC model-based comparison tests of the added compliance method with computer simulations and a standard method. Polish J. Med. Phys. Eng. **13**(1), 13–21 (2007)
14. Von Neergaard, K., Wirz, K.: Die Messung der Strömungswiderstände in den Atemwegen des Menschen, insbesondere bei Asthma und Emphysem. Z Klin Med 205, H. 1/2., 51–82 (1927). (in German)
15. Gólczewski, T., Pałko, K.J.: A method for quantification of lung resistive and compliant properties for spirometry interpretation support-tests on a virtual patient. Biocybern. Biomed. Eng. **33**(3), 136–144 (2013)
16. Gólczewski, T., Darowski, M.: Virtual respiratory system for education and research: simulation of expiratory flow limitation for spirometry. Int. J. Artif. Organs **29**(10), 961–972 (2006)
17. Gólczewski, T., Michnikowski, M., Darowski, M., Tomalak, W.: Tgol.e-spirometry: a system for e-learning of spirometry interpretation. http://virtual-spirometry.eu/en/index.html. Accessed 24 Mar 2019
18. Lubiński, W., Gólczewski, T.: Physiologically interpretable prediction equations for spirometric indexes. J. Appl. Physiol. **108**(5), 1440–1446 (2010)
19. Śliwiński, P., Górecka, D., Jassem, E., Pierzchała, W.: Guidelines for chronic obstructive pulmonary disease. Adv. Respir. Med. **82**(3), 227–263 (2014)
20. Pałko, K.J., Gólczewski, T., Zieliński, K., Kozarski, M., Darowski M.: Validation of the added compliance and resistance method for lung function tests on a population of artificial patients. In: XXXIX Congress of the European Society for Artificial Organs - ESAO 2012. http://www.esao2012.org/images/ESAO-Booklet.pdf. Accessed 24 Mar 2019
21. Global Initiative for Chronic Obstructive Lung Disease. Global strategy for the diagnosis, management, and prevention of chronic obstructive pulmonary disease. http://www.goldcopd.org. Accessed 24 Mar 2019
22. Gólczewski, T., Pałko, K.J.: Method and system for measuring the resistance coefficient and differential compliance of the respiratory system. Patent No. PL394157 (2011)

Image Analysis

Role of Let-7 Family miRNAs in Migration of Colorectal Cancer HCT 116 and Caco-2 Cells After Stimulation by the Adipokine Vaspin. Time-Lapse Live-Cell Microscopic Observations

Magdalena Skonieczna[1,2](✉), Dorota Hudy[1,2], Patryk Bil[1,2], Malgorzata Adamiec[1,2], Marta Stachowska[3], and Krzysztof Biernacki[4]

[1] System Engineering Group, Silesia University of Technology, Institute of Automatic Control, Akademicka 16, 44-100 Gliwice, Poland
Magdalena.Skonieczna@polsl.pl

[2] Biotechnology Center, Silesian University of Technology, Krzywoustego 8, 44-100 Gliwice, Poland

[3] District Hospital in Myszkow, Department of Internal Diseases, Aleja Wolności 29, 42-300 Myszków, Poland

[4] Department of Medical and Molecular Biology, Medical University of Silesia, School of Medicine with the Division of Dentistry, Jordana 19, 41-808 Zabrze, Poland

Abstract. Changes of transcript levels are among the first observable cellular responses to internal or external stimuli. Those changes may be caused by the well known phenomena of RNA interference, especially by directly acting, transcript-specific micro RNAs (miRNAs), capable of influencing gene expression by activation or silencing at the level of DNA or mRNA. It may therefore be crucial to estimate the expression levels of both a mRNA and of the miRNAs influencing it. Micro RNA level changes may also be relevant in changes of the influence of signalling pathways on specific genes. One of the greatest threats to developed societies nowadays is obesity, which is connected to various health problems including colon or breast cancer [9,34]. In many studies a connection between adipose tissue-derived hormonal activation and obesity can be observed. Adipokines are various regulators (growth factors, cytokines) of different signalling pathways, and could play a significant role in stimulation cell proliferation, cell cycle, angiogenesis, and also tumour growth and metastasis [21]. The let-7 miRNA family seems to be a good candidate to study adipose tissue-derived hormonal activation especially cancer migration and metastasis in cancer cells due to its involvement in conservative pathways with various nuclear factors like factor-B, Lin28/Let7 and TNF-alpha [15]. In this paper we have studied the regulatory role of miRNA let-7 in migration of HCT 116 colorectal cancer cells, and additionally in Caco-2 cells after hormonal stimulation with the adipokine vaspin. We found stimulatory effects of vaspin on Caco-2 cells at a dose of 3 ng/ml after 72 h of observations by time lapse live-cell microscopy in Wound Healing assays.

J. Korbicz et al. (Eds.): PCBEE 2019, AISC 1033, pp. 47–61, 2020.
https://doi.org/10.1007/978-3-030-29885-2_5

Keywords: Adipose tissue · Vaspin · Wound Healing assay · Live-cell time-lapse microscopic observations · miRNA let-7 family

1 Introduction

Contemporary biology has undergone a small revolution with the discovery of microRNAs (miRNAs), a class of around 22 nucleotide-long regulatory RNAs (Fig. 1). Micro RNAs are endogenously expressed, non-coding RNAs that inhibit the translation of target mRNAs by binding to complementary sites found in their 3' untranslated regions [36]. To date, thousands of miRNA genes have been identified by a combination of cloning, direct sequencing and bioinformatics techniques. Computational analysis suggests that the total number of miRNA genes may be greater than 1% of the total protein coding genes [20]. Experimental data and computational algorithms suggest that each miRNA may potentially target multiple (up to hundreds) mRNAs, which taken collectively indicates that over 30% of all genes may be regulated by miRNAs [20]. By targeting the mRNAs of protein-coding genes, miRNAs play a critical role in a variety of biological processes like development, cell growth, proliferation, migration, lineage determination and metabolism [1]. miRNAs have also been implicated in the etiology of a variety of complex diseases like Fragile X syndrome, Tourette's syndrome, Alzheimer's disease, neurodegenerative diseases and a large number of cancers with different origins [1,20,29].

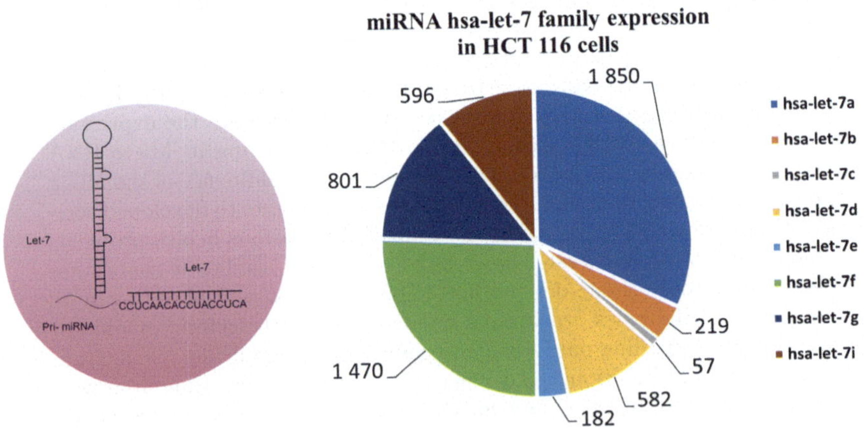

Fig. 1. Left, structure of pri-miRNA precursor for miRNA let-7 biogenesis (modified from [22]; left). Right, let-7 family expression in HCT 116 cells assayed by microarrays [2].

Examination of tumor-specific miRNA expression profiles has revealed widespread deregulations of these molecules in various cancers [19]. Moreover, it has been reported that over 50% of miRNAs can be mapped to regions in the genome associated with chromosomal locations linked to cancer [35]. Some miRNAs appear to act as oncogenes. For example, mir-21 is relatively overexpressed in glioblastoma multiform, cervical, and breast cancer [23] and increased expression of miR-21 appears to decrease apoptosis. Similarly, the locus containing seven miRNAs of the miR-17-92 polycistronic cluster is amplified in a variety of B-cell lymphomas [8]. Conversely, some miRNAs can act as tumor suppressors, genes whose deletion promotes the process of tumorigenesis. An important example is the let-7 miRNA and its target, the protooncogene Ras. Activating mutations in Ras result in increased expression, causing cellular transformation. MiRNAs from the let-7 family are down-regulated in human lung cancer, and it has been speculated that this leads to increased severity of the cancer due to misregulation of Ras [32].

The first cellular response, for any external stimuli or modification of internal equilibrium modification, could be changes on the transcriptional level. It is well known that short miRNAs are direct acting regulatory elements are short miRNAs, and play a crucial role in gene activation/silencing. Estimation of expression profiles of miRNA is important, as well as that of expression profiles of genes regulated by this miRNA. Relevant changes in miRNA level influence signaling pathways; the miRNA let-7 can act in cells through the evolutionarily conserved pathway, a feedback loop involving nuclear factor-B, Lin28/Let7 and TNF-alfa [15]. So, the let-7 miRNas family seems to be a good candidate to answer the question: is there a let-7 miRNA regulator in adipose tissue-derived hormonal activation of cancer migration and metastasis? We have estimated the level of let-7 miRNA in HCT 116 cells, with correlation to the migration potential. Additionally, in Caco-2 cells the migration was assayed after hormonal stimulation by the adipokine vaspin.

Numerous pathways targeted by let-7 can be connected to reduction of tumour severity, which may explain the fact that the let-7 miRNA group is considered a cancer metastasis stimulator, especially in breast cancer. [33] The let-7 family consists of 10 known members: let-7a, let-7b, let-7c, let-7d, let-7e, let-7f, let-7g, let-7i, miR-98, miR-202 and in normal physiological conditions they are primarily involved in gene regulation, cell adhesion and muscle formation. In case of cancer the let-7 family is associated with apoptosis, proliferation and invasiveness of cancer cells [33]. Let-7 expression is controlled at various stages of biogenesis, which involves numerous factors and signaling molecules, including adipokines. As chemoatractants, adipokines play a role in chemotaxis, which is believed to be a fundamental cause of metastasis in which external signals direct/orient and attract tumor cells. A set of adipokines, such as vaspin, visfatin, leptin or adiponektin, could be used as stimulators for miRNA expression and provide additional regulation of cancer cell migration (Fig. 2). Live single-cell observation with a time lapse system allowed for observation of microRNA-dependent microenvironment remodeling via exosomal vehicles production [17]. An understanding of the heterogeneity of tumor cells and communication among them via miRNA transfer may help in clinical treatments, and

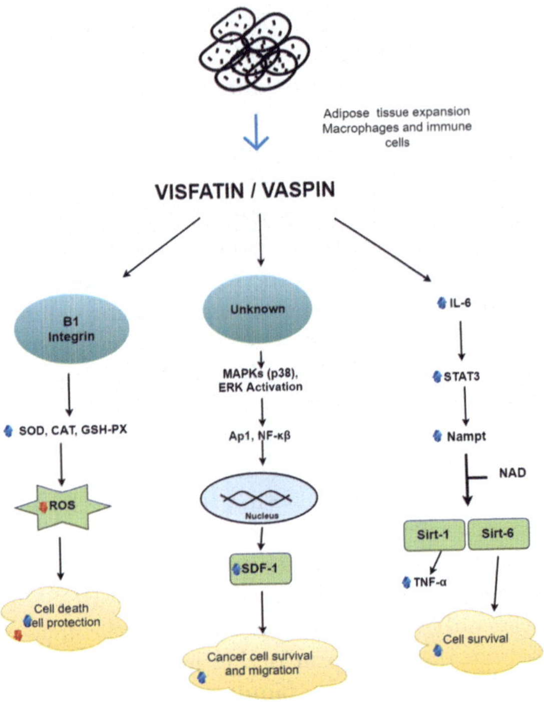

Fig. 2. Scheme of adipokine hormonal-dependent processes, e.g. migration and metastasis in cancer cells (modified from [4] and [3]).

miR-21 is already known as a regulator and promoter of migration and invasion processes in esophageal cancers [17]. It has been also reported that tumors which are able to secrete let-7-rich exosomes to the extracellular environment could maintain high carcinogenic and metastatic capacities [17]. This process is not yet fully understood, and varies in different cancer cell lines. Cancer metastasis could also be connected with adipose tissue miRNA production, such as reported for miR 1908 in mature human adipocytes [18]. Some of the adipokines, such as resistin and leptin, have been shown to down-regulate the expression of miR 1908 in human adipocytes, however there are still conflicting results with the up-regulation moieties [18]. Such inconsistent reports resulted in the investigation of regulatory models of cancer cell migration and invasiveness that includes evaluation of miRNAs after adipokines stimulation. Different miRNAs, together with specific adipokines, impact surrounding environmental niches and play a role in tissue remodeling; however in pathological situations with imbalanced physiological genes and miRNA expression it could result in disease progression like that reported in metastatic bone morrow cancers [11]. Due to these reasons a study of the influence of both adipokines stimulation and miRNA regulation in cancer cells seems to be necessary. Adipose tissue also influence s the inflammation process, which in chronic states could be a response for cancer development and progression [10]. Nucleic acids are a highly attractive class of biomarkers because of their easy and non-invasive extraction from samples (stool, blood, urine) and from tissues and cell lines models in vitro. DNA, RNA and nowadays miRNAs are widely used for prognosis and diagnosis [30].

Based on the current knowledge, a possible role of let-7 family miRNAs in migration and metastasis in cancer cells after adipokine stimulation was studied. Enhanced secretion of adipose-derived hormones like growth factors and proinflammatory cytokines allowed to focus on elements for studies of the pathogenesis of tumor growth, increased cell migration, and subsequently cancer metastasis (Fig. 2). After hormonal stimulation the phenotype of the cells changed radically, and only live-cell and time-lapse microscopy in vitro allowed to study the behavior of single cells or whole populations. Using a 72 h Wound Healing assay, vaspin [3 ng/ml] stimulated proliferation and migration of Caco-2 cells. In HCT116 cells the level of let-7 miRNA was estimated to study its role in cell migration. The results presented combine vaspin and miRNA let-7 cross-interactions in stimulating migration of cancer cells.

2 Materials and Methods

2.1 Cell Culture, Cytotoxicity and Migration Assays

Vaspin was obtained from Enzo Life Sciences (cat. No. ALX-201-360-C050) and dissolved in sterile PBS (PAA). Concentrated stock solutions (0.05 mg/ml) were stored at −20 °C and working solutions were prepared in fresh growth medium before assays. In vitro cytotoxicity assays (MTT; Sigma) based on mitochondrial dehydrogenase activity enabled assessment of the effect of vaspin at different doses [0; 1; 3; 5 and 10 ng/ml] on cells during 72 h. The IC_{50} (the concentration of drug causing 50% reduction in proliferation of cells i.e. growth or viability) was calculated from the dose-effect curve using CalcuSyn Version 2.0 (Dose-Effect Analyzer for Single and Multiple Drugs software; BIOSOFT). Before assays HCT116 (ATTC) cells and Caco-2 (ECACC) were plated 24 h before assay under standard conditions in DMEM-F12 (PAA) medium supplemented with 10% FBS (EURx) and penicillin/streptomycin (Sigma) on 96-well plates (Sarstedt), with 10 000 cells per well. For migration and Wound Healing assays cell monolayers at 80–90% confluence were scratched and vaspin was added to wells at appropriate concentrations. Microscopy observations were made manually every 24 h (HCT116), or automatically every 1 h (Caco-2).

2.2 Cell Transfection and Dual Luciferase Assay

HCT116 cells in 96-well plates were transfected with 100 ng pUC19-GFP control reporter plasmid and 100 ng Renilla luciferase reporter plasmids: psiCHECK2-let-7 8x (Addgene), termed here (+) 8 x let-7, and psiCHECK2 (Promega), termed here (−) 8 x let-7 using Lipofectamine 2000 (Invitrogen). Firefly and Renilla luciferase activities were measured 24 hours after transfection with the

Dual-luciferase assay (Promega). Renilla activity was estimated on the basis of visually verified presence of GFP (Green Fluorescence Protein) in cells transfected with pUC19-GFP control plasmid (Addgene).

2.3 miRNA Real Time PCR

Let-7a was chosen as a representative member of the let-7 group due to its prominent expression in HCT116 cells among members this family (Fig. 1). Quantification of miRNA let-7a was done using specific primers (has-let-7a; Applied Biosystems) for reverse transcription reactions. TaqMan MicroRNA Assays, with primer and probe sets designed to detect and quantify mature microRNAs (miRNAs) followed by producer's protocol (Applied Biosystems). These assays detected and quantified small RNA in 1 to 10 ng of total RNA with a dynamic range of greater than six logs. When the assays were used for microRNA analysis they enabled discrimination of mature miRNA sequences from their precursors (Fig. 1). For normalization of expression levels, reference U6 snRNA was used. Real time PCR reactions were performed using a 7900 HT Fast Real-time PCR System (Applied Biosystems) and the results were pre-analyzed with SDS 2.4 and RQ Manager 1.2.1 software. Final results are presented as threshold cycle Ct mean, normalized to U6, from 3 independent experiments ± SD using Microsoft Excel 2010.

2.4 Wound Healing Imaging

Standard Wound healing assays were performed to estimate migration and invasive potential after application of vaspin. Manual image acquisitions from scratched plates of HCT116 cells were made every 24 h using ZEISS MD1 Axiophot microscope, transit channel and 5x magnification. The scratched area was located using ImageJ software cells were counted. Results from each time point (0; 24; 48 and 72 h) are presented on chart as mean ± SD of time-depended confluency, mean from 3 separate image sets. An Olympus CellR microscope equipped with a temperature-controlled live-cell chamber was used for automated and time-lapse image acquisition at a predetermined position in control and vaspin treated Caco-2 cells. Images were automatically taken with 1 h acquisition intervals.

2.5 Images Pre-processing and Computational Analyses

Transparent light and phase-contrast images from time-lapse live-cell microscopy were read into Matlab R2016b software. The transparent light images were converted into negative images, and then merged with the phase-contrast images for better separation of the cells from background. In the next step the images were binarized, and through graphical operations the cell population was turned into a mask which was used for detection of the edge of cell population (border between cells and empty scratched area; Fig. 7; down). Pixels of the edge of the first image in the sequence were used as starting points of cell migration, and in the succeeding images migration pace was calculated in relation to these pixels in respective rows. The mean value of differences in positions of corresponding pixels in the mask edges was generated as a result. Final results are presented as plots for control and treated cells, displayed as mean ± SD from 10 separate image stacks.

3 Results

3.1 Transfection Efficiency

Twenty-four hours after transfection, plates ware evaluated visually for the transfection efficiency using fluorescence microscopy. Green signals emitted by HCT 116 cells (transfected by a pUC19-GFP control reporter plasmid) were counted, and expressed as the percent from 100 cells. The transfection efficiency of over 40% suggested that transfection with other plasmids, using lipofectamine would be on the same level, and this protocol was used for further experiments with measurements performed 24 h after transfection (Fig. 3).

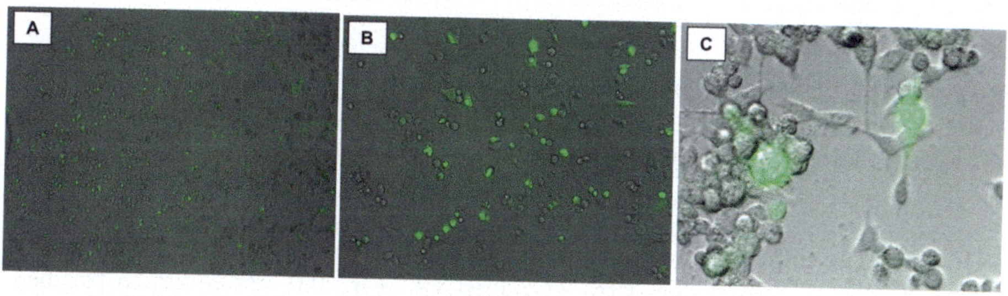

Fig. 3. Positive control transfected HCT 116 cells (green fluorescence), 24 h after transfection with plasmid pUC19-GFP. Images from FITC channel; magnification 20x (A), and 40x (B). Transmission and FITC channels merged, 100x (C).

3.2 Let-7 miRNA and Gene Expression Regulation

To monitor Renilla and/or Firefly luciferase signals in transfected cells we observed the luminescence of both luciferases using the Dual Luciferase System. The gene of interest was cloned into the multiple cloning region located 3' to the Renilla luciferase gene and its translational stop codon. After cloning, the vector was transfected into HCT 116 cells and the Renilla luciferase gene and the gene of interest was transcribed. If a specific miRNA let-7 bound to the targeted 3'-UTR region of mRNA for the gene of interest and cleaved and subsequently degraded it, a decrease of the Renilla luciferase signal was observed (Fig. 4A). The measurements were made 24 h after transfection of HCT116 cells, defined

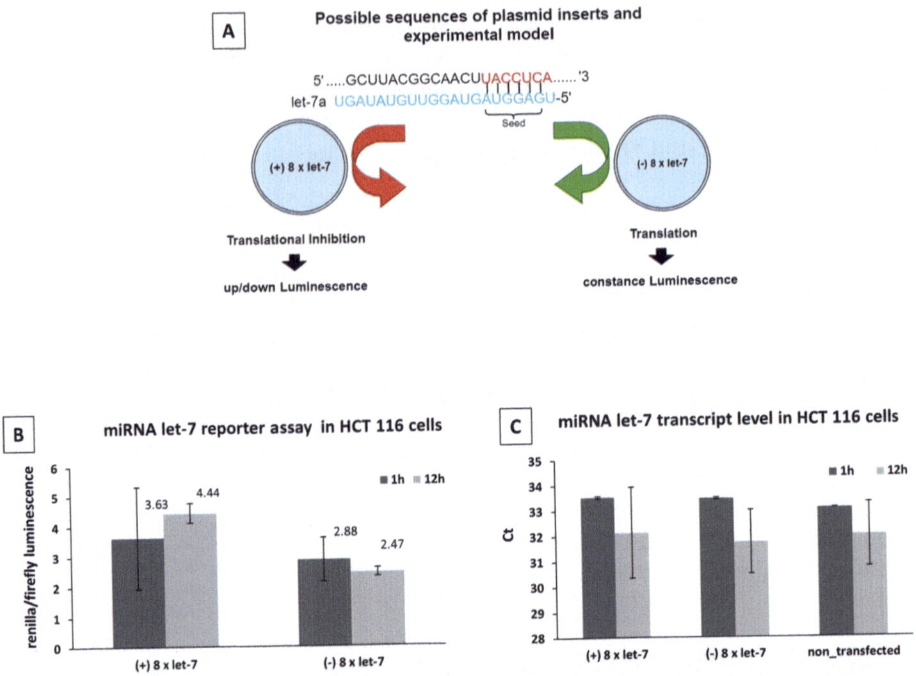

Fig. 4. (A) Let-7 miRNA interactions with reporter genes (modified from [25]); (B) Dual report assay for let-7 miRNA gene regulation at 1and 12 h after transfection; (C) let-7 miRNA level. Data show mean ± SD from 3 experiments.

as 1h, when the luminescence of reporter Renilla luciferase was similar to that in cells with plasmid unregulated by let-7 miRNA (Fig. 4B). However, in the next 12 h the let-7 regulated luciferase signal (+) 8 x let-7, increased from 3.63 at 1 h to 4.44 at 12 h (Fig. 4B). The signal for unregulated luciferase (−) 8 x let-7 was stable (Fig. 4C). Expression of let-7 miRNA decreased insignificantly after 12h, but for the whole populations of transfected with (+) 8 x let-7, (−) 8 x let-7 and untransfected cells let-7 miRNA expression levels at 1 and 12 h were similar (Fig. 4C). These findings suggested that in HCT 116 cells transfected with the reporter gene, after 12 h a lower level of miRNA let-7 did not decrease Renilla luciferase production. The constant level of let-7 miRNA suggests that let-7 miRNA may be used in regulation of other cellular migration process.

3.3 HCT 116 Migration Assay

Standard Wound Healing assays showed the natural potential of HCT 116 cells for proliferation and migration; an increased number of cells within scratched area was detected in long-time observations (Fig. 5, left). At confluence (over 60 cells/scratched area) cellular proliferation was slower at 48 h, contact inhibition and cellular death stopped migration and proliferation at 72 h (Fig. 5, right).

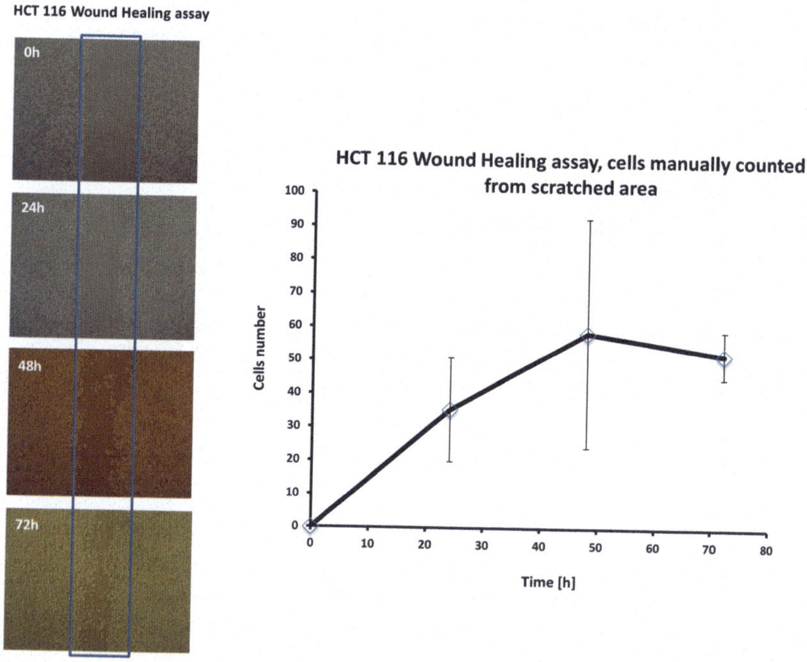

Fig. 5. Left, typical images from HCT116 Wound Healing assays; transmission microscopy; magnification 5x. Right, manually counted cells from the scratched area. Data show mean ± SD from 3 experiments.

Because of its stable level in HCT 116 cells (Fig. 4), miRNA let-7 played a possible role in such crowded population and could have inhibited cell motility [13].

3.4 Cytotoxicity Assay of Vaspin in Caco-2 Cells

Vaspin showed dose-dependend antiproliferative activity on Caco-2 cells with an $IC_{50} = 2.8$ ng/ml (Fig. 6). Higher doses started to be lethal and were not used for long-time microscopic observations.

3.5 Hormonally-Stimulated Migration

Wound Healing assays for observation of cell populations by time-lapse image acquisition were automised, so that with image pre-processing, cells movements were perfectly detected at each position every 1 h. The scratched area was monitored during 72 h and the confluency and position changes ware calculated for each image. Vaspin stimlulated the migration and proliferation of Caco-2 cells at dose of 3 ng/ml, but the differences between control and vaspin-treated populations were not statistically significant (Fig. 7).

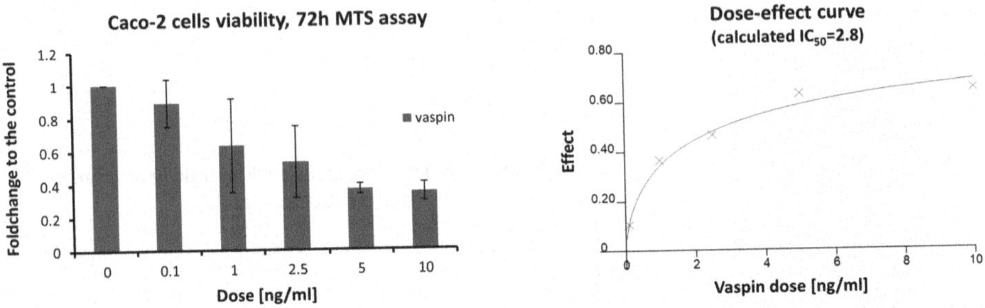

Fig. 6. Left, viability of Caco-2 cells after vaspin addition shown by MTT assays after 72 h. Right, IC_{50} calculated from the dose-effect curve using Dose-Effect Analyzer for Single and Multiple Drugs software. Values are means ± SD from 3 separate experiments

4 Discussion

One of developed countries problems is a population obesity, with accompanying health disorders, like an increased risk of colon cancer and breast cancer [9,34]. Adipose tissue delivers set of adipokines, which are regulators (growth factors, cytokines) of different signaling pathways. The adipokines could play a significant role in stimulating of cell growth, proliferation, cell cycle, angiogenesis and also tumor growth and metastasis [21]. Leptin is expressed in colorectal tumors with a greater expression in more aggressive tumors [3]. An adipose tissue, which is also a residuary of an adipose-derived stem cells (ASCs) could be a promising candidates for autologous cell-based regeneration therapies. [27] The existence of a feedback loop, involving nuclear factor-B, Lin28/Let7 and TNF-alfa has been described in ASCs transformation [15], also presumably in cancerogenesis, migration and cancer progression. Cell motility is one of the most important factors that can affect both, cell migration and invasion in cancer. A member of the let-7 miRNA family, let-7b has been reported as an inhibitor of mentioned above processes - it was observed that the motility of breast cancer cell was remarkably decreased after introducing an overexpression system for let-7b into the cells [13]. In opposite way, let-7 miRNAs are known as regulators of novel neurons in newborns and can regulate radial migration [28]. So, role of let-7 miRNAs family is wide and opposite in different molecular processes, that's why they were a good candidates to be either scavenged, or to be released via cancer cells in migration process.

Epigenetic factors, such as ionizing radiation, could modify primary structure of precursors and mature miRNAs (Fig. 1), what influence regulatory functions of these elements in gene expression [24]. We have already studied oxidative bases modifications in DNA/RNA [26] and we can conclude, that perfect matches or mismatches between nucleobases, formed by hydrogen bonds in secondary structure of nucleic acids, influence targeting 3' UTR mRNA of reporter genes by miRNA let-7. The changes are observed on the expression level [14,16,31]. After ionizing radiation, the nucleic acids interactions are modified and novel

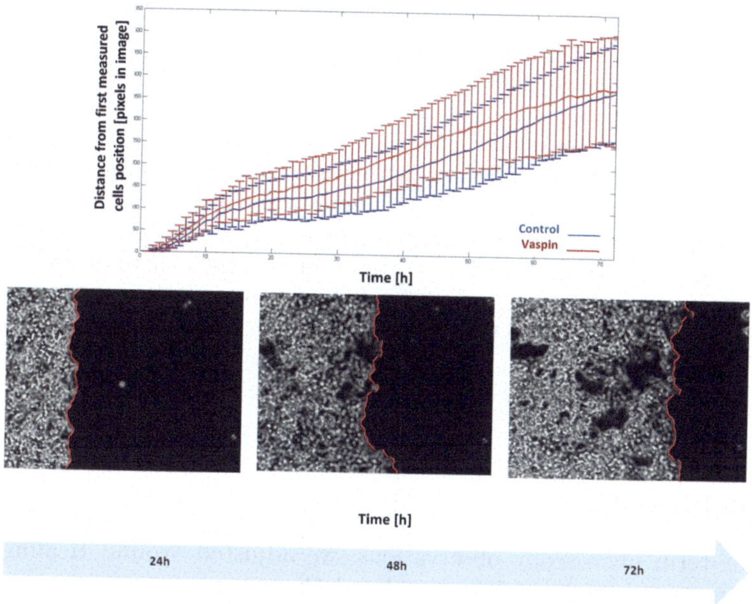

Fig. 7. Upper panel, effect of vaspin [3 ng/ml] in Wound Healing assays on Caco-2 cells. Migration rates for control and treated cells from analysis of images acquired every hour for 72 h at 10 positions. Data show mean ± SD from 3 experiments. Lower panel, images preprocessed from phase-contrast (transit channel) using ImageJ software (magnification 100x).

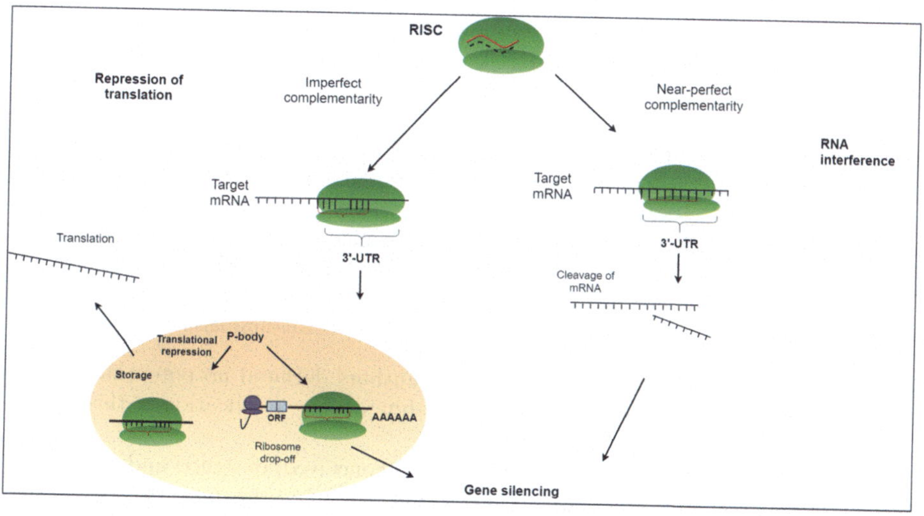

Fig. 8. Scheme of the degree of complementarity between the 3'- untranslated region (UTR) of the target mRNA and the seed region of the miRNA - mechanism of regulation (after [24]).

gene expression profile in cells occurred (Fig. 8), according to the near-perfect or perfect complementarity theory [24].

Previous findings also showed correlation, between adipokines level, visfatin, in the cells and activation of an antioxidative enzymes (catalazes, dismutases), after colorectal cancer HCT116 and melanoma Me45 cells with oxidative agents stimulation [5,6]. Cellular response, with apoptosis as main effect after adipokines-dependent stimulation, were observed and also nuclear co-localisation of visfatin, probably as an regulatory element in gene expression of treated HCT116 cells were found [7]. The correlation of previous findings with cancer migration and metastasis, corresponded with one of the main cancer cells hallmarks [12], where adipose tissue deliver to cancer cells oxidative agents, and released hormons played a role of nuclear regulators for gene expression. miRNA let-7 expressed at constant level in HCT 116 cells could be also connected to migration process.

5 Conclusions

Using long-term microscopy observations, we adjusted Wound Healing assays for multifunctional analysis. We stimulated Caco-2 colorectal cancer cells with vaspin at concentrations of 1, 3 and 5 ng/ml, and found increased migration at 3 ng/ml compared to untreated control cells (Fig. 7), confirming a correlation between adipokine concentration and cellular motility.

The results will allow for prediction of behavior of obesity-related tumor cells, their motility, migration and metastasis. The physiological aspect of such processes are still under investigation, and our novel findings improve knowledge in the area of pathogenesis of obesity and hormonal regulation of cancer. Adipose tissue, as a system for delivery of adipokines which are regulators (growth factors, cytokines), together with miRNAs, should be the object of consideration of different signaling pathways.

Acknowledgements. This work was supported by grant No. 02/010/BK_18/0102 (Malgorzata Adamiec) from Silesian University of Technology in Gliwice, Poland; and grant No. UMO-2015/19/B/ST7/02984 (Dorota Hudy and Patryk Bil) founded by the Polish National Science Center. Magdalena Skonieczna received financial support (grant No. 02/010/RGH17/0092) funded by the Rector of the Silesian University of Technology in Gliwice, Poland. Dr. Eng. Sebastian Student and MSc. Eng. Tomasz Hejmo are acknowledged for help at microscopic Wound Healing assays.

Declaration of Conflicting Interests. The authors declared no potential conflicts of interest with respect to the research, authorship, and/or publication of this article.

Author Contributions. Magdalena Skonieczna conceived, designed and carried out the experiments and wrote the paper. Dorota Hudy analyzed the data; Patryk Bil analyzed time-lapse images; Malgorzata Adamiec prepared Figures and Wound Healing assays; Krzysztof Biernacki help at experiments for gene reporter assays. Marta Stachowska was involved in revising the paper's important content. All authors read and approved the final manuscript.

References

1. Bartel, D.P.: MicroRNAs: target recognition and regulatory functions. Cell **136**, 215–233 (2009)
2. Biernacki, K., Hudy, D., Skonieczna, M., Cieslar-Pobuda, A., Gajda, K., Rzeszowska-Wolny, J.: Micro RNAS and modulation of gene expression in cells exposed to ionizing radiation. In: European Environmental Mutagen Society 42th Annual Meeting, p. 160 (2012)
3. Booth, A., Magnuson, A., Fouts, J., Foster, M.: Adipose tissue, obesity and adipokines: role in cancer promotion. Hormon. Mol. Biol. Clin. Invest. **21**(1), 57–74 (2015)
4. Bossert, N., de Bruin, D., Götz, M., Bouwmeester, D., Heinrich, D.: Fluorescence-tunable Ag-DNA biosensor with tailored cytotoxicity for live-cell applications. Sci. Rep. **6**, 37897 (2016)
5. Buldak, R., Gowarzewski, M., Buldak, L., Skonieczna, M., Kukla, M., Polaniak, R., Zwirska-Korczala, K.: Viability and oxidative response of human colorectal HCT-116 cancer cells treated with visfatin/enampt in vitro. J. Physiol. Pharmacol. **66**(4), 557–566 (2015)
6. Buldak, R., Polaniak, R., Buldak, L., Mielanczyk, L., Kukla, M., Skonieczna, M., Dulawa-Buldak, A., Matysiak, N., Zwirska-Korczala, K.: Exogenous administration of visfatin affects cytokine secretion and increases oxidative stress in human malignant melanoma Me45 cells. J. Physiol. Pharmacol. **64**(3), 377–385 (2013)
7. Buldak, R., Skonieczna, M., Buldak, L., Matysiak, N., Mielanczyk, L., Wyrobiec, G., Kukla, M., Michalski, M., Zwirska-Korczala, K.: Changes in subcellular localization of visfatin in human colorectal HCT-116 carcinoma cell line after cytochalasin B treatment. Eur. J. Histochem. **58**(2408), 239–246 (2014)
8. Calin, G.A., Liu, C.G., Sevignani, C., Ferracin, M., Felli, N., Dumitru, C.D., Shimizu, M., Cimmino, A., Zupo, S., Dono, M., Dell'Aquila, M.L., Alder, H., Rassenti, L., Kipps, T.J., Bullrich, F., Negrini, M., Croce, C.M.: MicroRNA profiling reveals distinct signatures in B cell chronic lymphocytic leukemias. Proc. Nat. Acad. Sci. U.S.A. **101**(32), 11755–11760 (2004)
9. Flegal, K.M., Kruszon-Moran, D., Carroll, M.D., Fryar, C.D., Ogden, C.L.: Trends in obesity among adults in the United States 2005 to 2014. J. Am. Med. Assoc. **315**(21), 2284–2291 (2016)
10. Ge, Q.: MicroRNAs and downstream adipokines regulated by adiponectin in vivo are novel targets for controlling adipose tissue inflammation. Thesis submitted in fulfilment of the requirements of the degree of Doctor in Biomedical Sciences Orientation: Endocrinology, Metabolism and Nutrition (2012)
11. Haider, M.T., Taipaleenmäki, H.: Targeting the metastatic bone microenvironment by microRNAs. Front. Endocrinol. **9**, 202 (2018)
12. Hanahan, D., Weinberg, R.A.: Hallmarks of cancer: the next generation. Cell **144**, 646–674 (2011)
13. Hu, X., Guo, J., Zheng, L., Li, C., Zheng, T.M., Tanyi, J.L., Liang, S., Benedetto, C., Mitidieri, M., Katsaros, D., Zhao, X., Zhang, Y., Huang, Q., Zhang, L.: The heterochronic microRNA let-7 inhibits cell motility by regulating the genes in the actin cytoskeleton pathway in breast cancer. Mol. Cancer Res. **11**(3), 240–250 (2013)
14. Hudy, D., Biernacki, K., Skonieczna, M., Gajda, K., Rzeszowska-Wolny, J.: Expression of a reporter gene containing let-7-targeted motives in cells exposed to ionizing radiation. In: 16th Gliwice Scientific Meetings, p. 131 (2012)

15. Iliopoulos, D., Hirsch, H.A., Struhl, K.: An epigenetic switch involving NF-kappaB, Lin28, let-7 microRNA, and IL6 links inflammation to cell transformation. Cell **139**, 693–706 (2009)
16. Jaksik, R., Lalik, A., Skonieczna, M., Cieslar-Pobuda, A., Student, S., Rzeszowska-Wolny, J.: MicroRNAs and reactive oxygen species: are they in the same regulatory circuit? Mutat. Res. Genet. Toxicol. Environ. Mutagen. **764**, 64–71 (2014)
17. Jiang, X., Song Hu, S., Liu, O., Qian, C., Liu, Z., Luo, D.: Exosomal microRNA remodels the tumor microenvironment. PeerJ **5**, e4196 (2017)
18. Jiang, X., Yang, L., Pang, L., Chen, L., Guo, X., Ji, X., Shi, C., Ni, Y.: Expression of obesity related miR 1908 in human adipocytes is regulated by adipokines, free fatty acids and hormones. Mol. Med. Rep. **10**, 1164–1169 (2014)
19. Kent, O.A., Mendell, J.T.: A small piece in the cancer puzzle: microRNAs as tumor suppressors and oncogenes. Oncogene **25**(46), 6188–6196 (2006)
20. Lewis, B.P., Shih, I., Jones-Rhoades, M.W., Bartel, D.P., Burgel, C.B.: Prediction of mammalian microRNA targets. Cell **115**, 787–798 (2003)
21. Li, H., Bai, E., Zhang, Y., Jia, Z., He, S., Fu, J.: Role of nampt and visceral adiposity in esophagogastric junction adenocarcinoma. J. Immunol. Res. (2017). Article ID 3970605
22. Libri, V., Miesen, P., van Rij, R.P., Buck, A.H.: Regulation of microRNA biogenesis and turnover by animals and their viruses. Cell. Mol. Life Sci. **70**(19), 3525–3544 (2013)
23. Liu, C.G., Calin, G.A., Volinia, S., Croce, C.M.: Microrna expression profiling using microarrays. Nat. Protoc. **3**(4), 563–578 (2008)
24. Lynam-Lennon, N., Maher, S.G., Reynolds, J.V.: The roles of microRNA in cancer and apoptosis. Biol. Rev. **84**, 55–71 (2009)
25. Mayr, C., Hemann, M.T., Bartel, D.P.: Disrupting the pairing between let-7 and Hmga2 enhances oncogenic transformation. Science **315**, 1576–1579 (2007)
26. Modrzejewska, M., Gawronski, M., Skonieczna, M., Zarakowska, E., Starczak, M., Foksinski, M., Rzeszowska-Wolny, J., Gackowski, D., Olinski, R.: Vitamin C enhances substantially formation of 5-hydroxymethyluracil in cellular DNA. Free Radical Biol. Med. **101**, 378–383 (2016)
27. Pérez, L.M., Berna, A., San Martín, N., Lorenzo, M., Fernández-Veledo, S., Gálvez, B.G.: Metabolic rescue of obese adipose-derived stem cells by lin28/let7 pathway. Diabetes **62**, 2368–2379 (2013)
28. Petri, R., Pircs, K., Jönsson, M.E., Åkerblom, M., Brattås, P.L., Klussendorf, T., Jakobsson, J.: let-7 regulates radial migration of new-born neurons through positive regulation of autophagy. EMBO J. e201695235 (2017)
29. Rybak, A., Fuchs, H., Hadian, K., Smirnova, L., Wulczyn, E.A., Michel, G., Nitsch, R., Krappmann, D., Wulczyn, F.G.: The let-7 target gene mouse lin-41 is a stem cell specific E3 ubiquitin ligase for the miRNA pathway protein ago2. Nat. Cell Biol. **11**(12), 1411–1420 (2009)
30. Shah, P., Cho, S.K., Thulstrup, P.W., Bjerrum, M.J., Lee, P.H., Kang, J.H., Bhang, Y.J., Yang, S.W.: Microrna biomarkers in neurodegenerative diseases and emerging nano- sensors technology. J. Mov. Disord. **10**(1), 18–28 (2017)
31. Skonieczna, M., Student, S., Cieslar-Pobuda, A., Herok, R., Jaksik, R., Rzeszowska-Wolny, J.: Oxidative RNA damage and its potential role in cellular responses to radiation. Acta Biochim. Pol. **57**(4), 131 (2010)
32. Takamizawa, J., Konishi, H., Yanagisawa, K., Tomida, S., Osada, H., Endoh, H., Harano, T., Yatabe, Y., Nagino, M., Nimura, Y., Mitsudomi, T., Takahashi, T.: Reduced expression of the let-7 microRNAs in human lung cancers in association with shortened postoperative survival. Cancer Res. **64**(11), 3753–3756 (2004)

33. Thammaiah, C.K., Jayaram, S.: Role of let-7 family microRNA in breast cancer. Non-coding RNA Res. **1**(1), 77–82 (2016)
34. Vucenik, I., Stains, J.P.: Obesity and cancer risk: evidence, mechanisms, and recommendations. Ann. New York Acad. Sci. **1271**, 37–43 (2012)
35. Woods, K., Hammond, J.M.T.S.M.: Direct regulation of an oncogenic micro-RNA cluster by E2F transcription factor. J. Biol. Chem. **282**(4), 2130 (2007)
36. Xie, X., Jun Lu, J., Kulbokas, E.J., Golub, T.R., Mootha, V., Lindblad-Toh, K., Lander, E.S., Kellis, M.: Systematic discovery of regulatory motifs in human promoters and 3' UTRs by comparison of several mammals. Nature **434**, 338–345 (2005)

Fractal Dimension Analysis of PET-MRI Liver Images for Various ROI's Sizes

Marta Borowska[1(✉)], Ewelina Bębas[1], Marcin Hładuński[2,3], Małgorzata Mojsak[2], and Edward Oczeretko[1]

[1] Department of Biocybernetics and Biomedical Engineering, Faculty of Mechanical Engineering, Bialystok University of Technology, Wiejska 45C, 15-351 Białystok, Poland
m.borowska@pb.edu.pl

[2] Laboratory of Molecular Imaging, Medical University of Białystok, Zurawia 71A, 15-540 Białystok, Poland

[3] Department of Nuclear Medicine, Medical University of Bialystok, Curie-Sklodowskiej 24a, 15-276 Białystok, Poland

Abstract. In this study, a fractal analysis is applied to Magnetic Resonance Images (MRI) of the liver simultaneously to PET imaging. The analysis was performed to examine the influence of the size of the region of interest (ROI) in calculating the fractal dimension. MRI from 31 healthy subjects were used for analysis, several or a dozen frames from each, giving a total of 426 images. Five different algorithms of fractal dimension calculation (box-counting, rectangular prism surface area, power spectral density, triangular prism surface area and the intensity difference scaling) are tested. Two sizes of ROI's were analyzed: 128×128 pixels and 64×64 pixels. The greatest similarities in the analyzed ROIs were obtained using the intensity difference scaling algorithm but others gave acceptable results.

Keywords: Fractal dimension · Region of interest · PET-MRI

1 Introduction

Fractal geometry is a powerful tool used in many fields of science and technology. Specific examples of its use can also be found in medicine. To make the correct diagnosis, in addition to the classical approach (e.g. interview), additional tests are used to provide quantitative information about the patient's condition [2,19, 20,23]. The examinations made using positron tomography are included in the imaging of the interior of the human body. Visual analysis of body PET-MRI scans, performed by an experienced radiologist, is often not sufficient for correct assessment of the subject's health. Doctors performing the analysis are able to identify only a small part of the information stored in the images, hence the need to look for the appropriate methods in the analysis of images that can help the correct diagnosis.

J. Korbicz et al. (Eds.): PCBEE 2019, AISC 1033, pp. 62–71, 2020.
https://doi.org/10.1007/978-3-030-29885-2_6

In most applications, the fractal dimension parameter (meaning the complexity of the object) of the analyzed object represented in the digital image is calculated. The most important feature of the fractal dimension is that it can be fractional. This distinguishes it from the topological dimension T that accepts integer values. The set of points forming the line has a topological dimension $T = 1$, in this case, it does not matter how complex the pattern is. Similarly, the set of points forming the surface has a topological dimension $T = 2$. This traditional approach does not provide any information about the degree of irregularity or fragmentation of a given figure. Regardless of whether the surface is flat or "rough", its topological dimension is 2. The straight line is characterized by the fractal dimension equal to the unity $D = 1$. However, the more complex the shape, the higher the fractal dimension value. For a very complex line drawn on the plane $D = 2$ and in the three-dimensional space $D = 3$. The fractal surface size is between 2 and 3. The fractal dimension is a measure of the degree to which a given object fills the space. The fractal dimension, directly related to the exponent appearing in the power relationship, is governed by a power-law [3]:

$$M(s) = const \cdot s^D$$

D (Hausdorff-Besicovitch dimension) is contained between the topological dimension of the fractal and the Euclidean dimension of the space, $M(s)$ - a measure, s - scale. Hausdorf-Besicovitch's definition is too complicated for practical estimation, but it could be used in the creation of other algorithms. Algorithms for estimation fractal dimension use geometrical or stochastic approach. The following are distinguished among geometric algorithms: planar triangles method [8,11], covering blanket method [1], at structuring element method [16], box dimension and lacunarity. Statistical algorithms include: variogram [4,7], power spectrum [12,21]. All of them were used in the medical image analysis [5,6,9,14,15,17]. One of the major problems encountered during the image analysis is a choice of parameters describing the image for the proper tissue characterization. There are no such parameters that fit each problem, regardless of imaging techniques, acquisition conditions, or the organ. The parameter characterizing the tissue can be evaluated taking into account its stability (instability). If changes in the region of interest (ROI) of a homogeneous image cause a significant change in the value of the parameter, then this parameter can be considered unstable and thus unusable in the description of a given type of image. This is important because the calculated ROI size is often different for different patients. The instability of a parameter and its dependence on the size of the ROI can be determined by the coefficient of variation, which is calculated from the obtained values for different ROI in the same homogeneous image. It expresses the ratio of coefficient variation (standard deviation) in comparison with the absolute value of its mean. The coefficient of variation increases with increasing instability of the parameter. It should not exceed the set threshold.

In this paper we analyzed fractal dimension of PET-MRI liver images calculated by five algorithms: box-counting method, rectangular prism surface area method, intensity difference scaling method, power spectral density method and triangular prism surface area method in the two sizes of ROI: 128 × 128 and 64 × 64 pixels in order to assess both the stability of the calculated fractal dimension and its dependence on the ROI size.

2 Materials and Methods

2.1 Experimental Setup

Magnetic resonance images from 31 patients without liver disease were used for analysis, several or a dozen frames from each, giving a total of 426 images. The data came from the Laboratory of Molecular Imaging, Medical University of Bialystok and were performed on a Magnetom Biograph mMR (Siemens Healthcare, Erlangen, Germany). MRI thorax imaging was performed simultaneously to PET imaging the following sequence protocol post contrast a transversal T1_vibe breath hold (TR 4.02 ms, TE 1.6 ms, 4 mm slice thickness, FoV 400 mm, matrix size 256, GRAPPA factor 2). The images for analysis were selected by two independent specialists. The region of interest cut from each frame was 128 × 128 pixels. Then, each of them was divided into four ROIs, 64 × 64 pixels each. The next step was to conduct the normalization process and to calculate the fractal dimension using the five methods mentioned. In the case of the box method, the binarization of the image by Otsu [18] was also performed before the calculation. Schema of the analysis process are presented on Fig. 1.

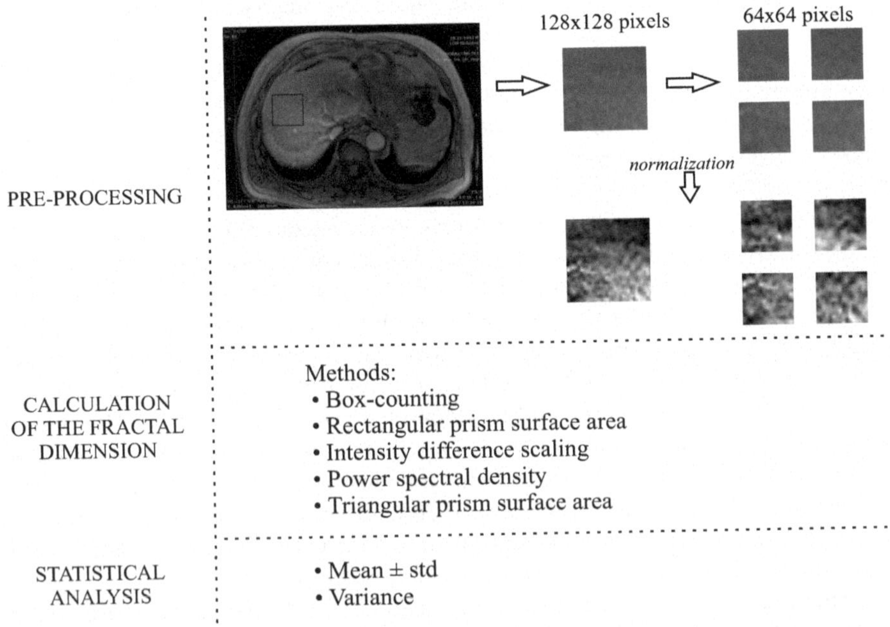

Fig. 1. Schema of analysis process

2.2 Methods of Estimation the Fractal Dimension

Box-Counting Method (BC). The box-counting dimension is based on the concept of counting the so-called boxes that cover the tested collection. This method requires binarization. It is determined by the formula [1]:

$$D_b = \lim_{\varepsilon \to 0} \frac{\ln(N(\varepsilon))}{\ln \frac{1}{\varepsilon}} \tag{1}$$

where: ε - side length of the grid element, $N(\varepsilon)$ - number of grid elements covering the image.

The algorithm for calculating a fractal dimension using the box-counting method [10] is as follows:

1. cover the object with a square grid with the side of the grid ε,
2. determining the number of meshes in which there are elements of the tested object - $N(\varepsilon)$,
3. remembering the above values,
4. repeating points 1–3 for a few or a dozen or so decreasing values of ε,
5. drawing the "fractal plot" putting on the axis of ordinates $\ln N(\varepsilon)$, and on the axis of abscissae $\ln \frac{1}{\varepsilon}$,
6. plotting a simple regression of the points obtained - its inclination is determined by the box dimension.

Rectangular Prism Surface Area Method "Skyscraper Method" (RPSA). The method is based on modeling the image using cuboids using pixel gray degrees. The fractal dimension D_r is calculated from the dependence of [8]:

$$S(\varepsilon) = c \cdot \varepsilon^{2-D_r} \tag{2}$$

where: $S(\varepsilon)$ - surface, which is a collection of cuboids with a base ε^2 and height $I(i,j)$, c - fixed scaling.

The algorithm for calculating a fractal dimension using the rectangular prism surface area method is as follows:

1. assigning a cuboid image of $\varepsilon \times \varepsilon \times I(i,j)$ to each matrix element $(i,j) - I(i,j)$ is the intensity (gray level) at the coordinates (i,j),
2. determination of the surface $S(\varepsilon)$ as a set of cuboids with base ε^2 and height $I(i,j)$, which is described by the formula:

$$S(\varepsilon) = \sum_{i,j} \varepsilon^2 + \sum_{i,j} \varepsilon \left[|I(i,j) - I(i+1,j)| + |I(i,j) - I(i,j+1)| \right] \tag{3}$$

3. the calculation of $S(\varepsilon)$ for $\varepsilon = 1$, which corresponds to the width of one element of the image matrix,
4. averaging gray levels in squares with $\varepsilon = 2, 4, 8, 16$, etc.,
5. calculation of $S(\varepsilon)$ for each ε.

Intensity Difference Scaling Method (IDS). The method of calculating a fractal dimension based on intensity differences uses a multiscale intensity difference vector ($MSID$):

$$MSID = [r_i(s_1), r_i(s_2), r_i(s_3), \ldots, r_i(s_k)] \tag{4}$$

where: $r_i(s_k)$ is the average intensity difference for all pairs of pixels lying at a given distance sk,

$$r_i(s_k) = \frac{\sum_{i_1=0}^{N-1}\sum_{j_1=0}^{N-1}\sum_{i_2=0}^{N-1}\sum_{j_2=0}^{N-1}[|I(i_2, j_2) - I(i_1, j_1)|]}{n_{s_k}} \tag{5}$$

where n_{s_k} is number of pixel pairs for the sk scale and the i_1, j_1, i_2, j_2 coordinates satisfy the dependence:

$$s_k = \sqrt{(i_2 - i_1)^2 + (j_2 - j_1)^2} \tag{6}$$

The algorithm for calculating a fractal dimension using the intensity difference scaling method [10] is as follows:

1. determining all possible distance scales between matrix elements in the whole image,
2. determining the total number of different pixel pairs of scales data,
3. reducing the number of $MSID$ components by using a normalized multiscale intensity difference vector ($NMSID$), in which scales take integer values,
4. calculation of the fractal dimension D_i from dependence $\log(r_i(k))$ from $\log(k)$, where $k = 1, 2, \ldots, s$.

Power Spectral Density Method (PSD). For the image matrix NxN the discrete Fourier transform pattern DFT looks like this:

$$F(u, v) = \sum_{x=0}^{N-1}\sum_{y=0}^{N-1} f(x, y) \exp\left(-\frac{2\pi i}{N}(xu + yv)\right) \tag{7}$$

where: u, v is the so-called spatial frequencies (horizontal and vertical).

The algorithm for calculating a fractal dimension using the power spectral density method [12] is as follows:

1. power spectrum calculation $P(u, v)$ for a given image matrix,
2. power spectrum sampling in many directions $\theta = \frac{1}{\tan(u,v)}$,
3. calculation of means $P(f)$ for each f,
4. calculating β from the slope of the regression line $\log(P(f))$ as a function of $\log(f)$,
5. calculation of the fractal dimension D_s from the formula:

$$D_s = 4 + \frac{\beta}{2} \tag{8}$$

Triangular Prism Surface Area Method (TPSA). This method is based on modeling the image using triangular prisms. The fractal dimension is calculated from the dependence of:

$$S(\varepsilon) = c \cdot \varepsilon^{2-D_t} \tag{9}$$

where: D_t - fractal dimension, $S(\varepsilon)$ - surface, which is a collection of cuboids with a base ε^2 and height $I(i, j)$, c - fixed scaling.

The algorithm for calculating a fractal dimension using the triangular prism surface method [11] is as follows:

1. creation of an image matrix model by assuming that a, b, c, d represent the gray levels of four adjacent elements, and e is the average of these values,
2. determination of four triangles for each element of the size matrix ε,
3. calculation of sums of surface areas $S(\varepsilon)$ of created triangles for all elements of the matrix,
4. repetition of the above steps for different values of ε,
5. determination of the fractal dimension D_t from the slope of the log regression line $\log(S(\varepsilon))$ as a function of $\log(\varepsilon)$.

3 Results

In order to assess the fractal dimension stability and its dependency on the ROI size, the various ROI sizes were used. The results of the mean values and standard deviations of fractal analysis of various ROI sizes and their coefficient of variation are presented in Table 1. The smaller the coefficient of variation, the more stable the method is - the value is not greater than 0.01 [13]. These five methods do not indicate significant differences in mean values of fractal dimension between the 128×128 sizes of ROI and 64×64 sizes of ROI. In Figs. 2, 3, 4, 5 and Fig. 6 visualizations of fractal dimension differences between the 128×128 sizes of ROI and sizes of ROI for various methods are shown. The values of the fractal dimension for box-counting method do not indicate statistical differences, but the values dispersion is significant (Fig. 2) in comparison to other methods. The values of the fractal dimension for intensity difference scaling method (Fig. 4) and rectangular prism surface area method (Fig. 3) give a very small dispersion of values. The values of fractal dimension for triangular prism surface area method (Fig. 6) and power spectral density method (Fig. 5) are acceptable with a relatively small variation of values. The intensity difference scaling algorithm (the correlation coefficient was not greater than a fixed threshold, 0.01) proved to be the most stable algorithm for calculating fractal dimension but the others are also in the accepted threshold. The results of the fractal dimension performed using different methods were different. Differences in values of fractal dimension estimated by various algorithms are inevitable, because the value of the exponent $M(s)$ can be bound with D [10,22] in different ways. The values of the fractal dimension of liver showed tissue homogeneity.

Table 1. The values of fractal dimension calculated using different methods (BC - box-counting, RPSA - rectangular prism surface area, IDS - intensity difference scaling, PSD - power spectral density, TPSA - triangular prism surface area) for two sizes of ROI.

Method	Dimension for size of ROI			
	128 × 128		64 × 64	
	mean ± std	Coefficient of variation	mean ± std	Coefficient of variation
BC	1.760 ± 0.148	0.0117	1.754 ± 1.148	0.0044
RPSA	2.220 ± 0.047	0.0010	2.262 ± 0.055	0.0011
IDS	2.224 ± 0.038	0.0006	2.225 ± 0.051	0.0007
PSD	2.251 ± 0.092	0.0038	2.127 ± 0.128	0.0046
TPSA	2.737 ± 0.115	0.0048	2.790 ± 0.112	0.0021

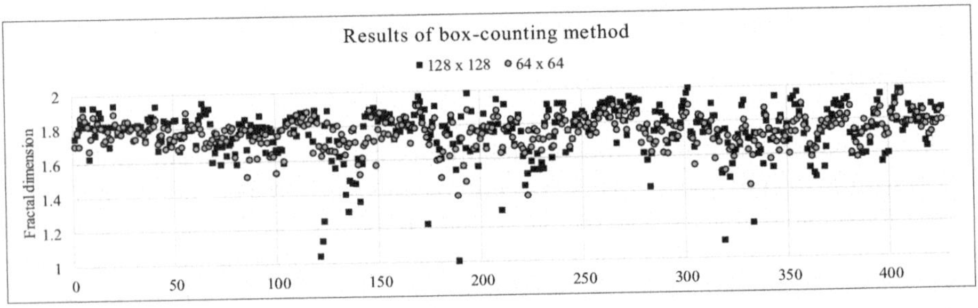

Fig. 2. Results of calculating the fractal dimension with the box-counting method

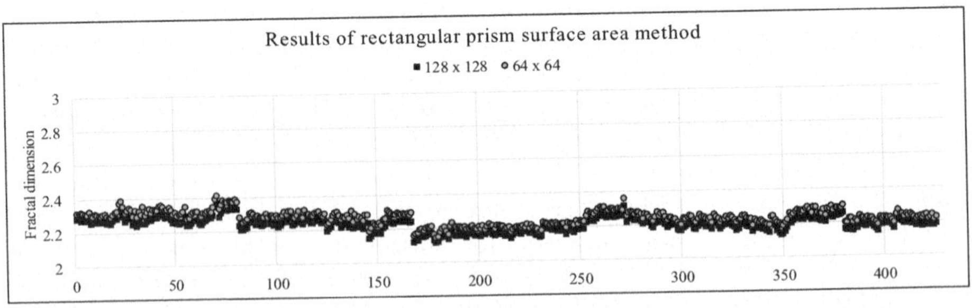

Fig. 3. Results of calculating the fractal dimension with the rectangular prism surface area method

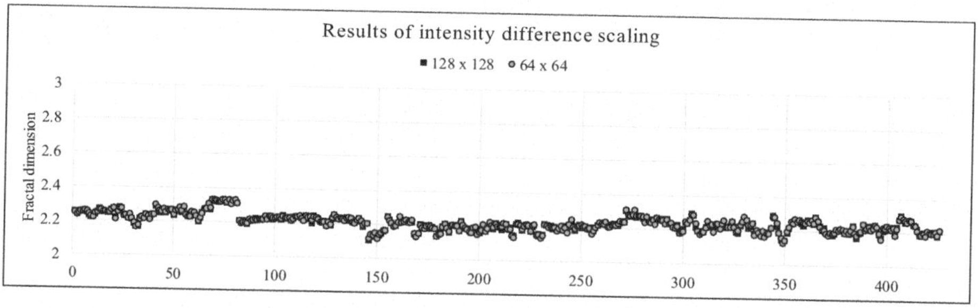

Fig. 4. Results of calculating the fractal dimension with the intensity difference scaling method

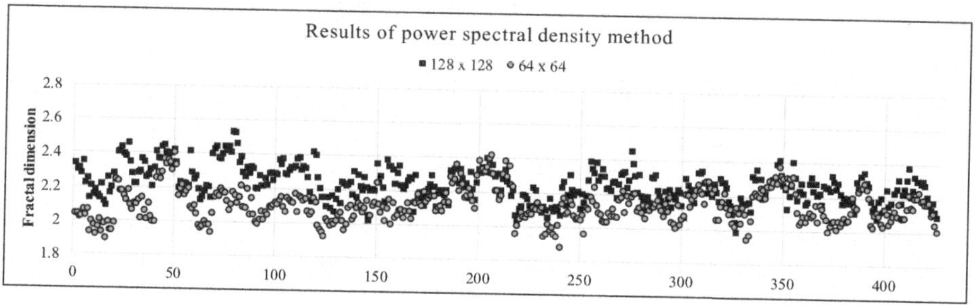

Fig. 5. Results of calculating the fractal dimension with the power spectral density method

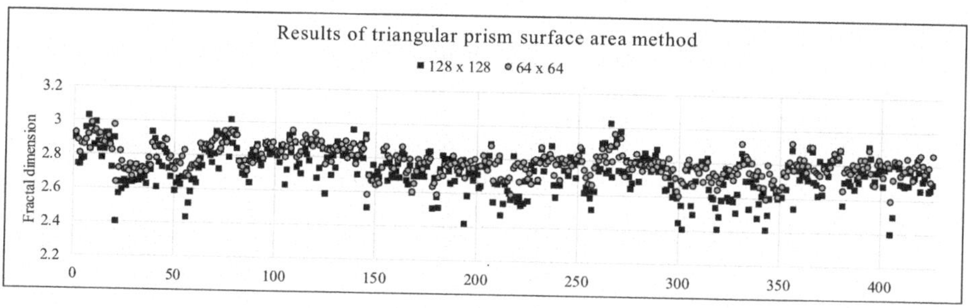

Fig. 6. Results of calculating the fractal dimension with the triangular prism surface area method

4 Conclusion

In this paper, five methods of calculating fractal dimension were presented. The advantage of these algorithms is its ability to analyze PET-MRI images regardless of the size of the analyzed area. The results obtained indicate parameters stability and independence of the ROI size. The results of calculation of the fractal dimension using different algorithms were different. This was related to different ways of combining the exponent $M(s)$ with D. Comparisons between images can be performed only when using the same method. The fractal dimension is a non-morphometric parameter, so very substantial differences can be obtained using various algorithms. The fractal dimension of medical images can be used for comparative statistics in comparing the two images or two regions of interest.

Acknowledgement. The research was performed as a part of the projects S/WM/1/2017 and was financed with the founds for science from the Polish Ministry of Science and Higher Education. This work was supported by Faculty of Mechanical Engineering, Bialystok University of Technology (MB/WM/3/2017). Research for patients was supported by Grant no. STRATEGMED2/266484/2/NCBR/2015.

Compliance with Ethical Standards. The study protocol was approved by the Local Ethical Committee of Medical University of Bialystok, Poland, and an informed consent was obtained from each patient.

Conflicts of Interest. The authors declare that there are no financial or personal relationships with other people or organizations that could inappropriately influence this study.

References

1. Ai, T., Zhang, R., Zhou, H., Pei, J.: Box-counting methods to directly estimate the fractal dimension of a rock surface. Appl. Surf. Sci. **314**, 610–621 (2014)
2. Al-Kadi, O.S., Chung, D.Y., Coussios, C.C., Noble, J.A.: Heterogeneous tissue characterization using ultrasound: a comparison of fractal analysis backscatter models on liver tumors. Ultrasound Med. Biol. **42**(7), 1612–1626 (2016)
3. Avnir, D., Biham, O., Lidar, D., Malcai, O.: Is the geometry of nature fractal? Science **279**(5347), 39–40 (1998)
4. Bianchi, F.D., Bonetto, R.D.: FERimage: an interactive program for fractal dimension, d_{per} and d_{min} calculation. Scanning **23**(3), 193–197 (2001)
5. Borowska, M., Borys, K., Szarmach, J., Oczeretko, E.: Fractal dimension in textures analysis of xenotransplants. Signal Image Video Process. **11**(8), 1461–1467 (2017)
6. Borowska, M., Oczeretko, E., Szarmach, J.: Fractal texture analysis in the irregular region of interest of the healing process using guided bone regeneration. In: Piętka, E., Kawa, J., Wieclawek, W. (eds.) Information Technologies in Biomedicine, vol. 3, pp. 103–114. Springer, Cham (2014)
7. Borowska, M., Szarmach, J., Oczeretko, E.: Fractal texture analysis of the healing process after bone loss. Comput. Med. Imaging Graph. **46**, 191–196 (2015)

8. Caldwell, C.B., Moran, E.L., Bogoch, E.R.: Fractal dimension as a measure of altered trabecular bone in experimental inflammatory arthritis. J. Bone Mineral Res. **13**(6), 978–985 (1998)
9. Chappard, D., Guggenbuhl, P., Legrand, E., Baslé, M.F., Audran, M.: Texture analysis of x-ray radiographs is correlated with bone histomorphometry. J. Bone Mineral Metab. **23**(1), 24–29 (2005)
10. Chen, S.S., Keller, J.M., Crownover, R.M.: On the calculation of fractal features from images. IEEE Trans. Pattern Anal. Mach. Intell. **15**(10), 1087–1090 (1993)
11. Clarke, K.C.: Computation of the fractal dimension of topographic surfaces using the triangular prism surface area method. Comput. Geosci. **12**(5), 713–722 (1986)
12. Dennis, T., Dessipris, N.: Fractal modelling in image texture analysis. In: IEE Proceedings F (Radar and Signal Processing), vol. 136, pp. 227–235. IET (1989)
13. Duda, D., Krętowski, M., Bézy-Wendling, J.: Computer-aided diagnosis of liver tumors based on multi-image texture analysis of contrast-enhanced CT. Selection of the most appropriate texture features. Stud. Logic Grammar Rhetoric **35**(1), 49–70 (2013)
14. Iftekharuddin, K.M., Jia, W., Marsh, R.: Fractal analysis of tumor in brain MR images. Mach. Vis. Appl. **13**(5–6), 352–362 (2003)
15. Losa, G.A., Merlini, D., Nonnenmacher, T.F., Weibel, E.R.: Fractals in Biology and Medicine. Birkhäuser, Basel (2012)
16. Maragos, P., Sun, F.K.: Measuring the fractal dimension of signals: morphological covers and iterative optimization. IEEE Trans. Signal Process. **41**(1), 108 (1993)
17. Mustonen, T., Koivisto, T., Vanninen, E., Vanninen, R., Kuikka, J.T.: Cerebral perfusion heterogeneity and complexity in patients with acute subarachnoid haemorrhage. Nuclear Med. Commun. **27**(2), 157–164 (2006)
18. Otsu, N.: A threshold selection method from gray-level histograms. IEEE Trans. Syst. Man Cybern. **9**(1), 62–66 (1979)
19. Piórkowski, A., Mazurek, P., Gronkowska-Serafin, J.: Comparison of assessment regularity methods dedicated to isotropic cells structures analysis. In: Choraś, R. (ed.) Image Processing & Communications Challenges, pp. 169–178. Springer, Cham (2015)
20. Stocker, D., Marquez, H.P., Wagner, M.W., Raptis, D.A., Clavien, P.A., Boss, A., Fischer, M.A., Wurnig, M.C.: MRI texture analysis for differentiation of malignant and benign hepatocellular tumors in the non-cirrhotic liver. Heliyon **4**(11), e00987 (2018)
21. Super, B.J., Bovik, A.C.: Localized measurement of image fractal dimension using gabor filters. J. Vis. Commun. Image Represent. **2**(2), 114–128 (1991)
22. Tolppanen, P., Stephansson, O., Stenlid, L.: 3-D degradation analysis of railroad ballast. Bull. Eng. Geol. Enviro. **61**(1), 35–42 (2002)
23. Xu, S.S.D., Chang, C.C., Su, C.T., Phu, P.Q.: Classification of liver diseases based on ultrasound image texture features. Appl. Sci. **9**(2), 342 (2019)

CNN Support to Diagnostics in Sjögren's Syndrome

Anna Korzynska[1(✉)], Jakub Zak[1], Krzysztof Siemion[1,2], Lukasz Roszkowiak[1], and Dorota Pijanowska[1]

[1] Nalecz Institute of Biocybernetics and Biomedical Engineering PAS, Ks. Trojdena 4, 02-109 Warsaw, Poland
akorzynska@ibib.waw.pl

[2] Department of Pathomorphology, Professor Witold Orlowski Public Clinical Hospital, Medical Center for Postgraduate Education, Czerniakowska 231, 00-416 Warsaw, Poland

Abstract. Sjögren's Syndrome is a systemic disease. Its diagnosis can be supported by histopathological examination of minor salivary glands. The value of the focus score equal or greater than one ensures the diagnosis of this disease. Human estimation is inherently subjective, which often leads to diverging results during the process of diagnosis. The paper proposes to use U-net CNN to find area of inflammation in WSI to support pathologist in the process of diagnosis by selecting foci for detailed inspection. Proposed neural network was trained based on tiles of size 256×256 pixels in magnification 400x from 13 digital slides stained with hematoxylin and eosin from patients with labial minor salivary gland biopsies. The ground truth was established by manual annotations done by a pathology resident. The accuracy and recall of proposed neural network model on testing dataset show the potential of machine learning for classification problem solving in this field.

Keywords: Digital microscopy · Deep learning · CNN · U-net · Sjögren's Syndrome

1 Introduction

Sjögren's Syndrome (SS) is a systemic disease, and its symptoms and signs are presented throughout the entire body. Those include a dry, gritty or burning sensation in the eyes, dry mouth, difficulty in talking, chewing or swallowing, a sore or cracked tongue, dry or burning throat, dry or peeling lips, a change in taste or smell, increased dental decay, joint pain, skin dryness, digestive problems, dry nose, debilitating fatigue [1]. The diagnosis is based on assessment of symptoms, signs, clinical tests, concentration of antibodies in blood and histopathological examination of minor salivary glands [2]. The last one mentioned is not obligatory, but may be extremely helpful [3,4]. The value of the focus score equal or greater than one strengthens the diagnosis of Sjögren's Syndrome.

J. Korbicz et al. (Eds.): PCBEE 2019, AISC 1033, pp. 72–81, 2020.
https://doi.org/10.1007/978-3-030-29885-2_7

Human estimation is inherently subjective, which often leads to diverging results during the process of diagnosis.

Machine learning approaches have been shown to improve the accuracy and automation of histopathologic slide analysis [5,6]. These types of approaches are reported to significantly reduce the tedious and laborious work of quantifications and reduce variability among pathologists [7,8].

Convolutional Neural Networks (CNNs) are being recognized as state-of-the-art in automated support in pathology domain [9]. They can be used in computer based decision support systems for digital pathology [10]. Using deep learning (DL) methods, CNNs can acquire complex histological patterns from microscopic images by finding salient features in images. The most meaningful features help solve the problems of classification, detection or segmentation of histological primitives such as nuclei, mitosis, epithelium and regions in tissue such as tumor margin [9,11].

The domain knowledge, in our cases pathology knowledge, is not necessarily implemented in the neural networks. It usually increases pathologists' mistrust for this types of method. But numerous papers show that well-trained CNNs allow to recognize patterns in images of the same type as those used during training, but which the network has never "seen" before [9]. CNN applications in digital pathology include recognition of the tumor regions [12] and its grows [10], nuclei detection or segmentation [9,12], epithelium or other structure of tissue segmentation [12], mitosis detection [9,12,13] and extraction of "hidden" tumor characteristics for biomarker development [14–17].

Our goal is to support pathological examination of minor salivary glands by finding regions of inflammatory infiltrate in Whole Slide Images (WSI) of tissue biopsy from patients presenting symptoms and signs indicating SS. In the next step of investigation we will develop an algorithm to detect lymphocytes' nuclei on the selected regions and count numbers of these nuclei. It will allow us to calculate focus score. According to PJ Waterhouse, it is defined as the number of lymphocytic infiltrate regions with more than 50 lymphocytes on the area of 4 mm^2 of tissue in the slide [18].

There are no available computer-assisted methods for the analysis of inflammation in SS patients' biopsy slides stained with hematoxylin and eosin (H&E). However, there are algorithms for finding the ratio of the number of T-lymphocytes to the number of B-lymphocytes in various stages of SS in salivary glands biopsy slides stained immunohistochemically against CD3 and CD20 using two color stains [4]. As in this investigation immunohistochemical process of lymphocytes staining is used, the quantification process is narrowed down to calculation of the ratio of number of pixels in brown and number of pixels in red colors in an image.

In the presented investigation tissue biopsies are stained with H&E. This causes lymphocytes' nuclei and other cell's nuclei present a blue tint because of the light absorption by the hematoxylin. Lymphocytes' nuclei can be detected from other cells' nuclei because of their smaller size, greater circularity in shape and greater chromatic density. As lymphocytes' cytoplasm is almost invisible around the cell's nuclei, they are mostly surrounded by background rather than pink color cytoplasm.

Currently, there are computer-assisted automatic or semiautomatic methods for the analysis of immune response in many forms of cancer e.g. breast or ovarian [9,10,12–14]. Results of proposed method should not be compared to results of this type of methods because each trained NN is adjusted to the texture of tissue of particular organ. Pattern of breast tissue with mammary glands and ducts and tissue patterns of the labial minor salivary gland with salivary glands affect clusters of lymphocytes' nuclei visible in H&E stained samples with various contents.

2 Materials and Methods

Firstly, the minor salivary gland is excised from the patient's lower lip and transferred to the pathology department as formalin-fixed tissue fragments. Based on these samples, microscopic slides of tissue fragments stained with H&E are prepared in a pathology department. These slides are examined by a pathologist at the next stage [2]. The doctor estimates the number of foci which contain at least 50 lymphocytes and the total surface of the salivary gland parenchyma. Afterwards, they calculate the number of foci per $4\,\text{mm}^2$ of area covered by parenchyma. This number is known as focus score [3,18].

For the purposes of this study, microscopic slides prepared for diagnostic purposes in previous years have been used. This study involved 13 digital slides (one slide for one patient). Prior to inclusion in the study, slides were manually reviewed to check tissue quality and confirm the presence of at least three salivary glands of appropriate size according to the guidelines [5]. After that all glass slides were digitalized using Pannoramic 250 Flash II (3DHISTECH, Budapest, Hungary) whole slide scanner. The scanner works with automatic setup which means with automatic brightness settings, 10 focus points, and with Plan Apochromat objective with magnification 40x. Digital slides were manually checked for blur artifacts, and affected slides were rescanned. As a result of extraction from all WSI, 107 fragments containing images of tissue (fragments with empty space and artifacts were removed) have been collected as tiff format images of average height 17843 ± 5494 and width 16473 ± 6037 pixels.

2.1 Ground Truth Annotations

A pathology resident with 4 years of experience provided ground truth annotations on a separate layer on these 107 images using GNU Image Manipulation Program 2 (GIMP2) software. Usually, pathologists first inspect the slides at lower magnification, then use higher one to identify areas of inflammation. On most digital slides, several chosen, not all areas of inflammation have been marked, if areas have been detected. Since accurate tracing of inflammation was very time consuming, the areas for annotation were coarse; a tool called a brush with a diameter several times larger than the average size of the lymphocyte nucleus has been used. The selected regions were saved as a separate map image specific for each slide and used during training and validation processes.

All annotated WSI fragments were randomly divided into two groups: training and testing sets of images. Training set consists of 103 images of which one is shown in Fig. 1 while testing set consists of four images each from different WSI. Two of testing imaging are presented in Fig. 2.

Based on annotated images prepared software exported regions of size 256×256 pixels in resolution 400x into separate images called tiles in semi-occluded manner with sliding windows moved by 64 pixels in x and y axis. This way, more than 150 000 images became the training set. More than 20 000 images became the testing set using the same tiling method with step equal to tile size. The tile size was chosen to ensure that nuclear features including cell organization and larger structures of minor salivary glands organization are observable in tiles, see examples of tiles in Fig. 1. In this investigation each tile from training set covers approximately 2.5×10^{-6} mm^2 area of tissue.

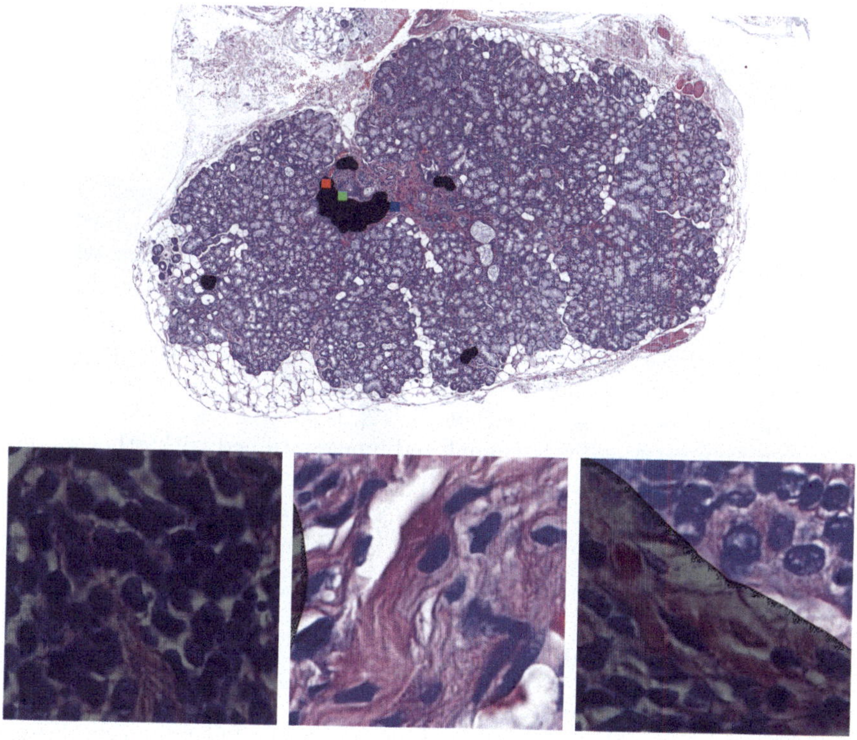

Fig. 1. Examples of tiles from training set and its location in fragment of WSI (image) with annotations. Top – full WSI fragment with annotations (in black) and with positions of presented below tiles (blue, red and green). Bottom – right tile extracted from place marked by red square; the left one – by blue square and middle by green one.

Training tiles were divided automatically into two sets: the first if they included wholly annotated region or the second if they included region partially or completely devoid of annotation. Automatically generated tiles in training sets were reviewed by the pathologist whether they contained clearly readable and high-quality tissue areas. Selected tiles have been used to train CNN for classification task.

2.2 Image Tiles Augmentation

Training of a CNN for a classification task requires thousands of training tiles but collecting image data originating from big number of WSI is costly and time consuming. To reduce the effort needed to acquire additional training data, improve the robustness and generalization and decrease the risk of overfitting, image augmentation is typically used. In this investigation the augmentation was done by manipulating a single tile's orientation because the orientation augmentation is justified in this type of images by the fact that the location of inflammatory areas have no inherent orientation with respect to imaging. In this study some tile image have been either rotated by 0, 90, 180 or 270° or sheared by 0.1 in either left or right direction (opencv nomenclature), thus changing the shape of the annotation.

2.3 CNN Training

Within neural networks, there is a group of architectures, called Convolutional Neural Networks. These are especially useful when analyzing images, which contain information with strong spatial correlation between pixels. The deep learning technologies abounds in numerous CNN architectures. In this investigation the U-net CNN [19] was chosen because it has been successfully used in tasks with classification and segmentation problems [11,20] and applied to other problems from not only the field of digital pathology [12]. It was decided to train NN from scratch because pre-trained CNNs analyzed images stained with other methods, most often using immunohistochemically stained tissue samples. This lead to CNNs focusing on different image features, than was required. The process of training from scratch was possible because of use two Tesla v100m cards with 32 GB RAM each.

2.4 Evaluation of Results

To evaluate proposed trained CNN suitable software has been written. To use trained CNN to detect the regions of inflammation, the following procedure has been prepared:

1. extraction of tissue area in a fragment of WSI and its division into separate sub-images of the size of training tile,
2. individual tile classification by proposed U-net CNN,
3. construction of the map of inflammatory foci based on all classification results.

To evaluate proposed trained CNN for classification task the quality of classification measures defined below have been used: accuracy, F-score and recall. The value of these measurements were calculated separately for each fragment (image) of WSI from testing set.

The measures are defined as:

- F-score:

$$F = \frac{2TP}{2TP + FP + FN} \tag{1}$$

- Accuracy:

$$A = \frac{TP + TN}{2TP + FP + FN} \tag{2}$$

- Recall:

$$R = \frac{TP}{TP + FN} \tag{3}$$

By superimposing the inflammatory infiltration map outputted by the CNN onto a corresponding ground truth mask it was possible to count true positive (TP), true negative (TN), false positive (FP) and false negative (FN) pixels.

Calculations, augmentation and training were performed using Python language (3.6+) and tensorflow and keras packages.

3 Results

3.1 Training

The process of training took 37 epochs (training through the entire dataset), each with 19 435 steps (training through the batch of images) with batch size (number of images used for training at once) 8. The model reached loss measured by binary crossentropy: 0.0187, and accuracy: 0.9886, as calculated by tensorflow.

3.2 Classification Performance Measured Against Pathologist Ground Truth

Visual assessment of maps constructed as the result of the procedure was performed by displaying it side-by-side with original H&E slides with superimposed pathologist annotations. Two fragments from testing set are shown with context of obtained results in Fig. 2. Subsequently, numerical results of classification performance for all fragments from testing set were calculated and presented in Table 1.

The best results have been achieved for second fragment shown in the first row in Fig. 2.

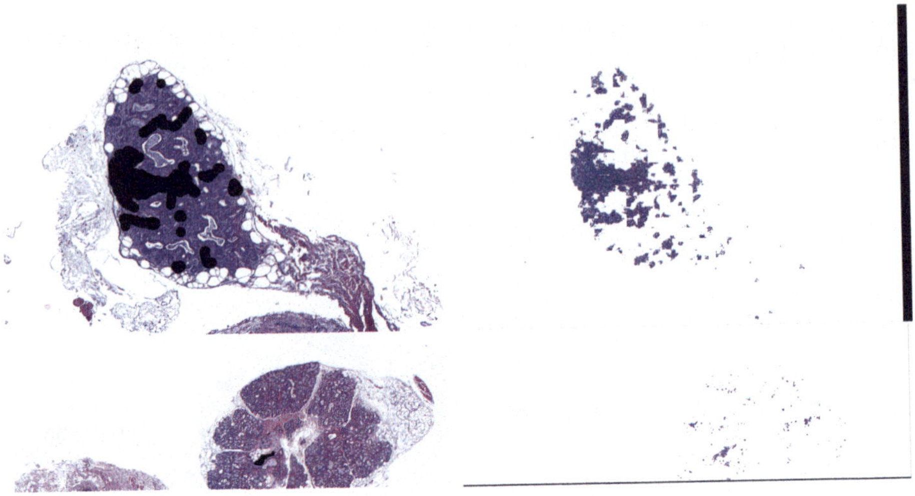

Fig. 2. WSI fragments from testing set with annotation (black area) in first column and results of inflammation area detection using proposed method in last column. The best result (upper row) and the worst result (bottom row) according to F-score.

Table 1. Comparison of F-score, accuracy and recall for validation set of 4 images (fragments of WSI).

Fragment of WSI	F-Score	Accuracy	Recall
1	0.19	0.99	0.7
2	0.66	0.96	0.82
3	0.53	0.94	0.79
4	0.37	0.97	0.39

4 Discussion

The neural network classification method presented in this paper achieves high accuracy of inflammation regions detection from 0.99 to 0.94 in comparison to the results of manual annotation done by pathologist on validation set. It means that almost all areas annotated by pathologist are detected by the trained CNN.

But results of F-score measurement are mediocre, ranging from 0.19 to 0.66. Visual assessment shows reasons why FP regions appears in results of classification by NN. The first is coarse style of annotations visible in avoiding the boundaries marked manually and automatically and in omitting some of lymphocyte nuclei clusters, particularly those small in size, by pathologist. The second is presence of classified pattern near nuclei of adipose tissue or other salivary glands structures other than parenchyma. All of these regions' texture is similar to the texture of annotated regions because there are also lymphocytes nuclei, but they were not marked as ground truth as they are not placed in parenchyma near blood vessel. So the validation by F-score is not suitable in this case because it based on FP which score high due to the selected method of annotation and due to the fact that pattern is met in structures other than parenchyma.

The next measurement of quality of classification results – recall has value from 0.79 to 0.82 as it shows the proportion of pixels classified correctly to all positive in ground true.

To increase quality of the results two types of improvements should be applied. The first one is introductory recognition if the area is parenchyma to avoid classification from other structures. The second is filtration of found regions according to the size.

5 Conclusions and Future Work

Study showing the potential of machine learning for classification problem solving in particular type of problems connected with digital images in pathology has been introduced. A simple problem of development of supporting tool for diagnosis of Sjögren's Syndrome was chosen. The first step of the study of inflammatory infiltration quantification based on digital images of stained tissue fragments obtained from minor salivary glands biopsy shows usefulness of deep neural networks.

Proposed method based on trained CNN should be developed to detect number of lymphocyte nuclei in detected inflammatory infiltration region in next step of investigation. It can be done using analogical AI method or typical image processing methods. Both steps of investigation used separately or together can help pathologists in quantitative analysis and become a base for supporting pathology through digital image analysis methods.

Because CNN model presented in this paper was trained and validated on slides prepared in one pathology laboratory it should be verified using digital slides of salivary glands from other laboratories. It is needed to ensure that proposed method is insensitive to the so called batch effect which is frequently observed among histological images from various hospitals because of non-biological experimental variations in various labs such as method of slide preparation, specifications of the imaging device, and their software and even an age of a sample.

Acknowledgements. This work has been supported by statutory founds of Nalecz Institute of Biocybernetics and Biomedical Engineering PAS. Authors are grateful to Comtegra S.A. for lending two Tesla v100m cards with 32 GB RAM each.

References

1. Hellquist, H.B., Skalova, A.: Histopathology of the Salivary Glands. Springer, Heidelberg (2014)
2. Varela-Centelles, P., Seoane-Romero, J., Sanchez-Sanchez, M., Gonzalez-Mosquera, A., Diz-Dios, P., Seoane, J.: Minor salivary gland biopsy in Sjogren's Syndrome: a review and introduction of a new tool to ease the procedure. Medicina Oral Patología Oral y Cirugia Bucal **19**, e20–e23 (2014)

3. Shiboski, C.H., Shiboski, S.C., Seror, R., Criswell, L.A., Labetoulle, M., Lietman, T.M., Rasmussen, A., Scofield, H., Vitali, C., Bowman, S.J., Mariette, X.: 2016 American college of rheumatology/European league against rheumatism classification criteria for primary sjögren's syndrome: a consensus and data-driven methodology involving three international patient cohorts. Arthritis Rheum. **69**(1), 35–45 (2016)
4. Costa, S., Schutz, S., Cornec, D., Uguen, A., Quintin-Roué, I., Lesourd, A., Berthelot, J.M., Hachulla, E., Hatron, P.Y., Goeb, V., Vittecoq, O., Pers, J.O., Marcorelles, P., Saraux, A., Devauchelle-Pensec, V.: B-cell and t-cell quantification in minor salivary glands in primary Sjögren's Syndrome: development and validation of a pixel-based digital procedure. Arthritis Res. Therapy **18**(1), 21 (2016)
5. Litjens, G., Sánchez, C.I., Timofeeva, N., Hermsen, M., Nagtegaal, I., Kovacs, I., van de Kaa, C.H., Bult, P., van Ginneken, B., van der Laak, J.: Deep learning as a tool for increased accuracy and efficiency of histopathological diagnosis. Sci. Rep. **6**(1), 26286 (2016)
6. Acs, B., Pelekanou, V., Bai, Y., Martinez-Morilla, S., Toki, M., Leung, S.C.Y., Nielsen, T.O., Rimm, D.L.: Ki67 reproducibility using digital image analysis: an inter-platform and inter-operator study. Lab. Invest. **99**(1), 107–117 (2018)
7. Veta, M., Pluim, J.P.W., van Diest, P.J., Viergever, M.A.: Breast cancer histopathology image analysis: a review. IEEE Trans. Biomed. Eng. **61**(5), 1400–1411 (2014)
8. Bhargava, R., Madabhushi, A.: Emerging themes in image informatics and molecular analysis for digital pathology. Ann. Rev. Biomed. Eng. **18**(1), 387–412 (2016)
9. Janowczyk, A., Madabhushi, A.: Deep learning for digital pathology image analysis: a comprehensive tutorial with selected use cases. J. Pathol. Inform. **7**(1), 29 (2016)
10. Gertych, A., Swiderska-Chadaj, Z., Ma, Z., Ing, N., Markiewicz, T., Cierniak, S., Salemi, H., Guzman, S., Walts, A.E., Knudsen, B.S.: Convolutional neural networks can accurately distinguish four histologic growth patterns of lung adenocarcinoma in digital slides. Sci. Rep. **9**(1), 1483 (2019)
11. Alom, M.Z., Yakopcic, C., Taha, T.M., Asari, V.K.: Microscopic nuclei classification, segmentation and detection with improved deep convolutional neural network (DCNN) approaches. arXiv e-prints (2018)
12. Litjens, G., Kooi, T., Bejnordi, B.E., Setio, A.A.A., Ciompi, F., Ghafoorian, M., van der Laak, J.A., van Ginneken, B., Sánchez, C.I.: A survey on deep learning in medical image analysis. Med. Image Anal. **42**, 60–88 (2017)
13. Saha, M., Chakraborty, C., Racoceanu, D.: Efficient deep learning model for mitosis detection using breast histopathology images. Comput. Med. Imaging Graph. **64**, 29–40 (2018)
14. Mobadersany, P., Yousefi, S., Amgad, M., Gutman, D.A., Barnholtz-Sloan, J.S., Vega, J.E.V., Brat, D.J., Cooper, L.A.D.: Predicting cancer outcomes from histology and genomics using convolutional networks. Proc. Natl. Acad. Sci. **115**(13), E2970–E2979 (2018)
15. Khosravi, P., Kazemi, E., Imielinski, M., Elemento, O., Hajirasouliha, I.: Deep convolutional neural networks enable discrimination of heterogeneous digital pathology images. EBioMedicine **27**, 317–328 (2018)
16. Bychkov, D., Linder, N., Turkki, R., Nordling, S., Kovanen, P.E., Verrill, C., Walliander, M., Lundin, M., Haglund, C., Lundin, J.: Deep learning based tissue analysis predicts outcome in colorectal cancer. Sci. Rep. **8**(1), 3395 (2018)

17. Pedraza, A., Bueno, G., Deniz, O., Cristóbal, G., Blanco, S., Borrego-Ramos, M.: Automated diatom classification (part b): a deep learning approach. Appl. Sci. **7**(5), 460 (2017)
18. Waterhouse, J.P.: Focal adenitis in salivary and lacrimal glands. Proc. R. Soc. Med. **56**, 911–918 (1963)
19. Ronneberger, O., Fischer, P., Brox, T.: U-net: convolutional networks for biomedical image segmentation. In: Navab, N., Hornegger, J., Wells, W., Frangi, A. (eds.) MICCAI 2015. Lecture Notes in Computer Science, pp. 234–241. Springer, Cham (2015)
20. Salvi, M., Molinari, F.: Multi-tissue and multi-scale approach for nuclei segmentation in H&E stained images. Biomed. Eng. OnLine **17**(1), 89 (2018)

On Stability of Feature Selection Based on MALDI Mass Spectrometry Imaging Data and Simulated Biopsy

Agata Wilk[1], Marta Gawin[2], Katarzyna Frątczak[1], Piotr Widłak[2], and Krzysztof Fujarewicz[1](✉)

[1] Institute of Automatic Control, Silesian University of Technology, Akademicka 16, 44-100 Gliwice, Poland
krzysztof.fujarewicz@polsl.pl
[2] Maria Sklodowska-Curie Institute — Oncology Centre, Gliwice Branch, Wybrzeze Armii Krajowej 15, 44-101 Gliwice, Poland

Abstract. In this work we analyse MALDI mass spectrometry imaging data for thyroid cancer samples. Such a data, containing information about spatial distribution of proteins/peptides, makes possible to make a virtual analysis how a technique of fine needle aspiration (FNA) biopsy, a routine diagnosis procedure for thyroid, influences the outcome i.e. a set of discriminative features between cancerous and normal tissue. We hypothesised that an impure dataset (consisting of normal cell contaminated cancer samples) would be beneficial in the terms of stable feature selection. We compared several methods of predictor selection on different datasets to perform an in-depth feature ranking stability analysis for thyroid cancer mass spectrometry data. Furthermore we examined the impact of sample contamination level on the selection.

Keywords: MALDI imaging mass spectrometry · Machine learning · Feature selection · Fine needle biopsy

1 Introduction

While it is universally agreed that a machine learning model will only be as good as its training set, and several studies have been conducted concerning the influence of the dataset on the feature selection process [12,23], little reference is available about the optimisation of the dataset preparation. In case of biomarker selection, the general direction is the highest possible purification of the samples. It can be achieved at the laboratory stage, for instance through laser microbeam microdissection, which results in almost pure cell populations [11], or computationally. Several interesting approaches have been proposed, including the ISOpure algorithm employing statistical regularization strategies [17], and a database-search enhanced method relying on Boolean operators [21].

The problem of sample heterogeneity is especially significant in diagnostic procedures such as fine needle aspiration biopsy. A number of machine learning

J. Korbicz et al. (Eds.): PCBEE 2019, AISC 1033, pp. 82–93, 2020.
https://doi.org/10.1007/978-3-030-29885-2_8

methods have been implemented for biopsy specimen feature selection and classification, ranging from image analysis techniques [3] to genomic and proteomic approaches. The studies based on omic data are usually preceded by sample purification [9,13].

We hypothesised that an impure dataset (consisting of normal cell contaminated cancer samples) would be beneficial in the terms of stable feature selection. We compared several methods of predictor selection on different datasets to perform an in-depth feature ranking stability analysis [8,19] for thyroid cancer mass spectrometry data. Furthermore we examined the impact of sample contamination level on the selection.

2 Data

2.1 MALDI-MSI Data

Tissue specimens were collected during thyroidectomy from 11 patients with papillary thyroid cancer (PTC) and stored as formalin-fixed paraffin-embedded material; the study was approved by the Institutional Ethics Committee (Approval No. KB/430-49/12). Tissue samples were sectioned (10 μm) and for each specimen cancer and normal (not malignant) tissue region was defined by a pathologist. Tissue specimens were processed and peptide spectra were registered in the m/z range of 600–4000 Da by MALDI-MSI with a 100 μm raster width as described in details elsewhere [14]. Spectral components (features) were identified in the normalized spectra by the Gaussian Mixture Model (GMM) as described in details elsewhere [15]. For every specific (x,y) point in every preparation and for every component in GMM model a convolution of a normalized spectrum and a component of Gaussian function was performed giving an integral of the pointwise multiplication of these two functions as the GMM component-related feature value as described previously [14]. In total, the dataset consisted of 75642 cancerous and 118791 non-malignant observations (spectra) with 2422 features (i.e., peptide ions) in each spectrum.

2.2 Dataset Preparation

In order to minimize the effect of possible errors during data acquisition as well as to counter the imbalances of sample numbers both between patients and tissue types, further data processing was required. A new dataset (later called sampling dataset) was generated by averaging the spectra of square neighbourhoods for randomly chosen points, under the constraints that the square contained more than 24 points and all of them belonged to the same type. That way a matrix was obtained, consisting of 50 observations of each type from every patient (for a total of 1100 records). In this dataset differences in the profile of imaged peptides between cancerous and healthy tissue was observed (Fig. 1) confirming the potential of registered peptides/proteins as hypothetical biomarkers of thyroid cancer.

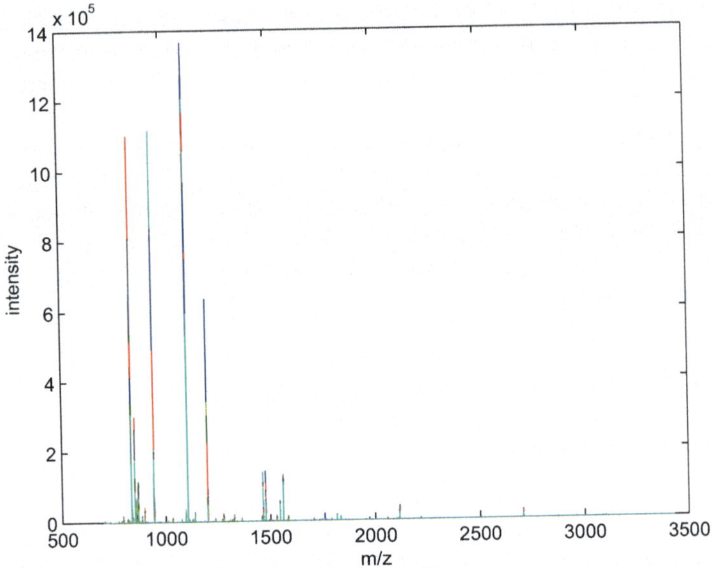

(a) cancerous tissue

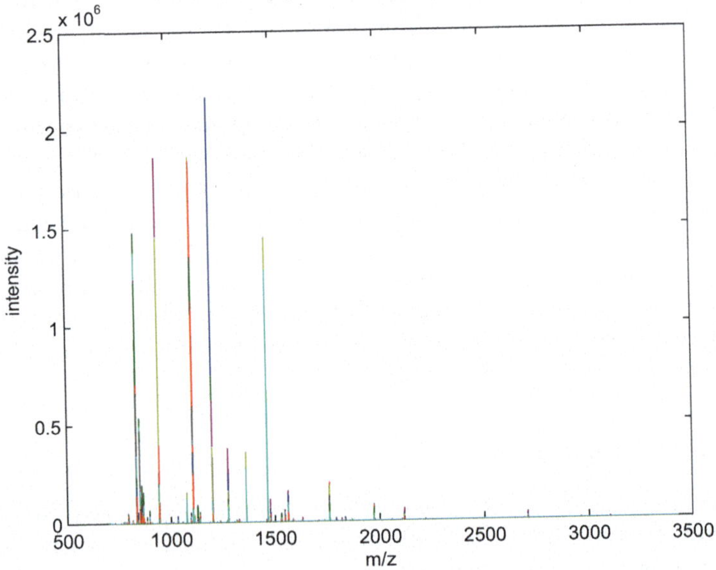

(b) non-cancerous tissue

Fig. 1. Spectra comparison between tissue types

2.3 Simulation of Biopsy Data

Due to the lack of available biopsy data, it had to be generated *in silico*. To construct a biopsy model, several assumptions had to be made:

1. A single biopsy procedure consists of five punctures with a common starting point anywhere outside of the tissue, and different end points within the thyroid.
2. The biopsy needle is rectangular in lateral cross-section.
3. All points 'covered' with the needle are included in a biopsy sample.
4. The spectrum of a biopsy sample is equal to a simple average of all the spectra of the included points.

Obviously, the above-described model is simplified, as it does not take into consideration punctures outside of the cross-section plane, or all the biological material extracted during a real-life fine needle aspiration (including, but not limited to blood, epithelial, and connective tissue). For a more realistic simulation of the biopsy procedure a dedicated interface was designed (Fig. 2), using which a second dataset (later referred to as biopsy dataset) was manually generated. It consisted of 15 samples of each type per patient. By subsetting the biopsy data, 30 sets were randomly created, each containing one sample of every type per patient.

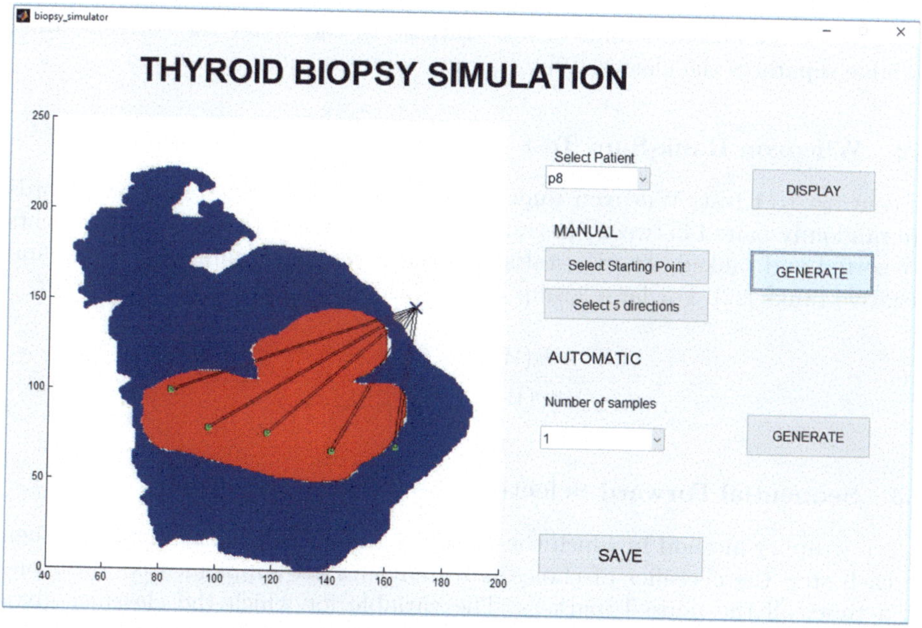

Fig. 2. GUI dedicated for biopsy simulation

3 Feature Selection

Feature selection algorithms can be divided into two main groups: filters and wrappers. In filter methods, predictors are chosen relative to a certain statistical measure which assigns a score to every feature. The main advantage of these kind of methods is their robustness and low computational cost. However, filters tend to ignore relationships between features which may lead to redundancy of resulting variable set. The second approach, wrapper methods, selects optimal features by employing the target model and assessing the contribution of each feature to the classification quality. These algorithms, while very unlikely to select correlated variables, are more prone to overfitting, and require significantly greater computation time [6,7,18].

In the experiment, three methods of feature selection were proposed: t test, Wilcoxon rank-sum test and sequential forward selection.

3.1 t Test

Let us assume that the observations come from groups of distributions $N(\mu_1, \sigma_1^2)$ and $N(\mu_2, \sigma_2^2)$, respectively, and cardinalities n_1 and n_2. The t statistics may thus be calculated as:

$$t = \frac{\mu_1 - \mu_2}{\sqrt{\frac{\sigma_1^2}{n_1} + \frac{\sigma_2^2}{n_2}}}$$

The greater the absolute value of test statistic is, the better the corresponding variable separates the classes [10].

3.2 Wilcoxon Rank-Sum Test

In contrast to t test, Wilcoxon rank sum is a non parametric test. The records are randomly paired between classes, then the differences between pair elements are sorted and ranked. The W statistic is the lower of the sums of positive and negative ranks [22]. For large group cardinalities a z-score can be calculated:

$$z = \frac{W - E(W)}{\sqrt{Var(W)}} = \frac{W - \frac{n(n+1)}{4}}{\sqrt{\frac{n(n+1)(2n+1)}{24}}}$$

3.3 Sequential Forward Selection

It is a wrapper method in which the feature set is initialized as empty and then in each step the classifier of choice is tested on a set supplemented with, one at a time, all the unused markers. The variable for which the classifier gives best results, according to a predefined loss function (usually one of the typical classification quality measures like error or accuracy), is added to the set [1].

4 Feature Ranking Stability

For a potential diagnostic application of the selected features, their choice must be stable, so that the variable retains its significance for separation of classes in the biopsy data. To test the feature ranking stability, selection was performed on the sampling set, an then on the 30 biopsy datasets. Additionally, the analysis was performed for datasets containing multiple biopsies from one patient, and for bootstrap sets of the sampling data. The importance of stability and sensitivity of feature rankings was also the subject of our previous works [5,20] (Fig. 3).

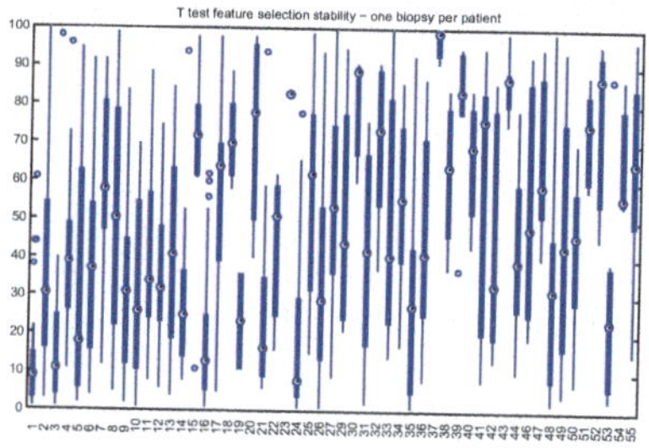

(a) one biopsy per patient

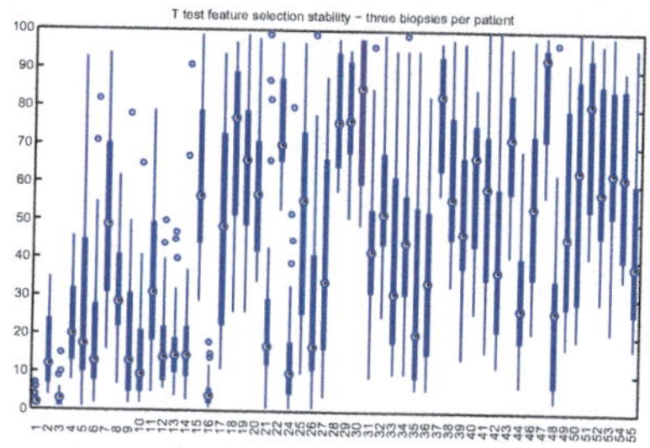

(b) three biopsies per patient

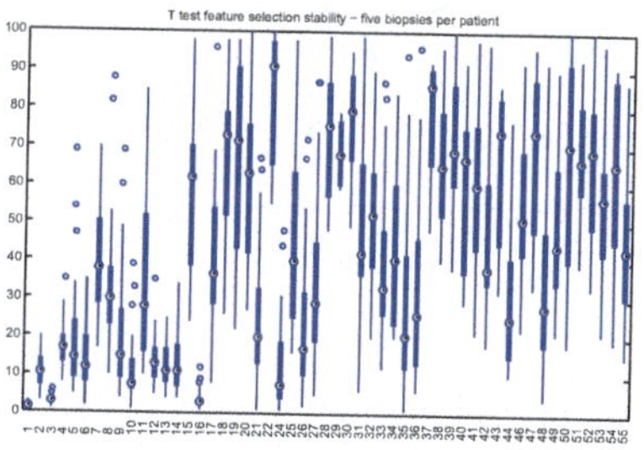

(c) five biopsies per patient

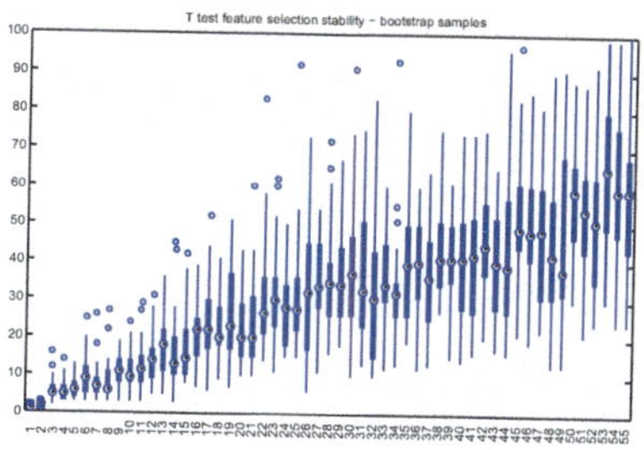

(d) Bootstrap samples

Fig. 3. Feature selection stability—t test. The x axis represents feature ranking in sampling data, the y axis—rankings in biopsy or bootstrap datasets

The obtained results clearly indicate, that the feature selection from the sampling set is highly unstable against biopsy data (containing cancerous samples with varying degrees of normal tissue contamination). Furthermore a certain stabilisation and decrease in ranking dispersion can be observed with the increase of samples taken from each patient. This realization may suggest the merits of multiple biopsy during the diagnostic process (other than the obvious benefit of increase in the probability of encountering a malignant nodule with the extra punctures). Further research is needed in this matter with emphasis on the aggregation of repeated biopsy data.

The apparent stabilisation for multiple biopsies raises a question whether the differences in marker selection are not caused by the small dataset size. To address this doubt one more stability test was conducted, this time utilising all the available biopsy data (Fig. 4).

While the trend first observed in bootstrap sampling is more pronounced, the differences in feature rankings are beyond any doubt, particularly for markers in further positions. This finding supports the approach to biomarker search in which, instead of focusing on purifying the dataset used for feature selection, a contaminated, non-homogenous dataset is favoured (as it is closer to the clinical data the target model will be used against).

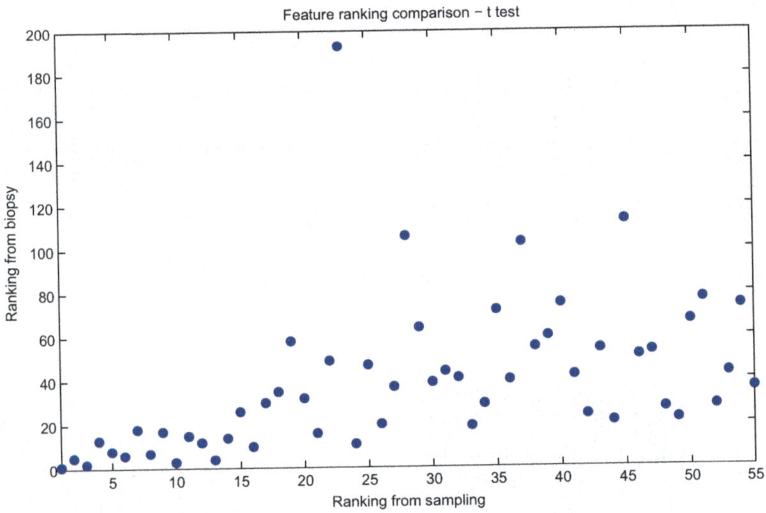

Fig. 4. Feature selection stability—t test. The x axis represents feature ranking for sampling data, the y axis—ranking for the entire biopsy set

For Wilcoxon rank-sum test the instability of feature selection for biopsy data is even more discernible (even though for bootstrap sampling the dispersion is lower than for t test). This might prove disadvantageous as the constructed model is likely to be less universal and irrelevant for new samples. Indeed, a simple model trained on features selected using this method proved worse than a model whose predictors were chosen with t test (Fig. 5).

The sequential forward selection algorithm was also applied to the datasets, the tested classifiers being SVM, LDA and decision tree. To reduce the computational cost of the method, it was preceded by a preliminary dimensionality reduction using the aforementioned t test filter, selecting 150 features.

Surprisingly enough, the SFS algorithm yields identical markers in the same order of selection as during the pre-filtering using t test. It could indicate the lack of correlation between features. Because the wrapper methods are much more computationally complex than filter ones, but do not result in noticeable performance improvement, they were abandoned in later analysis (Fig. 6).

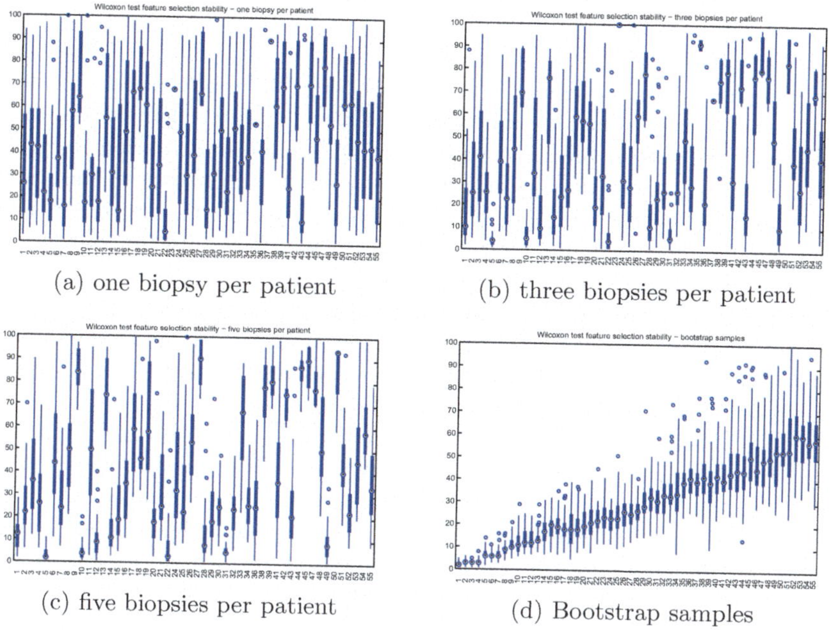

(a) one biopsy per patient

(b) three biopsies per patient

(c) five biopsies per patient

(d) Bootstrap samples

Fig. 5. Feature selection stability—Wilcoxon test. The x axis represents feature ranking in sampling data, the y axis—rankings in biopsy or bootstrap datasets

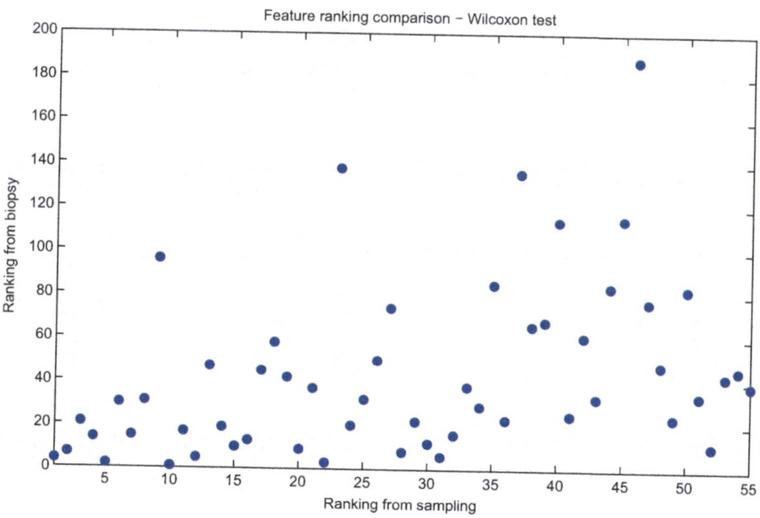

Fig. 6. Feature selection stability—t test. The x axis represents feature ranking for sampling data, the y axis—ranking for the entire biopsy set

5 Choosing a Dataset for Feature Selection

While the hitherto performed tests suggest that impure data is better to use for feature selection, the natural question concerns the existence of a preferable impurity level. For this purpose a new dataset was generated using the aforementioned biopsy simulator, modified to provide additional information referring to the percentage of cancer tissue in the biopsy specimen. Contrary to the previous data generation process, the biopsies intended as cancerous were not necessarily acquired by actively aiming for the tumour, which resulted in samples with practically full variability of cancer content.

The dataset was then divided into twenty groups according to the malignant tissue content. Due to both the nature of a biopsy procedure and the model used to simulate it, the groups with a very low or very high percentage of cancer cells contained fewer observations than the middle ones.

For each of the groups t test feature selection was performed (t test was chosen as the optimal method for its relative stability and simplicity). Then, rankings were determined of each selected marker in other groups.

Figure 7 confirms the hypothesis that the value of a feature as a marker for a model depends on the data subset based on which it was selected.

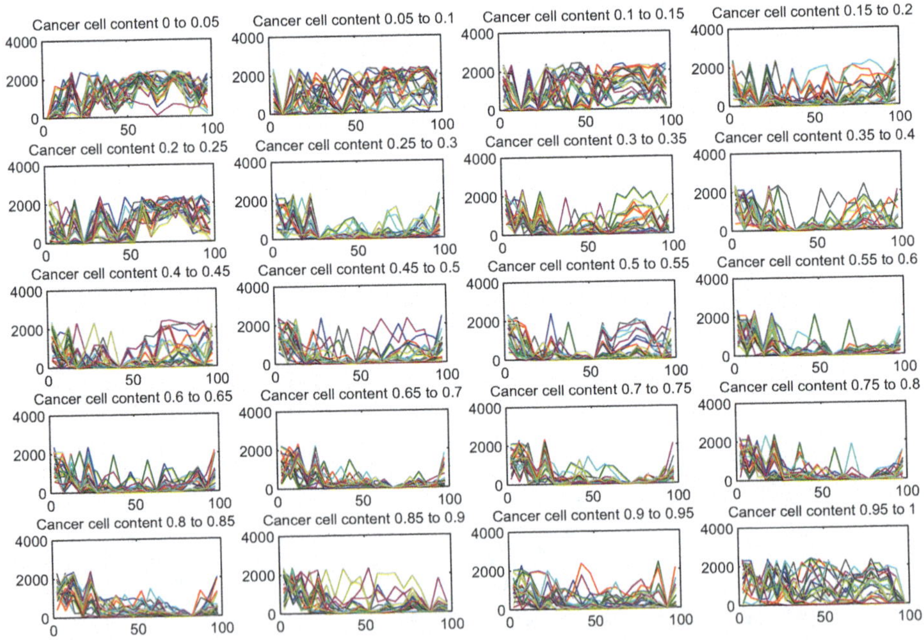

Fig. 7. Feature selection stability by sample content. Every plot corresponds to a range of cancer content in samples, for which the selection was performed. Each line describes a single feature. The x axes represent the percentage of cancer tissue, the y axes—feature's ranking position for the respective group

The most desired path of depicted lines is as flat as possible, ideally through the entire range of the x axis. Such features would be valid as biomarkers for biopsy diagnosis regardless of the specimen composition.

For very low malignant tissue percentages no features with the required properties emerge, which was to be expected. However, in certain groups (most notably for the percentages: 0.25–0.3, as well as all from 0.55 to 0.75), the obtained plots are very closed to the desired appearance for the rest of the scope. These subsets should therefore be favoured for mass spectrometry marker selection.

6 Future Research

The results suggest that by subsetting the data relative to the sample composition, more optimal and universally viable features might be selected. With this knowledge, further exploration of the approach proposed in this experiment might prove valuable in constructing a diagnostic model for clinical application, allowing for correct classification even for biopsies with high level of normal cell contamination.

Acknowledgement. This work was supported by Polish National Centre for Research and Development under Grant Strategmed2/267398/4/NCBR/2015 and Silesian University of Technology Grant 02/010/BK-18/0102. Data analysis was partially carried out using the Biotest Platform developed within Project n. PBS3/B3/32/2015 financed by the Polish National Centre of Research and Development (NCBiR) and described in [2,4,16]. Calculations were performed using the infrastructure supported by the computer cluster Ziemowit (www.ziemowit.hpc.polsl.pl) funded by the Silesian BIO-FARMA project No. POIG.02.01.00-00-166/08 and expanded in the POIG.02.03.01-00-040/13 in the Computational Biology and Bioinformatics Laboratory of the Biotechnology Centre at the Silesian University of Technology.

References

1. Aha, D.W., Bankert, R.L.: A Comparative Evaluation of Sequential Feature Selection Algorithms, pp. 199–206. Springer, New York (1996)
2. Bensz, W., Borys, D., Fujarewicz, K., Herok, K., Jaksik, R., Krasucki, M., Kurczyk, A., Matusik, K., Mrozek, D., Ochab, M., et al.: Integrated system supporting research on environment related cancers. In: Król, D., Madeyski, L., Nguyen, N. (eds.) Recent Developments in Intelligent Information and Database Systems, pp. 399–409. Springer, Cham (2016)
3. Filipczuk, P., Fevens, T., Krzyzak, A., Monczak, R.: Computer-aided breast cancer diagnosis based on the analysis of cytological images of fine needle biopsies. IEEE Trans. Med. Imaging **32**(12), 2169–2178 (2013)
4. Fujarewicz, K., Student, S., Zielański, T., Jakubczak, M., Pieter, J., Pojda, K., Świerniak, A.: Large-scale data classification system based on galaxy server and protected from information leak. In: ACIIDS 2017, pp. 765–773. Springer, Cham (2017)

5. Gaweł, D., Fujarewicz, K.: On the sensitivity of feature ranked lists for large-scale biological data. Math. Biosci. Eng. MBE **10**(3), 677–690 (2013)
6. Hand, D.J.: Data Mining Based in part on the article 'Data mining' by David Hand, which appeared in the Encyclopedia of Environmetrics. American Cancer Society (2013)
7. Haury, A.-C., Gestraud, P., Vert, J.-P.: The influence of feature selection methods on accuracy, stability and interpretability of molecular signatures. PLOS ONE **6**(12), 1–12 (2011)
8. Kalousis, A., Prados, J., Hilario, M.: Stability of feature selection algorithms. In: Fifth IEEE International Conference on Data Mining (ICDM 2005), p. 8, November 2005
9. Kim, Y., Jeon, J., Mejia, S., Yao, C.Q., Ignatchenko, V., Nyalwidhe, J.O., Gramolini, A.O., Lance, R.S., Troyer, D.A., Drake, R.R., Boutros, P.C., Semmes, O.J., Kislinger, T.: Targeted proteomics identifies liquid-biopsy signatures for extracapsular prostate cancer. Nat. Commun. **7**, 11906 (2016)
10. MathWorks. Two sample t-test, 23 March 2019
11. Nakamura, T., Furukawa, Y., Nakagawa, H., Tsunoda, T., Ohigashi, H., Murata, K., Ishikawa, O., Ohgaki, K., Kashimura, N., Miyamoto, M., Hirano, S., Kondo, S., Katoh, H., Nakamura, Y., Katagiri, T.: Genome-wide CDNA microarray analysis of gene expression profiles in pancreatic cancers using populations of tumor cells and normal ductal epithelial cells selected for purity by laser microdissection. Oncogene **23**(13), 2385–2400 (2004)
12. Oreski, D., Oreski, S., Klicek, B.: Effects of dataset characteristics on the performance of feature selection techniques. Appl. Soft Comput. **52**, 109–119 (2017)
13. Pankratz, D.G., Choi, Y., Imtiaz, U., Fedorowicz, G.M., Anderson, J.D., Colby, T.V., Myers, J.L., Lynch, D.A., Brown, K.K., Flaherty, K.R., Steele, M.P., Groshong, S.D., Raghu, G., Barth, N.M., Walsh, P.S., Huang, J., Kennedy, G.C., Martinez, F.J.: Usual interstitial pneumonia can be detected in transbronchial biopsies using machine learning. Ann. Am. Thoracic Soc. **14**(11), 1646–1654 (2017). PMID: 28640655
14. Pietrowska, M., Diehl, H.C., Mrukwa, G., Kalinowska-Herok, M., Gawin, M., Chekan, M., Elm, J., Drazek, G., Krawczyk, A., Lange, D., Meyer, H.E., Polanska, J., Henkel, C., Widlak, P.: Molecular profiles of thyroid cancer subtypes: classification based on features of tissue revealed by mass spectrometry imaging. Biochimica et Biophysica Acta (BBA) Proteins Proteomics **1865**(7), 837–845 (2017). MALDI Imaging
15. Polanski, A., Marczyk, M., Pietrowska, M., Widlak, P., Polanska, J.: Signal partitioning algorithm for highly efficient gaussian mixture modeling in mass spectrometry. PLOS ONE **10**(7), 1–19 (2015)
16. Psiuk-Maksymowicz, K., Płaczek, A., Jaksik, R., Student, S., Borys, D., Mrozek, D., Fujarewicz, K., Świerniak, A.: A holistic approach to testing biomedical hypotheses and analysis of biomedical data. In: Kozielski, S., Mrozek, D., Kasprowski, P., Małlysiak-Mrozek, B., Kostrzewa, D. (eds.) BDAS 2015, pp. 449–462. Springer, Cham (2015)
17. Quon, G., Haider, S., Deshwar, A.G., Cui, A., Boutros, P.C., Morris, Q.: Computational purification of individual tumor gene expression profiles leads to significant improvements in prognostic prediction. Genome Med. **5**(3), 29 (2013)
18. Saeys, Y., Inza, I., Larrañaga, P.: A review of feature selection techniques in bioinformatics. Bioinformatics **23**(19), 2507–2517 (2007)

19. Student, S., Fujarewicz, K.: Stable feature selection and classification algorithms for multiclass microarray data. Biol. Direct **7**, 33 (2012). 23031190[pmid], PMC3599581[pmcid], 1745-6150-7-33[PII]
20. Student, S., Fujarewicz, K.: Stable feature selection and classification algorithms for multiclass microarray data. Biol. Direct **7**(1), 33 (2012)
21. Türeci, Ö., Ding, J., Hilton, H., Bian, H., Ohkawa, H., Braxenthaler, M., Seitz, G., Raddrizzani, L., Friess, H., Buchler, M., Sahin, U., Hammer, J.: Computational dissection of tissue contamination for identification of colon cancer-specific expression profiles. FASEB J. **17**(3), 376–385 (2003). PMID: 12631577
22. Wilcoxon, F.: Individual comparisons by ranking methods. Biometrics Bull. **1**(6), 80–83 (1945)
23. Zhang, S., Zhang, C., Yang, Q.: Data preparation for data mining. Appl. Artif. Intell. **17**(5–6), 375–381 (2003)

Virtual Reality as a Tool for Ophthalmic Examination

Dorota Kamińska[1(✉)], Tomasz Sapiński[1], Grzegorz Zwoliński[1], Sławomir Wiak[1], Magdalena Kucharczyk-Pośpiech[2], and Michał Wilczyński[2]

[1] Institute of Mechatronics and Information Systems, Lodz University of Technology, Lodz, Poland
{dorota.kaminska,tomasz.sapinski,grzegorz.zwolinski}@p.lodz.pl

[2] I Departament of Eye Diseases, Medical University of Lodz, Lodz, Poland
michal.wilczynski@umed.lodz.pl

Abstract. In their daily practice ophthalmologists frequently face the problem of examining the vision abilities of immobilised patients. People affected by stroke or a demyelinating disease, who are often bedridden or struggle with movement disorders, are the best example of such a case. For such patients a relatively simple test becomes impossible to perform. The aim of this research is to transfer Ishihara colour test into virtual reality, thereby creating a mobile diagnostic tool for immobilised patients. Virtual reality headsets map the natural human perception of space and isolate the patient's eyes from any external light, which allows to control fully the colour and intensity of displayed image. Moreover, the headsets are light and portable, therefore, useful in evaluating the eye condition of immobilised patients. Performing eye examinations on aforementioned patients will improve the quality of their life. The application was tested among patients with and without colour blindness disorder. The achieved results suggest that the proposed solution can be used in ophthalmologists daily practice. Moreover, it shows that the application can be developed in the context of other ophthalmic examinations such as visual acuity or perimetry.

Keywords: Virtual reality · Ishihara color test · Ophthalmological examination

1 Introduction

Ophthalmologists, in their daily practice, frequently face the problem of examining immobilised patients. People affected by stroke or a demyelinating disease, who struggle with movement disorders, are the best example. In such cases a relatively simple test becomes impossible to perform. Virtual reality headsets are light and portable, therefore, can be useful in evaluating the eye condition of immobilised patients. Moreover, a VR headset maps the natural human perception of space and isolates the patient's eyes from any external light, which allows to control fully the colour and intensity of displayed image.

J. Korbicz et al. (Eds.): PCBEE 2019, AISC 1033, pp. 94–106, 2020.
https://doi.org/10.1007/978-3-030-29885-2_9

The aim of this research is to transfer ophthalmic examination, namely the colour perception test, into virtual reality, thereby creating a mobile diagnostic tool for immobilised patients. The research is conducted in a multidisciplinary team consisting of IT specialists and physicians working at the largest teaching hospital unit in Poland, which congregates specialists in the fields of, inter alia, ophthalmology and neurology, and what is the most important provides access to patients who are immobilised.

The concept of the application was established during brainstorming sessions among IT specialists and ophthalmologists. Additionally, during testing sessions feedback was gathered among physicians and patients in order to improve the tool and meet the expectations of final user groups.

The paper is organised as follows. In Sect. 2 we present virtual reality and its utilization in medicine and health care. Next section describes colour blindness and ophthalmic methods of its examination. Section 4 presents methodology: technology and software, research process, details of created application and conducted experiments. Section 5 presents discussion of the results and it is followed by a conclusion and a plan for the future.

2 Virtual Reality

Virtual Reality (VR) is a technology that provides an interactive computer generated environment in which one can see and move in a dynamically changing scenario [14]. VR simulates a user's physical presence in an artificially generated world and allows one to interact with it.

VR is defined as a set of computing technologies which provide a deep interaction and a natural experience of the world beyond reality [11]. It creates computer visualisation of objects, space and events. It can represent both, a real world as well as completely fictional elements.

Myron W. Krueger is considered to be the precursor of VR. He is an artist, researcher and computer scientist who developed and created a number of interactive video-installations. His designs and visions have become the prototypes of many VR systems, applicable in education, psychology and psychotherapy [20].

For the first time the term VR was specified by Jarona: *Virtual Reality is the use of computer technology to create the effect of an interactive three-dimensional world in which the objects have a sense of spatial presence.* In literature, one can also find another definition of virtual reality, which is $I^3 : Interaction + Immersion + Imagination$ [13]. Currently the I^3 paradigm is mainly achieved through the generation of visual, audio, less often tactile, smell or taste effects. In most cases this technology enables interaction with a computer-simulated environment through various manipulators.

Development of the first version of Oculus Rift [4] contributed to the popularisation of VR and the interest in VR devices is constantly growing. The business role of VR headsets is increasing as well, and the companies such as Facebook, Google, Microsoft, and Sony consistently invest in the development of this technology and find new uses for the hardware they manufacture. In 2016 the VR market was worth $ 1.6 billion, it is estimated that by 2020 it will increase to a range from 40 billion to 120 billion U.S. dollars [8].

Most of VR related content is focused on entertainment or commercial applications, as these areas provide the largest groups of VR headsets recipients. However, the possibilities of virtual reality do not end there. The dynamic growth and interest in the subject of virtual reality render it applicable in many other areas such as military simulations, psychological or medical research.

Virtual reality has been adopted by scientists and medical professionals from the University of Southern California Institute for Creative Technologies use VR as a therapy tool for soldiers with Post Traumatic Stress Disorder (PTSD). Currently it is used at clinical sites including hospitals, military bases and university centres. The use of this tool resulted in a meaningful reduction in PTSD symptoms [24].

Exposure therapy for patients with phobias is a similar VR application in treatment. At the University of Louisville patients face their fears, such as flying or claustrophobia in a controlled simulated environment [7]. An Argentinian psychologist, Fernando Tarnogol, created a software platform called Phobos to treat acrophobia and arachnophobia [16]. Deep, developed for Oculus Rift, might serves as another example of such application. It helps users dealing with anxiety, showing them how to take deep, meditative breaths in an underwater world [2].

Researchers from the University of Washington use VR as a distraction for patients with extensive burns to redirect their attention during an extremely painful dressings changing procedure. The studies have shown that this kind of therapy works better than morphine. In a similar vein, a study shows that virtual reality games may help patients with alleviate phantom limb pain [23].

VR interactive games are a great tool for socio-emotional therapy because of autistic children's intense interest in technology. Researchers from the University of Texas, Dallas put young adults in various social situations such as job interviews or blind dates to train their reading of social cues and expression of socially acceptable behaviour [18].

VR can also improve the quality of life of those who do not have the ability to get out in the real world, i.e. elderly or disabled people. Students from Stanford created a virtual reality experience for seniors confined at home, which allows them to experience the world outside their homes (a bike ride or a walk on the beach) [6].

ICT's Game Based Rehabilitation Lab developed interactive tools based on Kinect and Oculus Rift for rehabilitation of patients recovering from trauma, such as stroke, mild traumatic brain injury or spinal cord injury. In addition it is used for treating high-risk groups such as elderly adults, athletes and soldiers at risk of concussive head injuries [5].

Neurosurgeons at the Mayo Clinic, Ronald Reagan UCLA Medical Center, are planning complex brain surgeries using virtual reality technology. Wearing the Oculus Rift, they can navigate through 3D models based on MRI or CT scans to find the safest path to a hard-to-reach tumours [15].

Psychologists from Stanford shown how VR can change real-life behaviour. In their experiment participants were asked to chop a tree using a virtual mechanical saw. As their future studies shown, people who personally destroyed the virtual forest use less paper in the real world than the other group which was asked to merely imagine the same process [3].

Although virtual reality headsets are strongly associated with the sense of sight, only few research in the field of ophthalmology were found [12,22,27]. For example, in [26] the authors presented a visual field examination using VR glasses and evaluated the reliability of proposed method comparing obtained results with those of the Humphrey perimeter. They found comparable results in terms of field classification, which indicates the possibility of using VR in eye examinations.

3 Colour Blindness

The perception of colours is the ability to distinguish objects, based on sensitivity to the wavelength of light that these objects reflect, emit or absorb. The human nervous system distinguishes colour by comparing response of several types of eye cones to light. Cones are sensitive to different wavelengths of electromagnetic waves, for humans the visible light range is approximately 380–740 nm.

Colour blindness (also known as colour vision deficiency) is the inability to perceive differences between some or all colours that are normally perceived by others. It is usually a birth defect, however it can be a result of physical or chemical damage to the eye, optic nerve, or parts of the brain. The most common form of congenital colour blindness is red-green recognition deficiency, which is split into two different subtypes: protan (protanopia - complete, protanomalia - partial) and deutan (deuteranopia - complete, duteranomalia - partial). In protanopia the visible range of the spectrum is shorter, the part of the spectrum, which is perceived as blue-green by people without any deficiencies, appears grey. The whole visible range of the spectrum is divided in two areas separated by the grey part. Those areas appear as one colour system (different in each area) with different brightness and saturation. Red with a slight tinge of purple appears also as grey. In deuteranopia, that part of the spectrum, which normally appears as green, is perceived as grey, and the visible range of the spectrum is divided by that part into two, each appears as one system of colour. Similar to protanopia, purple also appears as grey, but in contrast the visible range of the spectrum is not contracted. Simulation of normal colour perception along with examples of protanopia and deuteranopia is presented on Fig. 1. In partial colour blindness the parts of spectrum which in protanopia appear as grey, in protanomalia appear as grayish indistinct colour, in deuteranomalia as a indistinct colour close to grey. Consequently, in red-green deficiencies blue and yellow colours appear to be remarkably clear compared with red and green. That can be the distinguishing feature for colour vision deficiencies test.

Colour vision deficiencies (CVD) affect about 10% of the population, mostly men. There is also a very rare group of people that are truly blind to colour,

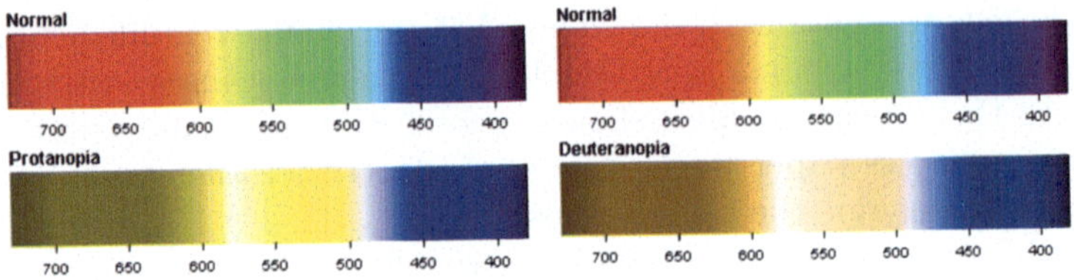

Fig. 1. Simulation of the normal, protanopia and deuteranopia perception of red and green color

it is usually associated with the impairment of central vision with photo-phobia and nystagmus. Colour blindness in blue and yellow is termed as tyritanomalia if partial, and tritanopia if complete, but it is extremely rare. In most cases except for the colour blindness, there is no abnormality in other visual functions. CVD is typically diagnosed with the Ishihara colour test or the Farnsworth-Munsell 100 hue test [19].

The Ishihara test, named after its designer Dr. Shinobu Ishihara, was created to give an opportunity to perform a quick and accurate test of colour vision disorder. It is the most common method used for establishing diagnosis on CVD cases and to distinguish possible causes and types of such disorders. It consists of a set of coloured plates, each contains dots forming a circle, appearing at random in colour and size. The part of the dots form a number or shape clearly visible to those with normal colour vision, and invisible, or difficult to see, to those with CVD.

The correct presentation of the plates, during a test, should be performed in a room lit adequately by daylight. Direct sunlight or electric light may produce some discrepancy in the results producing an alteration in the appearance of shades of colour. If it is impossible to use sunlight, artificial light should resemble the effect of natural one. During the examination the plate should be held 75 cm from the patient and tilted so that the plane of the paper is at right angles to the line of vision. The plates can be shown in any order. The patient has up to three seconds to answer in case of plates 1–17. Plates 18–24 are used when the patient is unable to read the previous series. In this case a line, composed of dots, connecting two points, should be traced within 10 s. The test is available as the original paper version and online. The paper version of plates should be kept in dark, except during use - exposure to sunlight causes a fading of the colour.

The physician has to assess the readings of plates from 1 to 15. If 13 or more numbers are identified properly, the colour vision is regarded as normal. In case of 9 or less correct recognised plates, the colour vision is regarded as deficient. Basing on the patients answer it is easy to distinguish the type of disorder between red-green deficiency and total colour blindness.

4 Method

4.1 Rendering Colours on Smartphone

Human eye contains two different groups of receptors - rods and cones. Colour vision is provided by three types of colour sensitive cones corresponding to red, green and blue. The response curves of these cones have been mapped by researchers - they have shown that colour samples could be matched by combinations of monochromatic colours: red (700 nm), green (546.1 nm) and blue (435.8 nm). By simply adding various amounts of these primary colours a large range of colours could be resembled. However, there are still some colours outside this range that could not be achieved by a pure addition. It was found, that by allowing negative values of red all colours could be matched. Those studies were summed up by identifying three colour matching functions, calculated for the primary colours in three monochromatic radiation wavelengths. On this basis the CIE RGB colour space was determined. Using CIE tabulated colour matching functions, one can calculate the values of RGB space coefficients for any colour with the distribution of the intensity of a given radiation. The colour of any spectral composition can be assigned to a point in the three-dimensional CIE space [21]. The most important colour spaces are included in international standards [25].

To assess the accuracy of the Ishihara colour vision test in VR, we had to make sure that the rendered colours (RGB) were as close as possible to those from the printout (CMYK). The colour calibration of standard monitors can be easily done with colour meter, however Android system does not allow for wide colour management. There are several colormeters, with apps supporting calibration external devices such as tablets and smartphones, on the market. However, these devices can only be calibrated through companion apps. For the purpose of this research ColorTRUE Aware SDK supported with X-Rite ColorMunki Display color meter was used. The X-Rite SDK allows to access the custom colour profile, enabling color management within apps [9].

4.2 Technology and Software

The ability to preform the test on patients who might be incapable of standing up or sitting, immobilised or unable to move on their own, was the main assumption for creating this tool. The test should not involve any action but the only two necessary ones: looking at the plate and orally reporting the number to the physician. In order to achieve those goals a mobile VR headset, namely Samsung Gear VR was selected for the task.

Samsung Gear VR is a headset developed by Samsung Electronics, in collaboration with Oculus. The Gear VR unit acts as the controller while a compatible Samsung Galaxy phone acts as the headset's display and processor. The application presented in this paper consists of VR headset equipped with a Samsung Galaxy S6 smart phone. The whole set is lightweight, portable, does not require any additional cabling while in use, allows to completely separate the user's eyes from any external lighting and fully control the what the user is viewing. It can be used for examining even bedridden patients.

4.3 Application Details

The application consist of two main components - control panel (Fig. 3) for the physicians and display module (Fig. 2) used by the patients. The earlier can be launched on any computer or laptop, it allows to create different sets of displayed plates in order to vary the examination or to change the set for different patients. The latter is an application for a mobile device compatible with a VR headset (a Samsung S6 mobile phone with Gear VR were used for this research), it presents Ishihara plates selected in the control panel. Moreover, the display module can be ported to almost any other VR headsets. A mobile solution was chosen due to its portability and the ability to operate without any additional cabling. This allows for the examination hardware to be used even by bedridden patients. The control panel allows the operator to choose a particular Ishihara plate to be displayed on the VR headset. The full test consists of 38 plates, however the operator can decide how many and which plates will be displayed. Both modules communicate with each other using a wireless network.

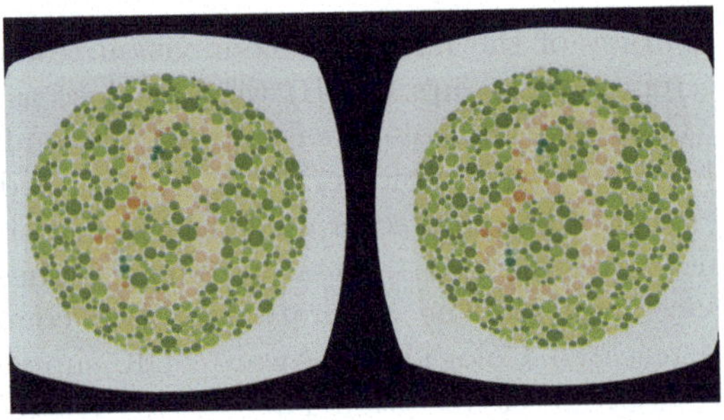

Fig. 2. Screen shot from the mobile application

For this research a test set of 17 plates was created according to Ishihara recommendations [17]. The plates which require tracing were omitted due to limited user interaction capabilities of the mobile VR headset and the assumption that during the examination the patient might be unable to move his, or her, head or arms. The application is implemented with the use of the C# programming language. This allows easy integration of all the software components, SDKs and libraries of this project. The application with 10 Ishihara plates can be made available upon request, for research purposes only.

4.4 Experiments

The application was tested on a group of 30 patients (20 men and 10 women) of different age ranging from 23 to 60. None of the patients reported neurological

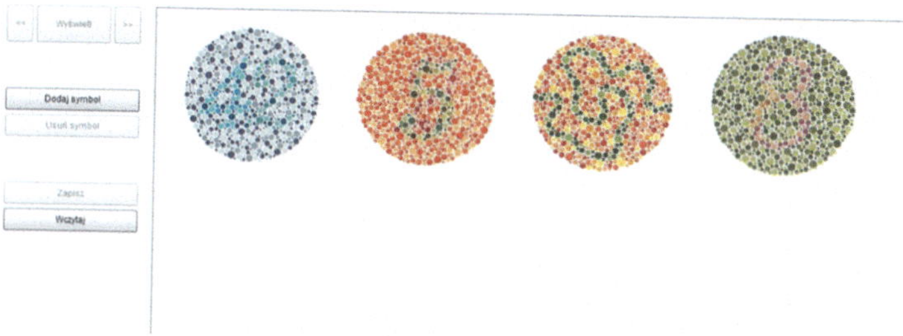

Fig. 3. The control panel of the colour perception test

diseases diagnosed in a medical examination, 4 of the patients reported a diagnosed CVD. More details about the patients are presented in Table 1.

All the testes were performed in a hospital room. Every examination was made during a single session. The patient was lying down on a bed and was asked to put on the VR headset, and during the whole test was supposed to utter the numbers displayed on the screen. First, the left eye was examined, followed by the right eye, and finally both simultaneously. During the test, the physician-operator presented the Ishihara plates in the headset using the control panel and noted the patient's responses. The process of examination is presented on Fig. 4.

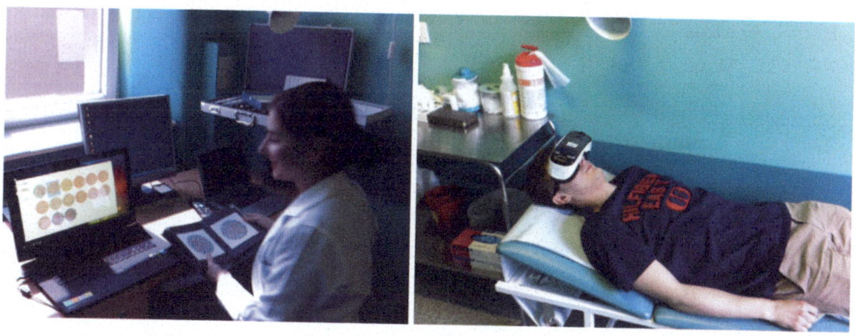

Fig. 4. Testing application process. On left: physician - operator. On right: the patient.

Next, the same patient was subjected to the regular Ishihara colour test with the use of paper plates, according to requirements described in Sect. 3. Similarly as in the case of the VR headset, the left eye was examined first, then the right, and finally both simultaneously. The results were compared with the previous examination and composed in Table 1.

Patients 1 to 25 (listed in Table 1) recognised all presented plates correctly, both in the headset and the paper version of the test. The results of patients with CVD (numbered 25 to 30) are presented in detail in Table 2, which shows

Table 1. Detailed information of patients undergoing the examination

Number	Sex	Age	Neurological diseases	Diagnosed colour blindness
1	Woman	23	NO	NO
2	Woman	23	NO	NO
3	Woman	23	NO	NO
4	Woman	22	NO	NO
5	Man	23	NO	NO
6	Man	23	NO	NO
7	Woman	32	NO	NO
8	Man	40	NO	NO
9	Woman	50	NO	NO
10	Woman	27	NO	NO
11	Man	26	NO	NO
12	Woman	27	NO	NO
13	Woman	23	NO	NO
14	Woman	23	NO	NO
15	Man	22	NO	NO
16	Woman	33	NO	NO
17	Man	32	NO	NO
18	Woman	60	NO	NO
19	Man	60	NO	NO
20	Woman	58	NO	NO
21	Man	59	NO	NO
22	Woman	32	YES	NO
23	Woman	34	NO	NO
24	Woman	23	NO	NO
25	Woman	25	NO	NO
26	Man	23	NO	YES
27	Woman	43	NO	NO
28	Man	33	NO	YES
29	Man	40	NO	YES
30	Man	25	NO	YES

what they recognised on printed Ishihara plates in comparison with their representatives in VR.

In our study patients with or without a CVD responded identically to examination using the printouts and the headset. The diagnosis was made for patients no. 26, 27, 28, 29 was mild protanomalia and mild deuteranomalia (mildly affecting red–green hue discrimination) and mild protanomalia and strong

Table 2. Analysis of answers to each plate: X - plate cannot be read, Blank space - the reading is indefinite, The numerals in parenthesis - they can be read but they are comparatively unclear

Plate	No CVD	Patient no. 26	Patient no. 27	Patient no. 28	Patient no. 29	Patient no. 30
1	12	12/12	12/12	12/12	12/12	12/12
2	8	3/3	8/8	8/8	3/3	3/3
3	29	70/70	29/29	20/20	70/70	20/20
4	5	2/2	5/5	9/9	2/2	2/2
5	3	5/5	3/3	8/8	5/5	5/5
6	15	17/17	15/15	15/15	17/17	17/17
7	74	21/21	21/21	24/24	21/21	21/21
8	6	X/X	6/6	X/X	X/X	X/X
9	45	X/X	15/15	X/X	X/X	X/X
10	5	X/X	5/5	5/5	X/X	X/X
11	7	X/X	7/7	1/1	X/X	X/X
12	16	X/X	16/16	X/X	X/X	X/X
13	73	X/X	23/23	X/X	X/X	X/X
14	X	45/45	X	X/X	45/45	X/X
15	X	5/5	X	X/X	5/5	X/X
16	26	26/26	26/26	26/26	26/26	2
17	42	42/42	42/42	42/42	4/4	4

deuteranomalia for patient 30. The Samsung Gear VR - Virtual Reality Headset colour vision test using pseudoisochromatic color plates can deliver test results comparable to those yielded by conventional colour vision tests. Further investigations on larger groups of patients are required to confirm our results.

5 Conclusion and Future Works

Early diagnosis of colour blindness is crucial to support an individual affected by this visual perception alteration. However, in their daily practice ophthalmologists frequently face the problem of examining vision abilities of immobilised patients. The best example of such a case are people affected by stroke or demyelinating disease, who struggle with movement disorders. Even relatively simple examinations become impossible to perform.

According to the World Health Organization, 15 million people suffer stroke worldwide each year: 5 million die and another 5 million are permanently disabled [10]. A stroke affects the brain, therefore, it can produce visual defects often accompanied by physical disability associated with difficulty or inability to perform active movement, problems with balance and coordination and postural disorders.

Multiple Sclerosis (MS), which is just one example of demyelinating disease, can affect the brain and/or spinal cord, causing a wide range of potential symptoms, including problems with vision, arm or leg movement, sensation or balance. Nowadays, this is the most widespread disabling neurological condition of young adults. Multiple Sclerosis Foundation estimates that about 2.5 million people around the world have MS [1]. The main symptoms of MS include difficulty walking, problems with balance, coordination and vision.

These are just two examples. There are many more diseases causing physical disability and may require ophthalmology consultation (i.e. head and spinal trauma). However, many cases are geriatric patients living alone or in rural areas with no possibility to get to a hospital for a proper examination.

Computer systems and new technology such as virtual reality goggles are becoming an essential part of physician's work. It is a rapidly growing field of research, however at the moment there is no available VR application supporting ophthalmologists. Such a tool would be useful for diagnosis and treatment of immobilised patients.

Our idea is to create a low-cost, intuitive and transportable device which simulates various ophthalmological examinations for immobilised patients. We propose the interactive tool based on virtual reality, which will be developed to improve the quality of life of such patients. The main objective of the proposed tool, as well as the examination, is to substantially improve the examination process and as a result the quality of life of people with impaired movement and to allow them to better function in the society. The solution could be used outside of hospitals, anywhere, even in the developing world.

It would also take off some pressure from the physicians who had to make important decisions without the possibility of patient's examination due to his or her immobilisation. With the proposed tool we intend to support physicians, facilitate their work with immobilised patients and facilitate treatment documenting process, making their work easier, more controllable and less stressful. The tool is based on the actual needs of patients and physicians, through their continuous feedback during development process.

Promising results of this research confirm our assumptions that VR can be useful in ophthalmology. We have just started with developing other types of ophthalmic examinations such as: visual acuity, visual field, contrast sensitivity and other colour perception tests. All above mentioned examinations are the valuable source of information which can provide a basis for subsequent research concerning diagnosis of various types of disorders and the degree of impairment.

The tool described in this paper is the first successful attempt to port ophthalmological examinations to a VR headset. The authors intend to continue developing the application and supplement the applications with other examinations.

References

1. Bohemia interactive simulations. http://www.healthline.com/health/multiple-sclerosis/facts-statistics-infographic
2. Deep - a meditative and psychoactive VR game. http://owenllharris.com/deep/
3. New virtual reality research–and a new lab–at stanford. http://news.stanford.edu/news/2011/april/virtual-reality-trees-040811.html
4. Oculus rift. https://www.oculus.com/
5. Video games for rehabilitation. http://ict.usc.edu/news/video-games-for-rehabilitation/
6. Virtual reality helps home bound seniors enjoy the outdoor. http://www.paloaltoonline.com/news/2014/05/04/virtual-reality-helps-home-bound-seniors-enjoy-the-outdoor
7. Virtual reality therapy for phobias. http://psychiatry.duke.edu/divisions/general-psychiatry/virtual-reality-therapy-phobias
8. Virtual reality (VR): a billion dollar niche. https://www2.deloitte.com/global/en/pages/technology-media-and-telecommunications/articles/tmt-pred16-media-virtual-reality-billion-dollar-niche.html
9. X-Rite colorTRUE aware SDK. http://blog.xritephoto.com/
10. World Health Report. World Health Organization (2002)
11. Berg, L., Vance, J.: Industry use of virtual reality in product design and manufacturing: a survey. Virtual Reality **21**(1), 1408–1411 (2017)
12. Borgersen, N.J., Skou Thomsen, A.S., Konge, L., Sørensen, T.L., Subhi, Y.: Virtual reality-based proficiency test in direct ophthalmoscopy. Acta ophthalmol. **96**(2), e259–e261 (2018)
13. Burdea, G., Coiffet, P.: Virtual Reality Technology. Wiley, New York (2003)
14. Cameirão, M.S., i Badia, S.B., Verschure, P.F.M.J.: Virtual reality based upper extremity rehabilitation following stroke: a review. J. Cyberther. Rehabil. **1**(1), 63–74 (2008)
15. Gardner, J.A.: Augmented and Virtual Reality in Medicine: 6 Applications We're Keeping Our Eye on. MedTech, Boston (2016)
16. Iozzio, C.: Virtually revolutionary. Sci. Am. **311**, 26 (2014)
17. Ishihara, S.: Test for Color Blindness: 24 Plates. Kanehara Trading Inc., Tokyo (1972)
18. Kandalaft, M., Didehbani, N., Krawczyk, D., Allen, T., Chapman, S.: Virtual reality social cognition training for young adults with high-functioning autism. J. Autism Dev. Disord. **43**(1), 34–44 (2013)
19. Kinnear, P.R., Sahraie, A.: New farnsworth-munsell 100 hue test norms of normal observers for each year of age 5–22 and for age decades 30–70. Br. J. Ophthalmol. **86**(12), 1408–1411 (2002)
20. Krueger, M.: Artificial Reality 2. Addison-Wesley Professional, Reading (1991)
21. Lindon, J., Tranter, G.E., Koppenaa, D.: Encyclopedia of Spectroscopy and Spectrometry. Elsevier, Amsterdam (2016)
22. Nesaratnam, N., Thomas, P., Vivian, A.: Stepping into the virtual unknown: feasibility study of a virtual reality-based test of ocular misalignment. Eye **31**(10), 1503 (2017)
23. Ortiz-Catalan, M., Sander, N., Kristoffersen, M.B., Håkansson, B., Brånemark, R.: Treatment of phantom limb pain (PLP) based on augmented reality and gaming controlled by myoelectric pattern recognition: a case study of a chronic PLP patient. Front. Neurosci. **8**, 24 (2014)

24. Rizzo, A., Pair, J., Graap, K., Manson, B., McNerney, P.J., Wiederhold, B., Wiederhold, M., Spira, J.: A virtual reality exposure therapy application for Iraq war military personnel with post traumatic stress disorder: from training to toy to treatment. In: Roy, M. (ed.) NATO Advanced Research Workshop on Novel Approaches to the Diagnosis and Treatment of Posttraumatic Stress Disorder, pp. 235–250. IOS Press, Washington D.C. (2006)
25. Tkalcic, M., Tasic, J.F.: Colour spaces - perceptual, historical and applicational background. In: The IEEE Region 8 EUROCON 2003: Computer as a Tool (2003)
26. Tsapakis, S., Papaconstantinou, D., Diagourtas, A., Droutsas, K., Andreanos, K., Moschos, M.M., Brouzas, D.: Visual field examination method using virtual reality glasses compared with the humphrey perimeter. Clin. Ophthalmol. (Auckland, NZ) **11**, 1431–1443 (2017)
27. Xu, L., Huang, M., Lan, J., Huang, W., Wang, X., Zhang, G., Li, X., Shasha, P., Chu, H., Wiederhold, B.K., et al.: Assessment of binocular imbalance with an augmented virtual reality platform in a normal population. Cyberpsychol. Behav. Soc. Netw. **22**(2), 127–131 (2019)

Texture Description for Classification of Fine Needle Aspirates

Łukasz Jeleń(✉)

Department of Computer Engineering, Wrocław University of Science and Technology, wyb. Wyspiańskiego 27, 50-370 Wrocław, Poland
lukasz.jelen@pwr.edu.pl

Abstract. In breast cancer diagnosis, fine needle aspiration biopsy is an important diagnostic tool. It is used to estimate cancer malignancy grade that is further required for treatment determination. In this paper we describe a scheme based on pattern recognition and image processing techniques for automatic breast cancer malignancy grading from cytological slides of fine needle aspiration biopsies. To determine a malignancy classification of the slide we propose to extract textural features of nuclei with an application of local binary patterns. Based on texture determination, we present an improved classification system for cancer malignancy grading.

Keywords: Breast cancer classification · Local binary patterns · Texture classification · Fine needle biopsy

1 Introduction

In recent years computerized cancer diagnosis plays a crucial role in cancer examination procedure. According to the Polish National Cancer Registry [1], in Poland alone, the number of reported cancer cases doubled in the last three decades. According to statistics, around 300 000 middle-aged women in European Union will be diagnosed with breast cancer. This makes it the highest oncological problem affecting developed countries. About 89 000 of cases will be fatal. To reduce this high ratio, a number of computer aided techniques were developed to make the process more reliable and faster. The aim here is to diagnose a cancer is an early development stage. Cancers in their early stages are more vulnerable to treatment and we can assume that most of the diagnosed cases will lead to a successful recovery. Conversely, most advanced cancers stages are usually almost impossible to treat.

To overcome the problem of late diagnosis, screening mammographic tests were introduced and when a suspicious region in the image is noted, a fine needle aspiration biopsy (FNA) is taken. FNA is a minimally invasive method to extract a small tissue sample of the questionable breast tissue. This procedure allows for the description of the type and malignancy grade of the cancer. Depiction of malignancy plays a crucial role in the determination of patient's treatment.

J. Korbicz et al. (Eds.): PCBEE 2019, AISC 1033, pp. 107–116, 2020.
https://doi.org/10.1007/978-3-030-29885-2_10

Cancer malignancy is graded according to the three point scale that was first proposed by Bloom and Richardson in 1957 [4]. The proposed scheme was originally derived for assessment of malignancy from histopathological slides and today is very popular among pathologists. They use it for grading not only histological but also cytological tissue.

In literature we can find a numerous approaches to computerized cancer diagnosis [11]. Most of these approaches deal with the classification between benign and malignant cases [12] of histological tissue. To the best of authors knowledge, the first description of the computerized breast cytology classification problem was provided by Wolberg *et al.* in 1990 [21]. In this work, authors presented an application of a multi-surface pattern separation method applied to cancer diagnosis. Their idea was wildly propagated among researchers what led to the description of other breast cancer computer aided systems. In 2009, Malek *et al.* [15] described an application of active contours to nuclei segmentation and a fuzzy c-means classifier for classification of 80 malignant and 120 benign cases and in 2010, Niwas *et al.* provided a description of texture features for cancer classification [16]. In 2011, Jeleń *et al.* [13] described a neural network approach to breast cancer malignancy classification and in 2016, Jeleń *et al.* [12] provided a wide description of morphological features used for breast cancer classification. In that same work they provided an extended research on feature classification power and selection providing information about the best performing features.

Most recent approaches include a deep learning and convolutional neural network approaches. In 2017, Araújo *et al.* [3] used convolutional neural networks for feature extraction. In 2018, Kowal *et al.* [14] presented a similar study to this described in [12] on feature selection problem. Authors used convolutional neural networks for segmentation and have performed feature selection test on 500 images of benign and malignant cases.

For complete cancer diagnosis it is necessary to also determine the malignancy stage of cancer which is called a malignancy grading, and this is the problem that we are focusing on in this paper. The pre-screening process before taking an FNA results in the situation in which the biopsy slide being classified is nearly always malignant. Henceforth, in this work, we deal with a malignancy grading problem instead of malignancy diagnosis. In 2017, Alsaedi *et al.* [2] described six computer-aided grading frameworks assigning malignancy grades to cytological images of FNA biopsies of breast cancer.

In this paper we address an important issue of texture features that can be used for computer-aided grading of cancer malignancy. Texture is a very valuable source of information when analyzing structures in an image [9,19] and therefore, a textural description of the nucleus is an important feature of the computerized breast cancer malignancy grading system. According to Bloom and Richardson textural analysis allows for the determination of the chromatin distribution and detection of mitotic cells.

The aim of the described study is to present an improved classification scheme that takes into consideration a textural description of nuclei in the slide and will determine the malignancy of breast cancer tissue. Additionally we will describe a set of morphological features that are used for comparison purposes.

2 Database Description

The database used in this study includes images of fine needle aspiration biopsy that were collected by dr. Artur Lipiński during an FNA examination at the Department of Pathology and Oncological Cytology of the Medical University of Wrocław, Poland. Slide preparation technique includes staining with Haematoxylin and Eosin, known as HE technique. The choice of the staining agents allows for visualizing a nuclei with purple and black dyes, cytoplasm with shades of pink and red blood cells with orange/red dyes. The setup consisted of an Olympus BX 50 microscope with CCD–IRIS camera mounted on the head of the microscope and a MultiScan Base 08.98 software. Such a system allowed for recording of images with the resolution of 96 dots per inch (dpi) and a size of 764×572 pixels.

There are 218 images in the database that represent two classes of cancer malignancy, namely intermediate (G2) and high (G3) malignancy grades (see Fig. 1). The lack of low malignancy images is caused by the fact that these cases very rarely require FNA and in recent years there was only a few of them at the Medical University of Wrocław. There are 134 cases of intermediate and 84 of high malignancy. For all cases, breast tissue was surgically removed during follow-up biopsy and it was histopathologically graded using the Bloom–Richardson [4] grading scale which confirmed the FNA grading. Therefore all the cases in our database were histopathologically validated.

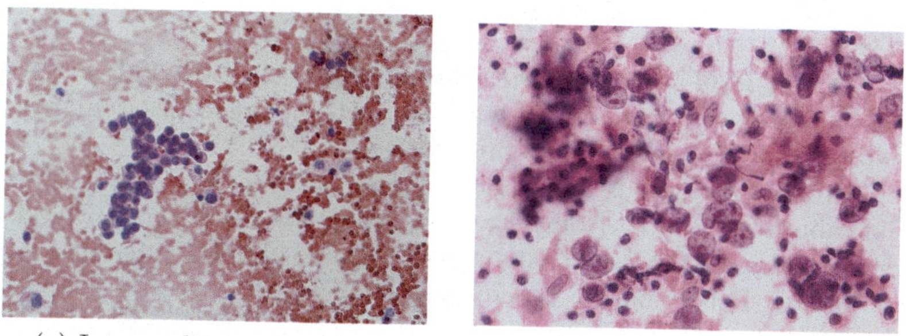

(a) Intermediate malignancy (b) High malignancy

Fig. 1. Example of images for one case in the database.

3 Methods

Recently a number of classification frameworks were presented [2]. All of them use shape descriptors to represent relevant Bloom-Richardson features. In [12] except for features definition, authors preformed a comparison of their discriminatory powers to choose a subset of the best performing features. According to the authors, there are two types of features required for proper malignancy classification. These features are described in Sect. 3.2. To correctly classify breast cancer malignancy we need to calculate features at two different magnification of the slide. Lower magnification will allow for the determination of cells behavior and their ability to form groups. This simply means if there are more loosely spread cells in the image, the more malignant the case should be. On higher magnification we will be able to determine cells morphology as well as textural differences between nuclei.

In this paper we propose to use only texture description to extract features from images of higher magnification images. In Sect. 3.1 we will describe Local Binary Patterns that were used as a texture measure. Last part of this section is devoted to the description of classification scheme that was used for the evaluation of the textural features performance.

3.1 Texture Description

Textural features are used to measure the texture information of the image [9,19]. Here, the texture of the nucleus is taken into consideration. To extract textural features, a local binary patterns method was applied. This method was first proposed by He and Wang in 1990 [20]. Here, we use a variation of the method that was described by Ojala *et al.* in 2002 [17]. They described an efficient approach based on the local binary patterns and nonparametric discrimination of sample and prototype distributions to rotation invariant texture classification. Authors defined a texture Tx of a graylevel image in a local neighborhood as a joint distribution of gray levels as represented in Eq. 1.

$$Tx = t(g_c, g_0, ..., g_{P-1}), \tag{1}$$

where g_c is a gray level associated with the center of neighborhood, g_p for $p \in \{0, ..., \mathrm{P}-1\}$ is a gray level associated with P pixels arranged in circular manner, equally spaced on a circle with a defined radius.

To obtain the Gray–Scale invariance authors subtracted the gray level value of the neighborhood center from gray level values of the neighborhood pixels and further scaling of the gray scale using only signs of the differences and not the exact values. This situation is represented by Eq. 2.

$$Tx \approx t(s(g_0 - g_c), s(g_1 - g_c), ..., s(g_{P-1} - g_c)), \tag{2}$$

where

$$s(x) = \begin{cases} 1 & \text{if } x > 0 \\ 0 & \text{if } x < 0 \end{cases}.$$

According Ojala *et al.* adding a 2^p binomial factor to each scaling term, the characterization of the local spatial image texture structure can be rewritten by Eq. 4.

$$LBP_{P,R} = \sum_{p=0}^{P-1} s(g_p - g_c)2^p. \tag{3}$$

To further remove the effect of rotation, authors defined a following relation:

$$LBP^{ri}_{P,R} = \min\{ROR(LBP_{P,R}, i) | i = 0, 1, ..., P-1\}, \quad (4)$$

where $ROR(x, i)$ is a circular bit-wise right shift on the P-bit number x i times, which means clockwise rotation of the neighborhood until maximal number of the most significant bits will be 0.

3.2 Morphological Features

As mentioned previously, there are two image magnifications that are used for breast cancer malignancy grading. Morphological feature extraction in this paper takes both of these types into consideration. The magnifications of images corresponds to the magnifications used during a cytological examination of a breast tissue. For the calculation of the low magnification features, images recorded with 100x magnification are used and for the calculation of high magnification features, images obtained with 400x magnification are used.

Low Magnification Features (LM) – these features are defined based on the number of groups and their area (A_{100}). The first feature is calculated as a number of groups in the image that weren't removed during the segmentation process and the area is calculated as the average number of nuclei pixels. The third feature describes nuclei dispersion within an image and is defined as a variation of cluster areas (A_c) which is determined by the following equation [12]:

$$\frac{1}{D} = \frac{1}{n-1} \sum_{i=1}^{n} (A_c - A_{100})^2, \quad (5)$$

High Magnification Features (HM) – the feature vector constructed by extraction of high magnification features includes 30 features calculated according to [12]. These features are divided into binary, momentum, histogram, textural and color based features.

Binary features that were calculated for set of nuclei $N = \{N_1, N_2, ..., N_n\}$ from a binary image ($\mathcal{I}$), where N is defined as a collection of all connected components and the nucleus N_i. Using this definition, the following features are extracted [12]:

- **Area** (A_{400}) – A_i is defined as the sum of all nuclei pixels of the nucleus N_i.
- **Perimeter** ($Perim$) – is a length of the nuclear boundary of a nucleus N_i that is approximated by a length of the polygonal approximation of the boundary.
- **Convexity** ($Conv$) – is defined as a ratio of the nucleus area and the area of the minimal convex polygon that contains the nucleus, called a convex hull.

- **Eccentricity** (Ecc) – calculated as a ratio of the distance between focal points of an ellipse matched with a nucleus having the same second–moments as the segmented nuclei, and its major axis length.
- **Centroid** ($(x - Ctr, y - Ctr)$) – For each nucleus, the centroid (Ctr_i) is a point ($\overline{x}_i, \overline{y}_i$) called a center of mass of the extracted nucleus along each row (X) and column (Y).
- **Orientation** (Or) – For the coordinate system placed at the centroid ($\overline{x}_i, \overline{y}_i$) of the nucleus the orientation is defined as

$$Or_i = tan(2\theta_i), \tag{6}$$

where the angle θ_i is measured counterclockwise from the x–axis.
- **Projection** ((x-Prj, y-Prj)) – to calculate the feature Prj_i for each nucleus, projections along rows (x-Prj_i) and columns (x-Prj_i) are performed.

Momentum–based features use normalized central moments, η_{ij} to calculate rotation, scaling and translation invariant features as described in [19]. Using these η_{ij} values, seven momentum–based features, were calculated.

Another set of features is histogram based. The histogram is treated as a probability distribution function of grey level values from the red channel in the image and describes their frequencies in that image [12]. Based on this definition, five histogram-based features were calculated. Additionally statistical features of a texture were calculated based on the gray level co–occurrence matrix that describes the relationships between a pair of pixels and their grey levels [19]. Assuming that the distance between the pixels and the directions are given four textural features were extracted.

Last set of features is based on the spherical coordinate transform applied to the RGB image. Determining a histogram of the converted image 5 color–based features were computed.

3.3 Malignancy Classification

To classify the cytological FNA tissue we build four classifiers that take a feature vector (see Sect. 3.2) as an input and respond with a two element output vector $(1, 0)^T$ for intermediate malignancy and $(0, 1)^T$ for high malignancy. In the remainder of this section, classification methods are presented and, in the following section, their ability to classify malignancies is studied.

Decision Trees (Tree) – most of the traditional pattern recognition algorithms are based on the feature vectors that are real–valued and some kind of metric can be applied to them [6]. Tree classifiers on the other hand are able to solve classification problems that involve nominal data such as a list of attributes like fruit colors and sizes.

Decision trees are constructed in a way where the classes are held in the leaves of the tree and the decision rules are kept in the internal nodes including the root [18]. Classification with decision trees seeks a path from the root to the correct leaf creating a decision path. Here, we make use of the CART (Classification And Regression Trees) method described by Breiman *et al.* [5] that provides a general framework for decision tree construction. In general, the tree–growing process declares the node to be a leaf or finds another property that can be used to split the data represented at the node into subsets creating new nodes. This is process is run recursively until all the data is represented by the constructed tree.

Linear Discriminant Analysis (LDA) – is one the simplest classification algorithms that requires a construction of a, so called, decision boundary. This boundary is constructed as discriminant function of a form presented by [6]:

$$g(x) = w^t x + \omega_0, \tag{7}$$

where w is a weight vector and ω_0 is a threshold weight.

In a general way we can rewrite the the decision boundary as:

$$g(x) = \omega_0 + \sum_{i=1}^{d} \omega_i x_i. \tag{8}$$

Here, ω_i represents the components of w. If we add to Eq. 8 terms related to products of pair of x, a definition of a quadratic discriminant function (Eq. 9) will be determined.

$$g(x) = \omega_0 + \sum_{i=1}^{d} \omega_i x_i + \sum_{i=1}^{d} \sum_{j=1}^{d} \omega_i x_i. \tag{9}$$

To train an LDA based classifier we need to calculate a decision boundary according to Eq. 9.

Support Vector Machines (SVM) – are used to separate two or more classes of patterns or data points by constructing a boundary between them [10]. An unknown point will be classified according to its orientation with respect to the boundary. To estimate the boundary between classes we support vectors from each class. This is performed as an iterative approach that minimizes an error function.

$$\frac{1}{2} w^T w + C \sum_{i=1}^{N} \varepsilon_i \tag{10}$$

with the following restrictions:

$$y_i(w^T \phi(x_i) + b) \geq 1 - \varepsilon_i \quad and \quad \varepsilon_i \geq 0, \quad i = 1, ..., N \tag{11}$$

where C and b are constants, w is the weight vector, ε_i is a bias value that deals with overlapping cases and ϕ is a kernel function that transforms input data into the feature space. In this work we have used a linear kernel as described by Huang *et al.* in [10].

Classifiers Ensembles (Ensemble) – enhance a single–model classifiers offering hybrid approaches [8] for a better classification of two-class classification problems. Here, we use an AdaBoost. M2 approach, which is defined for a multi-class problems. The algorithm seeks an approximation of the Bayes classifier $B(x)$, where x is a set of observations, based on a combination of other classifiers such as decision trees [7]. During training, weights of misclassified points are increased and a second classifier in the ensemble will use these new weights. This approach is called boosting. Such a procedure is iterated and can lead to a creation of numerous classifiers [22]. After training, each built classifier will be assigned with a score. Such a combination of classifiers will define an AdaBoost classifier as a classification model. All the classifiers in the final model are linearly connected.

4 Results and Discussion

In this paper we discuss a problem of malignancy classification of fine needle aspirates. For this purpose we have calculated textural features based on local binary patterns as described in Sect. 3.1. For our calculations we have used 8 pixels neighborhood. Additionally a set of morphological features proposed by Jeleń *et al.* [12] was extracted as described in Sect. 3.2.

Obtained feature vectors were presented as an input to four different classifiers to check their ability to distinguish between two malignancy classes: G2 and G3. Tests were performed using a 5–fold cross–validation method. Presented Table 1 classification results are averages over the 5 folds used for testing.

To perform a complete analysis of the texture description, we have calculated local binary patterns not only to a graylevel image (LBP features) but additionally for a combination of R, G and B channels. Table 1 summarizes obtained classification results for these features also. Further analysis shows that red and blue channel together provided the best classification of malignancy cases. Additionally we included morphological features of low magnification images to show the performance of a complete malignancy classification scheme.

From the results presented in this section we can see that classification based only on the morphological features from high magnification images provided the worst results. Textural features alone outperformed them but when low magnification features were taken into consideration, the proposed scheme was able to classify malignancy with accuracy around 90% or better. Inclusion of low magnification features to the local binary patterns calculated for a red and blue channel provided the best classification with an accuracy of 97.6% for an SVM classifier. From Table 1 we can also see that this classifier outperforms all other tested classifiers for textural features. Classification results obtained for morphological features clearly show that classifier ensembles perform much better.

Table 1. Classification accuracies for different feature vectors and classifiers.

Classifier	morphological features		LBP features		LBP for RB channels	
	without LM	with LM	without LM	with LM	without LM	with LM
Tree	66.4%	92.2%	82.1%	92.1%	82.6%	92.6%
LDA	67.3%	94.1%	79.8%	89.8%	79.4%	88.9%
SVM	64.4%	89.4%	87.1%	**97.1%**	88.1%	**97.6%**
Ensemble	77.2%	*96.0%*	83.0%	93.0%	83.5%	93.2%

	LBP for RG channels		LBP for GB channels	
	without LM	with LM	without LM	with LM
Tree	81.2%	89.8%	80.3%	89.7%
LDA	83.5%	91.4%	79.8%	85.5%
SVM	86.3%	**95.9%**	87.2%	**97.2%**
Ensemble	84.7%	93.6%	84.4%	94.4%

5 Conclusions and Further Work

In this paper we conducted a research on malignancy grade classification based on textural features calculated with local binary patters. We have compared classification accuracies of texture features extracted for different color channels and graylevel image against accuracies obtained for classification of morphological features. As described in the previous section, the best performing scheme included local binary patterns with low magnification features yielding an accuracy of 97.6% when SVM classifier was used. The worst performing setup included only morphological features extracted form high magnification features. Considering the above, we can conclude that textural representation of fine needle aspirates would be very good choice for breast cancer malignancy classification.

In the further research the classification framework that includes images for all three malignancy classes, should be tested. Introducing the low malignancy class will make the classification more difficult and therefore even a more interesting problem. Also, it would be necessary to perform per nuclei analysis but the problem of occluding cells needs to be addressed and taken into consideration.

References

1. Polish National Cancer Registry, December 2019
2. Alsaedi, M., Fevens, T., Krzyżak, A., Jeleń, Ł.: Cytological malignancy grading systems for fine needle aspiration biopsies of breast cancer. In: 2017 IEEE International Conference on Bioinformatics and Biomedicine (BIBM), pp. 705–709 (2017)
3. Araújo, T., Aresta, G., Castro, E., Rouco, J., Aguiar, P., Eloy, C., Polónia, A., Campilho, A.: Classification of breast cancer histology images using convolutional neural networks. PLOS ONE **12**(6), 1–14 (2017)
4. Bloom, H.J.G., Richardson, W.W.: Histological grading and prognosis in breast cancer. Br. J. Cancer **11**, 359–377 (1957)

5. Breiman, L., Friedman, J., Stone, C.J., Olshen, R.A. (eds.): Classification and Regression Trees. Chapman & Hall, Boca Raton (1993)
6. Duda, R.O., Hart, P.E., Stork, D.G.: Pattern Classification, 2nd edn. Wiley Interscience Publishers, New York (2000)
7. Freund, Y., Schapire, R.E.: A decision-theoretic generalization of on-line learning and an application to boosting. J. Comput. Syst. Sci. **55**(1), 119–139 (1997)
8. Galar, M., Fernandez, A., Barrenechea, E., Bustince, H., Herrera, F.: A review on ensembles for the class imbalance problem: bagging-, boosting-, and hybrid-based approaches. IEEE Trans. Syst. Man Cybern. Part C Appl. Rev. **42**(4), 463–484 (2012)
9. Haralick, R.M.: Statistical and structural approaches to texture. IEEE Trans. Syst. Man Cybern. **67**(5), 786–804 (1979)
10. Huang, T.M., Kecman, V., Kopriva, I.: Kernel based algorithms for mining huge data sets: supervised, semi-supervised, and unsupervised learning. Springer (2006)
11. Irshad, H., Veillard, A., Roux, L., Racoceanu, D.: Methods for nuclei detection, segmentation, and classification in digital histopathology: a review-current status and future potential. IEEE Rev. Biomed. Eng. **7**, 97–114 (2014)
12. Jeleń, Ł., Krzyżak, A., Fevens, T., Jeleń, M.: Influence of feature set reduction on breast cancer malignancy classification of fine needle aspiration biopsies. Comput. Biol. Med. **79**, 80–91 (2016)
13. Jeleń, Ł., Lipiński, A., Detyna, J., Jeleń, M.: Grading breast cancer malignancy with neural networks. Bio-Algorithms Med. Syst. **2**(14), 47–54 (2011)
14. Kowal, M., Skobel, M., Nowicki, N.: The feature selection problem in computer assisted cytology. Int. J. Appl. Math. Comput. Sci. **28**(4), 759–770 (2018)
15. Malek, J., Sebri, A., Mabrouk, S., Torki, K., Tourki, R.: Automated breast cancer diagnosis based on GVF-snake segmentation, wavelet features extraction and fuzzy classification. J. Signal Process. Syst. **55**(1–3), 49–66 (2009)
16. Niwas, S.I., Palanisamy, P., Sujathan, K.: Complex wavelet based texture features of cancer cytology images. In: 2010 International Conference on Industrial and Information Systems (ICIIS), pp. 348–353 (2010)
17. Ojala, T., Pietikainen, M., Maenpaa, T.: Multiresolution gray-scale and rotation invariant texture classification with local binary patterns. IEEE Trans. Pattern Anal. Mach. Intell. **24**(7), 971–987 (2002)
18. Safavian, S.R., Landgrebe, D.: A survey of decision tree classifier methodology. IEEE Trans. Syst. Man Cybern. **21**(3), 660–674 (1991)
19. Umbaugh, S.E.: Digital Image Processing and Analysis: Human and Computer Vision Applications with CVIPTools, 2nd edn. CRC Press, New York (2011)
20. Wang, L., He, D.-C.: Texture classification using texture spectrum. Pattern Recognit. **23**(8), 905–910 (1990)
21. Wolberg, W.H., Mangasarian, O.L.: Multisurface method of pattern separation for medical diagnosis applied to breast cytology. Proc. Natl. Acad. Sci. U.S.A. **87**, 9193–9196 (1990)
22. Zhu, J., Rosset, S., Zou, H., Hastie, T.: Multi-class adaboost. Stat. Interface **2**, 349–360 (2006)

Computer-Aided Medical Diagnosis

Novel Digital Pathology Method for Computer-Aided Analysis of Histopathological Images Obtained from Dystrophic Muscle Biopsies

Wlodzimierz Klonowski[1(✉)], Bozenna Kuraszkiewicz[1], Anna M. Kaminska[2], and Anna Kostera-Pruszczyk[2]

[1] Nalecz Institute of Biocybernetics and Biomedical Engineering PAS, Warsaw, Poland
wklon@ibib.waw.pl
[2] Department of Neurology, Medical University of Warsaw, Warsaw, Poland

Abstract. Background: Despite the introduction of a full range of genetic diagnostic tests and sophisticated techniques in modern pathology, interpretation of histopathological images obtained from muscle biopsies remains important in the daily practice of neuropathology since it can give indications of the severity and the rate of progression of neuromuscular disease. In this paper, we propose a simple and time saving method for quantitative assessment of severity of Duchenne Muscular Dystrophy (DMD) based on computer-aided analysis of histopathological images obtained from biopsies of dystrophic muscles.

Methods: The method that we propose, colour filtration pixel-by-pixel of the whole virtual slides (*CFPP method*), enables semi-quantitative evaluation of morphological structure of the muscular tissue. We retrospectively analyzed digital microscopic images of DMD muscle tissue from patients and healthy persons. The images were acquired with x400 magnification, from original microscopy slides coming from biopsies. For comparison, we constructed histograms of muscle fiber diameters with the areas of measurements selected randomly from the center of each section.

Results: CFPP method allows to distinguish DMD tissue from normal muscle tissue, as well as it makes possible to assess quantitatively the severity of DMD through assessment of connective tissue index, a proliferation index dedicated to the grading of DMD. The method required only choosing of small representative regions of muscle tissue, of fatty and connective tissue, and of nuclei.

Conclusions: Results demonstrate usefulness of the proposed method in neuropathological assessment of DMD severity. It is a promising technique that can assist tissue-based diagnosis and can be used for virtual slides evaluation.

Keywords: Digital pathology · Biomedical engineering · Histochemistry · Image analysis · Computer-assisted analysis · Biopsy · Dystrophy · DMD · ILF method · CFPP method

J. Korbicz et al. (Eds.): PCBEE 2019, AISC 1033, pp. 119–130, 2020.
https://doi.org/10.1007/978-3-030-29885-2_11

1 Introduction

Victor Dubowitz in 1970's revolutionized the role of muscle biopsies in the diagnosis of muscular dystrophies [1] making muscle biopsy then a gold standard in diagnosis of these diseases. Recent developments of automated systems for analysis of genetic material have caused that nowadays histopathological methods based on images from muscle biopsies [2] are applied not so often as in previous decades, since they are more time-consuming. Nevertheless, histopathological evaluation of muscle biopsies are still clinically applied [3, 4] along with other diagnostic methods as muscle MRI or 6 min walk test. Immunohistochemistry analysis is particularly essential for the diagnosis of patients with a strong clinical suspicion of muscular dystrophy but in the absence of mutations detected by molecular genetics. The aim of this study is to develop simple and quick semi-automatic method for analysis of histopathological images to assist evaluation of muscle dystrophies [5], especially Duchenne Muscular Dystrophy (DMD) [6]. The method is based on computer-aided analysis of images - **c**olour **f**iltration **p**ixel-by-**p**ixel (*CFPP method*) of the whole histopathological virtual slides. Our algorithm permits the evaluation of diagnosis-associated image features obtained from texture of hematoxylin-eosin-stained histopathologic slides. CFPP method was successfully applied in digital pathology for grading lymphomas [7].

DMD is a severe genetic disease characterized by a progressive muscle weakness and loss in structural integrity of muscle fiber membranes, caused by deficiency of the dystrophin protein [8]. Respiratory complications and cardiomyopathies are common causes of DMD patients premature death.

Healthy muscle has a characteristic appearance, and is made up of closely-packed fibers, which are more or less evenly sized. Dystrophic changes on muscle biopsy include variation in fiber size, enlarged internal nuclei, grown fatty and connective tissue [9], and presence of regenerating and degenerating fibers [10]. As the disease progresses the microscopic images will show a greater proportion of fatty and connective tissue. In later pathology the image shows an increase in internal nuclei and variability of myofiber size [11]. Hence, digital pathology helps in neuropathological diagnostics [12].

2 Materials and Methods

Among many features, severity of dystrophic muscles is assessed based on the amount and distribution of irregular myocytes, and the amount of fatty and connective tissue in the microscopic images obtained from biopsy (*whole specimen*). When grading the severity of dystrophy, pathologist concentrates his/her attention on some fragments of the image. Muscle pathologists usually show histograms of the "lesser fiber diameter", which document very well the degree of muscle wasting (cf. [1]). The evaluation based on the whole specimen is frequently compared to the evaluation in the manually chosen and magnified *regions of interests (ROIs)* containing *objects* (cells' nuclei) that are hematoxilin-stained (H) and so turned blue, and *objects* (myocytes) that are eosin-stained (E) and so turned pink.

Manual grading, based on 'eye-ball' (visual) assessment, is frequently time consuming. Besides, it is usually focused on up to 100–200 myofibers being assessed along with surrounding connective tissue. This kind of visual evaluation requires patience and accuracy in choosing *ROIs*. Hence, our objective is to develop method for quick and global assessment of microscopic images of dystrophic muscles.

We retrospectively analyzed 13 digital microscopic images of DMD muscle tissue from 4 patients (average±SD age 6.5 ± 1.3 years). The images were obtained from 2 subjects in moderate (8 slides) and 2 subjects in advanced DMD stage (5 slides) at the Department of Neurology, Medical University of Warsaw. The images were acquired with x400 magnification, from original microscopy slides coming from biopsies. The biopsies were performed as a part of medical diagnostic procedure, so no IRB approval was required. Additionally, we have analyzed 3 microscopic slides of normal muscle tissue; since obtaining a muscle biopsy from a patient is an invasive procedure the biopsies were not obtained from healthy persons but from patients that had been suspected of DMD but those suspicions turned out to be wrong tracks.

To construct histograms of muscle fiber diameters the areas of measurements were selected randomly from the center of the sections. To overcome the possible distortion of the muscle, the lesser diameter of the muscle fibers were measured [1]. Total of 200 fibers were measured for each histogram.

Among the full range of methods used in the muscle dystrophy analysis earlier we choose Image Landscapes' Fractal Dimension (*ILF*) method [13] that is based on Higuchi's fractal dimension algorithm. However, in analysis of biopsies of DMD muscles the results of such fractal analysis are not satisfactory - the method does not differentiate between moderate and advanced stages of the illness (Fig. 1).

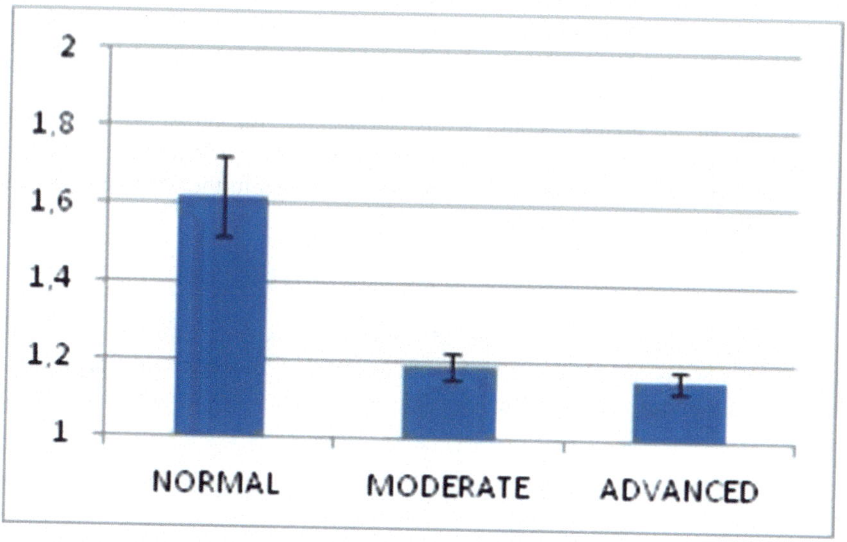

Fig. 1. Mean±SD values of Higuchi fractal dimension calculated from microscopic images of biopsies of normal muscle tissue (n = 3) and of moderate (n = 8) and advanced DMD (n = 5).

Herein, we propose a new dedicated method for quick semi-automatic assessment of severity (stage) of DMD from dyes-stained microscopic images of muscle biopsies. Our method is based on **c**olor **f**iltration **p**ixel-by-**p**ixel of the whole virtual slides (*CFPP method*) and semi-quantitative evaluation of morphological structure of the muscular tissue. The method is simple and quick – it does not require either time consuming manual marking of artifact-free representative *ROIs* or complicated detection of myocytes to grade the dystrophic tissue. Instead, we propose to count the numbers of pixels belonging to the H-stained nuclei (L_H), to the E-stained myocytes (L_E), and to the E-stained connective tissue including sarcolemma and not stained regions belonging to adipocytes (L_C) in the given whole specimen. The images were analyzed on a standard PC with Intel i7 dual core processor and 8 GB RAM. All analyses were performed in MATLAB 2016b.

We define two quantities characterizing dystrophic changes that will be called the *connective tissue indices*:

$$CI_H = (L_C + L_H)/(L_C + L_E) \quad (1)$$

$$CI_C = L_C/(L_C + L_E) \quad (2)$$

The calculations are easily done with MATLAB. Each whole specimen is appropriately color filtered pixel by pixel. Color-coding of computer images most often uses RGB system - the color of a pixel is expressed as a triplet, *(r, g, b)* i.e. *(read, green, blue)*, each component of which can vary from *0* for the darkest one to *255* for the brightest one (Fig. 2). Such a texture-based information calculated from three colors values per each pixel is independent from any segmentation procedure.

The whole RGB-space has nearly 17 millions of points (256 × 256 × 256). To characterize biopsies of DMD muscles we consider cardinality of two subspaces – one classified as consisting of pixels corresponding to H-stained nuclei and another classified as consisting of pixels corresponding to E-stained nuclei.

So, for each of .tiff image to count pixels belonging to the nuclei that are H-stained we choose pixels with *(r, g, b)* components greater than minimum values further denoted with index *Hn,* but smaller than maximum values further denoted with index *HN, i.e.* pixels fulfilling the condition

$$(r_{Hn} < r < r_{Hn})\, AND\, (g_{Hn} < g < g_{Hn})\, AND\, (b_{Hn} < b < b_{Hn}) \quad (3)$$

To count pixels belonging to the E-stained connective tissue we choose pixels with *(r, g, b)* components greater than minimum values further denoted with index C*m* but smaller than maximum values further denoted with index *CM, i.e.* pixels fulfilling the condition

$$(r_{Cm} < r < r_{Cm})\, AND\, (g_{Cm} < g < g_{Cm})\, AND\, (b_{Cmc} < b < b_{Cm}) \quad (4)$$

Pixels belonging to adipocytes are omitted or partially included in the count of connective tissue pixels, as fatty tissue is usually completely washed out from the microscopy specimens.

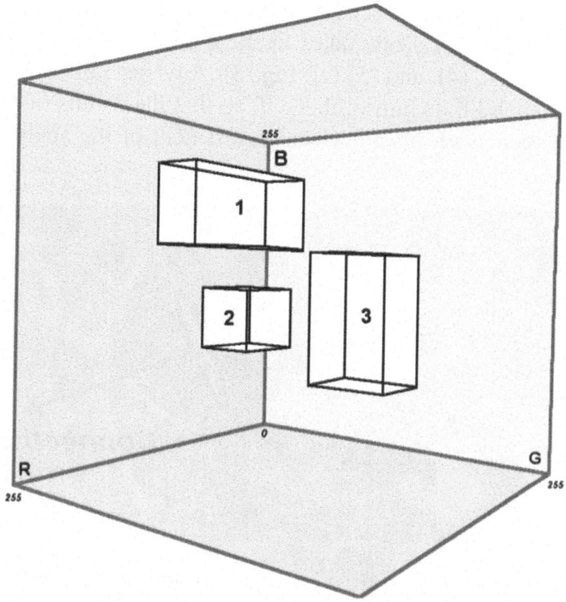

Fig. 2. RGB-space and its subspaces corresponding to different colours of pixels.

In general, to count pixels belonging to the E-stained myocytes we choose pixels with *(r, g, b)* components greater than minimum further denoted with index *Em* but smaller than maximum values further denoted with index *EM, i.e.* pixels fulfilling the condition

$$(r_{EM} < r < r_{EM})\,AND\,(g_{EM} < g < g_{EM})\,AND\,(b_{EM} < b < b_{EM}) \tag{5}$$

But quite often, like in the presented case, for the number of pixels belonging to E-stained myocytes, L_E, one can take all the remaining pixels in the image

$$L_E = L - L_H - L_C \tag{6}$$

where L denotes the total number of pixels in the given specimen.

It is important that for at least one component, *r* or *g* or *b,* the intervals of values for pixels belonging to H-stained (2) and to E-stained (3, 4) structures should be completely disjoint. Such color filtration also filters out the remaining of the empty background and majority of artefacts.

When a fragment of a slide is viewed in MATLAB then clicking on a pixel shows *(r, g, b)* components of this pixel. To choose *Hn* values one clicks on several dark-blue pixels belonging to the nuclei, one writes down the shown *(r, g, b)* values, and for calculation of L_H one takes average value of the corresponding component as the limits in (2); similarly, to choose *HN* values one clicks on several light-blue pixels. To choose *Em* or *Cm* values one clicks on dark pixels belonging to myocytes or connective tissue, respectively; to choose *EM* or *CM* values one clicks on some light pixels belonging to myocytes or connective tissue, respectively. One writes down the shown *(r, g, b)* values and for calculation of L_E or L_C one takes average values of the corresponding component as the limits in (3), (4), and (5) (cf. Fig. 3). Any specialist who is analyzing the images may choose such limits himself/herself, so that the results of computations will show the best agreement with his/her visual assessment of the analyzed images.

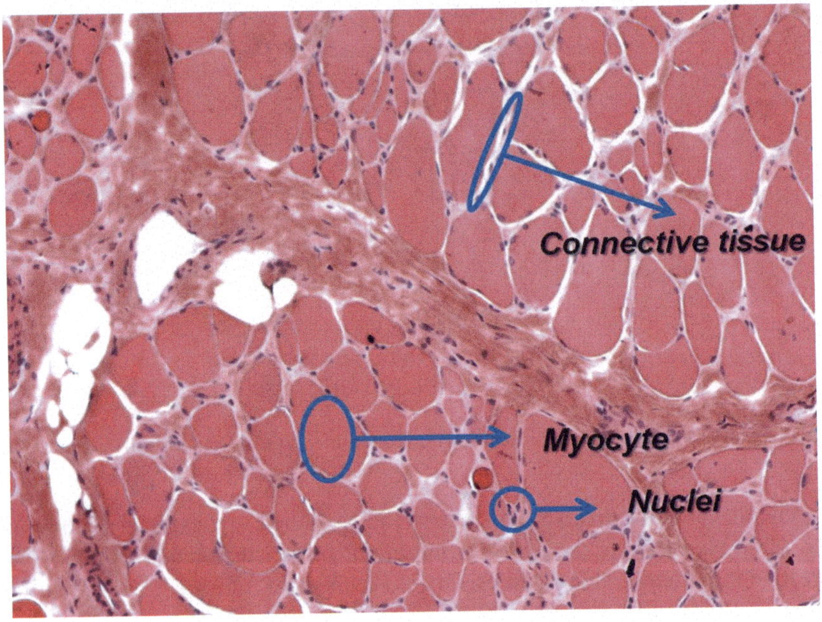

Fig. 3. Example of regions: Nuclei – H-stained, counted into L_H; Myocyte – E-stained, counted into L_M; Connective tissue – partially E-stained sarcolemma with not stained adipocytes, counted into L_C. These regions were used to establish the limits of (r, g, b) in Eqs. (7)–(8).

In the presented case, to count the number of pixels belonging to H-stained nuclei, L_H, we choose the following *(r, g, b)* intervals:

$$(<105, 205>, <50, 80>, <120, 160>) \tag{7}$$

To count the number of pixels belonging to E-stained connective tissue, L_C, we choose the following *(r, g, b)* intervals:

$$(<200, 255>, <105, 255>, <170, 255>) \tag{8}$$

To calculate L_H, L_E and L_C the whole specimen may be subdivided into any number of disjoint parts or even read into RAM pixel by pixel. It is important since it makes possible to analyze specimens even on a PC with 8 GB of RAM. L_H, L_E and L_C are used to calculate the connective tissue indices, CI_H and CI_C (Eqs. (1)–(2)).

3 Results

We have calculated connective tissue indices CI_H and CI_C for 12 slides of 4 patients in two grades of DMD and for 3 images of normal muscles and compared with the assessment made by trained pathologist who used histograms of fiber diameters.

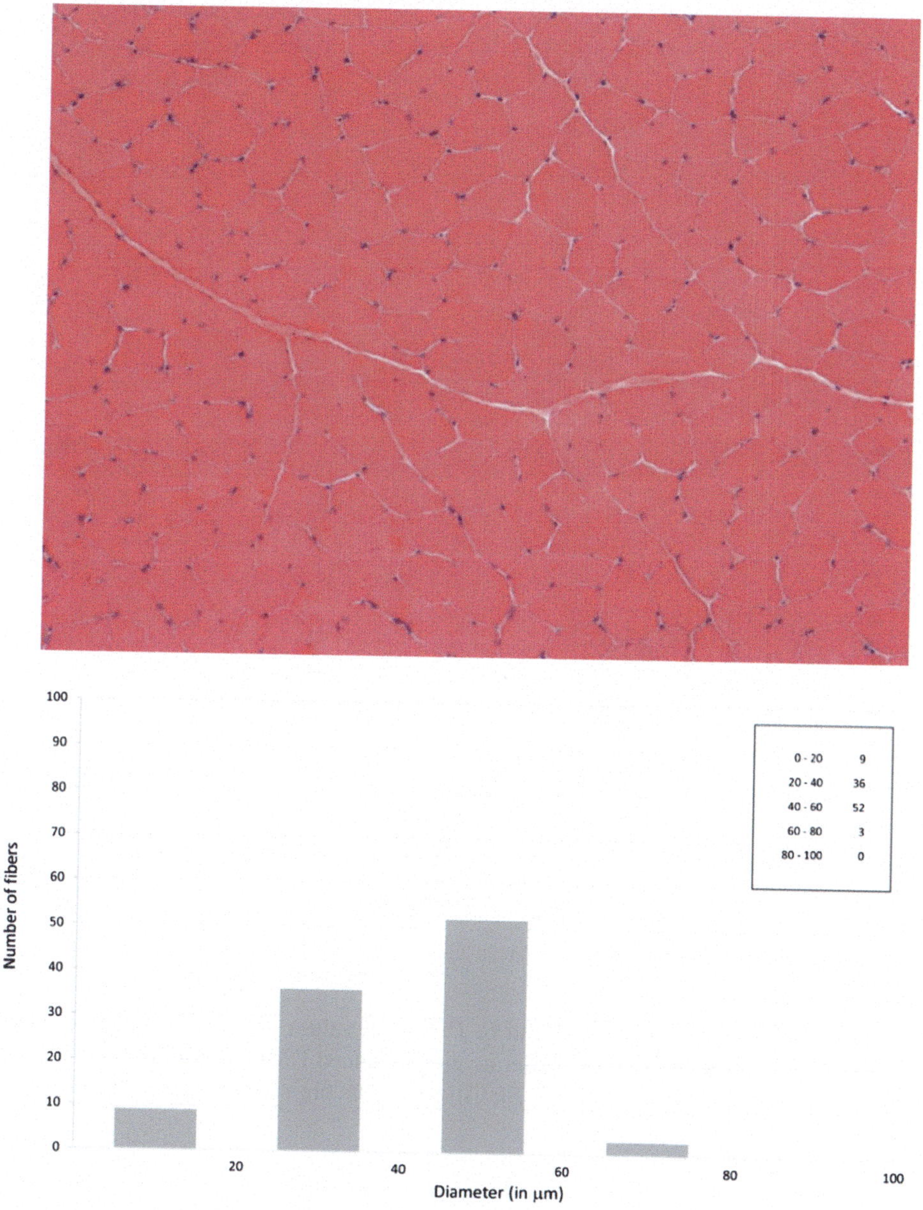

Fig. 4. Normal muscle. Mean fiber diameter 48 μm; $CI_H = 0.032 \pm 0.001$.

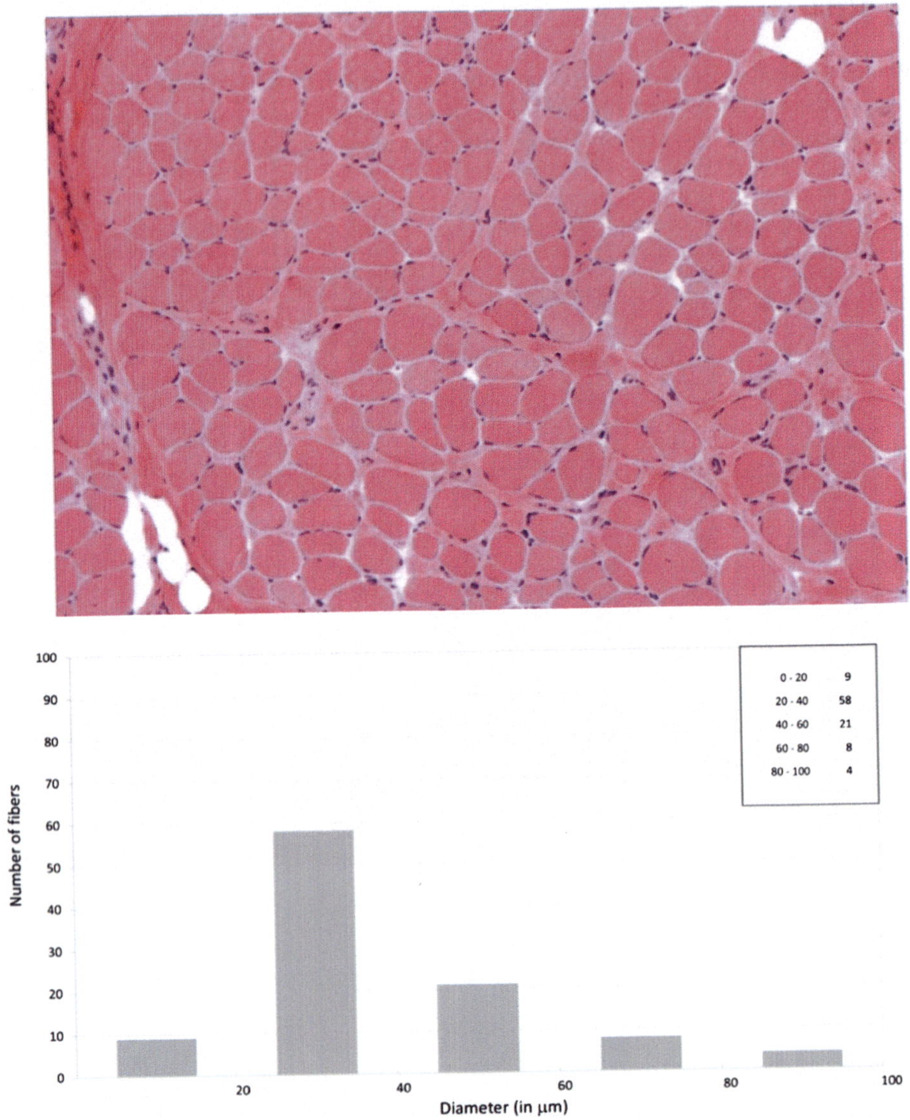

Fig. 5. Moderate stage of DMD. Mean fiber diameter 39 μm. $CI_H = 0.18 \pm 0.03$.

Examples are shown on Figs. 4, 5 and 6. The mean values and standard deviations of the connective tissue indices for moderate and advanced DMD as compared to normal muscle are shown on Figs. 7 and 8. Statistical results are summarized in Table 1.

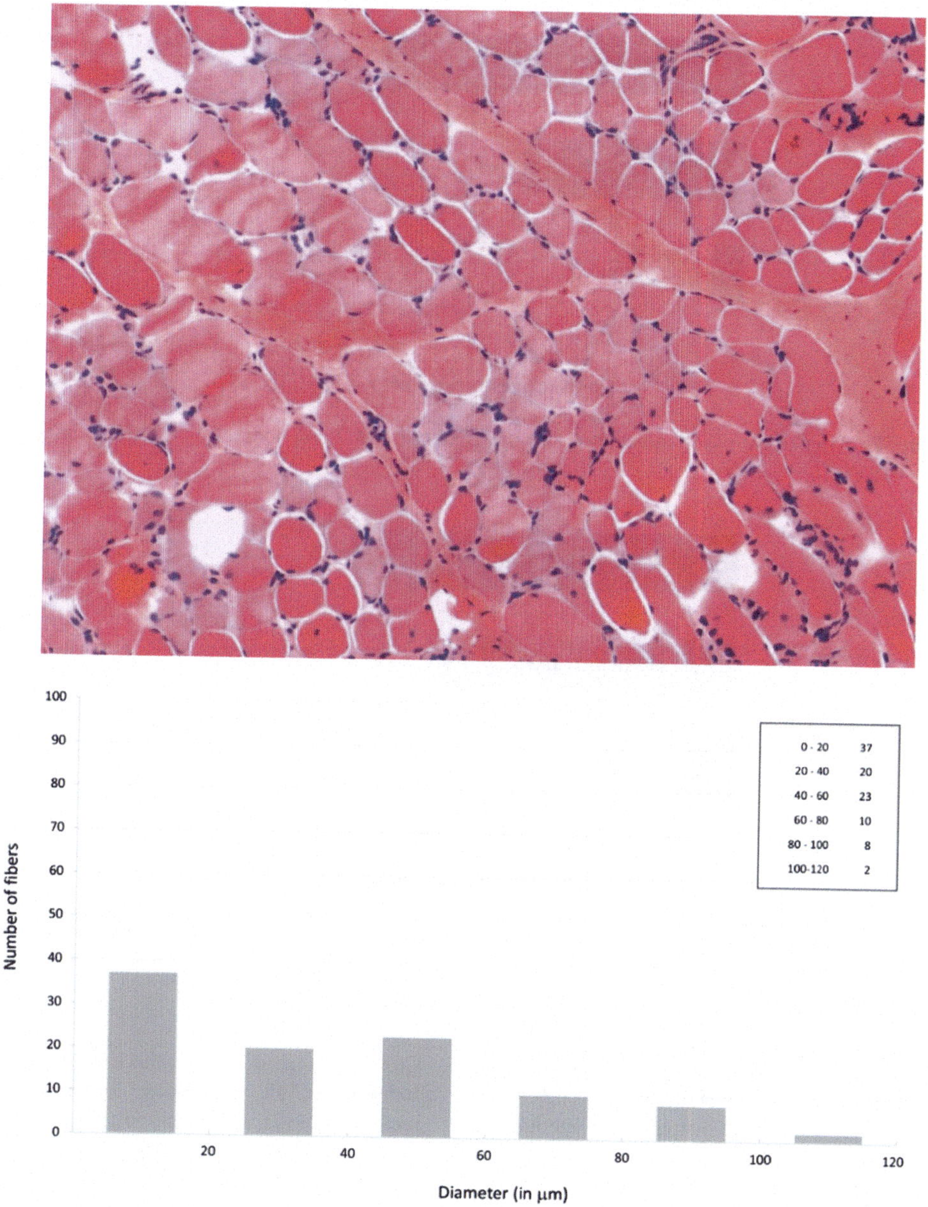

Fig. 6. Advanced stage of DMD. Mean fiber diameter 26 µm. $CI_H = 0.31 \pm 0.09$.

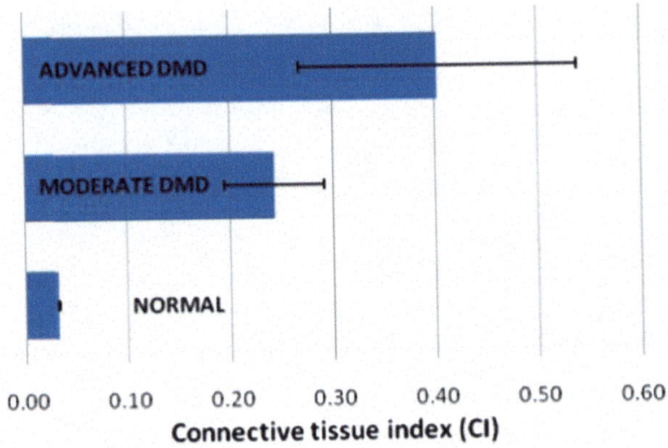

Fig. 7. Connective tissue indices CI_H for the analyzed DMD and normal muscle tissue images.

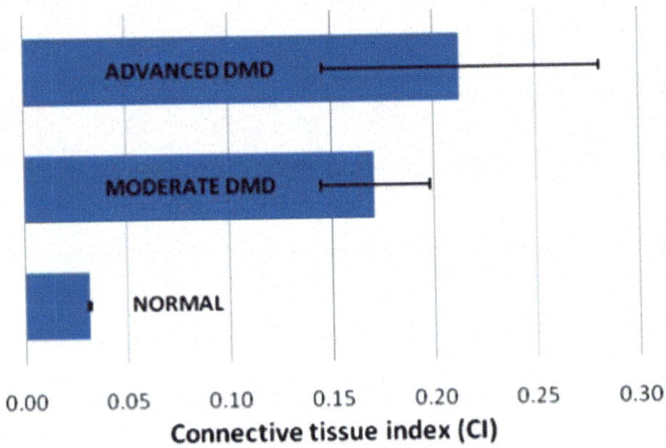

Fig. 8. Connective tissue indices CI_C for the analyzed DMD and normal muscle tissue images.

Table 1. Comparison of methods of grading DMD

'Eye-ball' assessment	Mean Fiber Diameter [μm] from histogram	Higuchi FD (ILF method)	Connective tissue Index CI_C CFPP method (Eq. (2))	Connective tissue index CI_H CFPP method (Eq. (1))
Normal muscle	48	1.61 ± 0.11	0.036 ± 0.002	0.032 ± 0.001
Moderate stage of DMD	39	1.19 ± 0.02	0.17 ± 0.03	0.24 ± 0.05
Advanced stage of DMD	26	1.17 ± 0.01	0.21 ± 0.07	0.40 ± 0.13

4 Conclusions

There is still a need for digitized quantitative histopathology image analysis of biopsy specimens [2, 14, 15]. We have presented a fast semi-automatic method for evaluation of muscle tissue microscopy images. CFPP method is simple, reliable, and rapid compared to conventional methods. It allows to distinguish DMD tissue from normal muscle tissue, as well as it makes possible to assess quantitatively the severity of DMD through assessment of connective tissue index, a proliferation index dedicated to the grading of DMD. From the operator's point, the method requires only choosing of a small representative regions of muscle tissue, fatty and connective tissue, and of nuclei. As the method is applicable globally to a single image, it does not require colour scale standardization or brightness and contrast among sets of microscopic images. Its general principle is applicable in digital pathology with whole slide imaging (WSI) and virtual microscopy (VM) [7].

It is a promising technique that can assist tissue-based diagnosis and be used for virtual slides evaluation. CFPP method is much simpler and much less time consuming than classical 'lesser fiber diameter' method. Table 1 demonstrates that proliferation index in the form of connective tissue index CI_H (1) well differentiates stages of DMD and well corresponds to classical DMD staging that uses fiber diameter – the more advanced stage of DMD the lower is the mean fiber diameter and the greater is connective tissue index CI_H. Higuchi Fractal Dimension calculated using ILF method and CI_C index calculated using CFPP method do not differentiate between moderate and advanced DMD stages. Further research on large data sets are needed.

Acknowledgements. This work was partially supported by Nalecz Institute of Biocybernetics and Biomedical Engineering Polish Academy of Sciences and by Warsaw Medical University statutory activities. It was also supported by COST Action BM1304 MYO-MRI.

List of Abbreviations

CFPP method: Colour Filtration Pixel-by-Pixel method
CI_H, CI_C: connective tissue indices
DMD: Duchenne Muscular Dystrophy
ILF method: Image Landscapes' Fractal Dimension method
ROIs: regions of interests
VM: virtual microscopy
WSI: whole slide image

References

1. Dubowitz, V., Sewry, C.A., Fitzsimons, R.B.: Muscle Biopsy: A Practical Approach, 2nd edn. Balliére-Tindall, London (1985)
2. Gurcan, M.N., et al.: Histopathological image analysis: a review. IEEE Rev. Biomed. Eng. **2**, 147–171 (2009). https://doi.org/10.1109/RBME.2009.2034865
3. Waddell, L.B., Evesson, F.J., North, K.N., Cooper, S.T., Clarke, N.F.: Diagnosis of the muscular dystrophies. In: Hegde, M., Ankala, A. (eds.) Muscular Dystrophy. IntechOpen, London (2012). https://doi.org/10.5772/1242
4. Challa, S., Uppin, M.S.: Approach to the interpretation of muscle biopsy. In: Sundaram, C. (ed.) Muscle Biopsy. IntechOpen, London (2012). https://doi.org/10.5772/10.5772/1241
5. Kayser, K., Hoshang, S.A., Metze, K., Goldmann, T., Vollmer, E., Radziszowski, D., Kosjerina, Z., Mireskandari, M., Kayse, G.: Texture- and object-related automated information analysis in histological still images of various organs. Anal. Quant. Cytol. Histol. **30**(6), 323–335 (2008)
6. Klonowski, W., Gomolka, R., Kuraszkiewicz, B., Kaminska, A., Stepien, P.: Simple semi-automatic method for quantitative analysis of dystrophic muscle biopsies. J. Neuromuscul. Dis. **4**(S1), 27 (2017). First International Conference on Imaging in Neuromuscular Disease, Berlin, Germany, 19–21 November
7. Klonowski, W., Korzynska, A., Gomolka, R.: Computer analysis of histopathological images for tumor grading. Physiol. Meas. **39**(3) (2018). Focus issue on the new field of Network Physiology. http://iopscience.iop.org/article/10.1088/1361-6579/aaa82c/pdf
8. Manzur, A.Y., Muntoni, F.: Diagnosis and new treatments in muscular dystrophies. J. Neurol. Neurosurg. Psychiatry **80**, 706–714 (2009)
9. Norwood, F.M., et al.: EFNS guideline on diagnosis and management of limb girdle muscular dystrophies. Eur. J. Neurol. **14**(12), 1305–1312 (2007)
10. McNally, E.M., Pytel, P.: Muscle diseases: the muscular dystrophies. Annu. Rev. Pathol. **2**, 87–109 (2007)
11. Wren, T.A., et al.: Three-point technique of fat quantification of muscle tissue as a marker of disease progression in Duchenne muscular dystrophy: preliminary study. AJR Am. J. Roentgenol. **190**, W8–W12 (2008)
12. Klonowski, W.: Applications of Chaos theory methods in clinical digital pathology. In: Skiadas, Ch.H., Skiadas, C. (eds.) Handbook of Applications of Chaos Theory, pp. 681–690. CRC Press, Boca Raton (2016)
13. Lerski, R.A., de Certaines, J.D., Duda, D., Klonowski, W., Yang, G., Coatrieux, J.L., Azzabou, N., Eliat, P.-A.: Application of texture analysis to muscle MRI: 2 – technical recommendations. EPJ Nonlinear Biomed. Phys. **3**(2) (2015). https://epjnonlinearbiomedphys.springeropen.com/articles/10.1140/epjnbp/s40366-015-0018-0
14. Suriyonplengsaeng, C., et al.: Immunohistochemistry of sarcolemmal membrane-associated proteins in formalin fixed and paraffin-embedded skeletal muscle tissue: a promising tool for the diagnostic evaluation of common muscular dystrophies. Diagn. Pathol. **12**, 19 (2017). https://doi.org/10.1186/s13000-017-0610-y
15. Vido, J.R., Adam, R.L., Lorand-Metze, I.G.H., Metze, K.: Computerized texture analysis of atypical immature myeloid precursors in patients with myelodysplastic syndromes: an entity between blasts and promyelocytes. Diagn. Pathol. **6**, 93 (2011)

Deep Neural Networks for Breast Cancer Diagnosis: Fine Needle Biopsy Scenario

Bartosz Miselis[1(✉)], Thomas Fevens[1], Adam Krzyżak[1], Marek Kowal[2], and Roman Monczak[3]

[1] Department of Computer Science and Software Engineering, Concordia University, Montréal, Canada
{b_miseli,fevens,krzyzak}@encs.concordia.ca
[2] Institute of Control and Computation Engineering, University of Zielona Góra, Zielona Góra, Poland
m.kowal@issi.uz.zgora.pl
[3] Department of Pathomorphology, University Hospital in Zielona Góra, Zielona Góra, Poland
r.monczak@issi.uz.zgora.pl

Abstract. In this study, we focus on the problem of computer-aided diagnosis of breast cancer using cytological images of fine needle biopsies. We explore the potential of modern deep neural network architectures by comparing five different convolutional neural networks trained to classify the specimen as either benign or malignant. For experimentation, we use 550 cytological images of fine needle biopsies from 50 patients, balanced between benign and malignant cases, acquired at the University Hospital in Zielona Góra, Poland. We found that the convolutional neural network Inception-v3 is the best model, reaching 91.86% accuracy and 0.97 value for area under the curve (AUC).

Keywords: Computer-aided diagnosis · Breast cancer · Fine needle biopsy · Convolutional neural networks

1 Introduction

Breast cancer is the most common cancer in women and second most common cancer type in the world. In 2018, there were over 2 million new cases estimated worldwide, of which more than 620,000 resulted in death [2]. The key to reduce its toll is not only to develop more effective treatments, but also to detect it as early as possible, increasing the probability of full recovery—99% of women with the breast cancer diagnosed at the earliest stage survive for at least 5 years compared to ~27% for those diagnosed with the most advanced stage [10].

A number of examinations can help detect changes in breast tissue: palpation[1], ultrasound, mammography, magnetic resonance imaging (MRI) and

[1] There is a discussion whether palpation should or should not be recommended.

J. Korbicz et al. (Eds.): PCBEE 2019, AISC 1033, pp. 131–142, 2020.
https://doi.org/10.1007/978-3-030-29885-2_12

biopsy. Excluding self-examination, the remaining four methods are nearly always analyzed by an expert who has to spend a significant amount of her time interpreting the data, whether it is a mammogram, an MRI scan or a tissue sample viewed under a microscope. There are different methods to help make the diagnosis as well, for instance, tests on cells' proteins, RNA or DNA. In the majority of cases, breast changes are either found by palpation or during a routine annual mammography. Often though, this is not enough to suggest a proper treatment—if the doctor is not sure whether the change is cancerous or not, the next step is to perform a biopsy.

During biopsy, a piece of tissue from the suspicious area is removed so that it can be analyzed in the lab for a presence of cancerous cells. Two most common types of biopsies are: fine needle biopsy (FNB)[2] [24] and a core needle biopsy (CNB). FNB is less invasive (no need to surgically cut the skin of the breast), can be done in doctor's office and in many cases it is possible to make the diagnosis on the day of the visit [26]. On the other hand, CNB is more invasive (requires an incision) and takes longer to process, but allows the doctor to make much more confident and detailed diagnosis [1,3].

In the era of Deep Learning [17] being the state-of-the-art approach for a wide range of computer vision tasks [19,27], there is a great potential to adopt some of its methods for cancer diagnosis. Empowering the doctors with such tools not only decreases the time spent on samples analyses, but also increases the confidence in the diagnosis [5,18,25]. The efficacy of computer-aided diagnosis used as the assisting method for breast cytology screening has been evaluated in numerous research, with variable results depending on the methodology applied to the specific problem [7,8,20]. In this work, we focus on FNB as the efficient, reliable and minimally invasive procedure and analyze how the existing deep learning methods perform on the task of cancer diagnosis using the data from this examination.

2 Related Work

In the previous studies on classifying FNB specimens as either benign (non-cancerous) or malignant (cancerous), various approaches have been proposed [6,14,15]. A common thread for all of them is that they were all based on nuclei segmentation, followed by feature extraction and feature classification.

Żejmo et al. [28] proposed an alternative solution to FNB breast cancer classification. They suggested a novel approach based on convolutional neural networks (CNNs). The authors compared two different model architectures that were trained to classify breast cancer from raw image patches (without manual feature extraction)—AlexNet [16] and GoogLeNet [23]. The best accuracy rate of 83% was obtained by GoogleNet model. In this work, we build upon their results and explore how modern deep architectures compare to their predecessors.

The paper is structured as follows. Section 3 describes the dataset used in this study. Section 4 contains a detailed description of the method used to train the

[2] Although FNB is a type of biopsy, it is also classified as a cytology examination.

models. Section 5 presents the analysis of the results obtained from the experiments. Section 6 summarizes the outcomes of this paper.

3 Dataset

Data Collection. The dataset of cytological images of breast cancer FNB used in this study were acquired at University Hospital in Zielona Góra, Poland. Tissue samples were collected from affected tissue using a 0.5-mm-diameter needle under the control of an ultrasonograph. Next, the material was fixed with Cellfix (Shadon) fixative spray and dyed with hematoxylin and eosin (H&E). The time between preparation of smears and their preservation in fixative spray did not exceed 3 s. Overall, 67 specimens were prepared from 67 patients with breast changes. Furthermore, all cases were histologically confirmed, 25 of them being benign and 42 malignant. Moreover, every patient classified as a benign case was either biopsied or followed for a year.

Digitization of glass slides was done using Olympus VS120 Virtual Microscopy System. The system consists of a 2/3″ CCD camera and 40x lens, resulting in 0.172 μm resolution per pixel. The average size of the slide from the system is approximately 200,000 × 100,000 pixels. All the scans were prepared using Extended Focal Imaging (EFI) system. After digitization, pathologists from University Hospital in Zielona Góra were asked to identify smaller regions of digital slides that, according to them, contain tissue fragments that are useful for further analysis. For each whole slide, 11 regions of interests (ROI) were selected. Each ROI was converted to an RGB image of size 1583 × 828 pixels. Overall, 737 ROIs were selected: 275 benign (25 slides × 11 ROIs) + 462 malignant (42 slides × 11 ROIs).

Data Preparation. The raw dataset consisting of 25 benign and 42 malignant cases required further preparation and preprocessing before classification. First, to balance the data and be consistent with [28] (to obtain a smaller version of the same datasets—25 benign and 25 malignant cases), we randomly sampled from malignant patients (resulting in 25 patients, instead of 42). Next, we independently split benign and malignant patients into training (60%—15 patients) and validation (40%—10 patients) sets.

Since the CNNs used in this work require input image dimensions not to be too large (for instance, when training neural networks on ImageNet dataset [4], the golden standard is 224 × 224 pixels), we decided to divide each 1583 × 828 pixels ROI into overlapping patches of size 256 × 256. We did this by sliding a 256 × 256 window with a 32-pixel step (both in horizontal and in vertical dimensions). Each ROI resulted in 697 patches.

For further consistency with [28], we followed their approach of forming final training and validation data not from all the available patches, but from those that contain a high enough ratio of cellular material with respect to background. For each patch, *cellular coverage ratio* (the ratio of number of pixels with cellular material divided by the number of all pixels in the patch) needed to be calculated.

Table 1. Number of patches resulting from SVM filtering. Discrepancy between benign and malignant cases gradually increases with higher cellular coverage ratio thresholds.

	Training		Validation	
	Benign	Malignant	Benign	Malignant
50%	51211	53797	43841	36091
75%	33774	24269	30715	10198
90%	20838	9894	18356	2266

To do that, we trained a simple SVM classifier on 20 manually labeled ROIs (10 benign and 10 malignant, each coming from a different patient). Each label was a binary image, separating the content into *cellular material* (positive class) and *background* (negative class). Pairs of pixels from the image and its label were extracted and fed into the classifier. To speed up SVM training, we randomly sampled equal numbers of positive and negative pixels from 20 ROIs used in the procedure. Furthermore, we applied zero-mean, unit variance standardization on pixels used to train the SVM. The classifier was evaluated on all pixels coming from an additional 6 ROIs (3 benign and 3 malignant, manually labeled). To counterbalance the difficulty of discriminating cellular material from background we introduced an additional weight that enabled us to control what fraction of the pixels used to train the classifier originates from a positive class. Our experiments showed that the SVM classifier trained on barely 1000 pixels with the positive class weight equal to 0.9 (900 pixels with cellular material and 100 pixels with background) is good enough in our task—it got a 0.92 balanced accuracy value on pixels from 6 test ROIs, understood as the average of true positive and true negative ratios.

The trained SVM was used to calculate cellular coverage ratio for all the patches in the dataset. Ratio values were used to create 3 different datasets: first one containing all the patches that had at least 50% of positive pixels, second one that had at least 75% and the third one containing all patches with at least 90% of pixels with cellular material. From now on, we will denote these datasets simply as 50%, 75% and 90% datasets, respectively. Exact results of SVM filtering can be found in Table 1. One can easily observe that benign and malignant classes are not represented uniformly across the datasets. Even though for 50% datasets imbalance is not serious enough to cause trouble, with 75% and 90% datasets it could disrupt metrics like accuracy in evaluating the model's performance. That is the reason we decided to perform one last step before training CNNs: patches sampling.

The final step of datasets generation was patches sampling. We randomly selected patches from either a benign or malignant class (depending which one was more frequent) to make them equal. Datasets prepared in the way described in this section could finally be fed into neural networks.

4 Experiments

For the experiments presented in this work, five different CNN architectures have been evaluated: AlexNet [16], GoogleNet [23], SqueezeNet [12], DenseNet [11] and Inception-v3 [22]. The choice of AlexNet and GoogleNet was inspired by the previous work on the same datasets [28]. SqueezeNet (700K parameters) appeared to us as a lightweight network with the potential to reach the level of accuracy of much more computationally demanding architectures, like AlexNet (60M parameters) or VGG (140M parameters) [21]. We had the initial intuition that our problem, when compared with, for instance, ImageNet Large-Scale Visual Recognition Challenge (ILSVRC) (2 class problem versus 1000 classes) is a grain of sand and using the architectures achieving state-of-the-art results would result in an enormous overfitting. To mitigate this issue, we chose the models with fewer parameters like—the aforementioned—SqueezeNet, DenseNet (8M parameters), followed by one larger architecture—Inception-v3 (23M parameters). Note that unless otherwise indicated, the implementation is from the `keras.applications` library.

The AlexNet architecture [16] won ILSVRC in 2012. This network is simply a series of convolutions, interlacing with several max-pooling layers. In our implementation of AlexNet, we used the one-GPU variant. Additionally, we decided to reduce number of parameters of the network by replacing two final fully-connected (FC) layers with 4096 units by a single global average pooling operation, followed by a FC layer with 2 units. This reduced number of parameters of the architecture from 60 million to 1 million (60× less). We also introduced a batch normalization layer [13] after each convolution (but before the activation) to help train the network. The decision to reduce the number of parameters in AlexNet was based on the observation that initial runs of the original architecture resulted in significant overfitting.

The Inception architecture [23] achieved state-of-the-art results in the ILSVRC in 2014. Its 22 layers deep variant—known as GoogleNet—was evaluated in our work. Inception introduces the idea of the *inception module* which is the key concept that helped to improve model's computational efficiency. We built a GoogleNet variant using `keras`, with several modifications to reduce the number of parameters of the architecture: for each auxiliary classifier, we replaced each FC layer with 1024 units + dropout with a single global average pooling layer, followed by a FC layer with 2 units (since we work on 2 classes). We also replaced the linear layer at the very end of the network with a FC layer with 2 units, followed by a softmax activation. When compiling the model, we assigned different weights to network output (1.0) and auxiliary classifiers (0.3). Final loss was a weighted sum of all three losses.

Iandola et al. [12] took on a challenge to create a lightweight neural network architecture that would be able to reach an accuracy comparable to large models on the ImageNet dataset, but with significantly fewer parameters. They accomplished that goal by creating the SqueezeNet network with AlexNet-level accuracy yet with 50× fewer parameters. We implemented SqueezeNet v1.1 in `keras`, based on the original `Caffe` version.

The Densely Connected Convolutional Network (DenseNet) paper [11] embraces the observation that introducing shorter connections between layers that are close to the input and those that are close to the output enables training of substantially deeper and more accurate models. They build upon this idea and present an architecture that connects each layer to every other layer in a feed-forward way. DenseNets have several strong advantages: they help with vanishing gradient problem, significantly reduce the number of the network's parameters, amplify feature propagation and feature reuse. The DenseNet121 variant was used in this work.

Szegedy et al. [22] took a slightly different approach from other ILSVRC 2012 competitors and instead of blindly increasing both network depth and number of parameters, they followed more structured approach: scale up the network, but in a way that utilizes added computation as efficiently as possible. The outcome of their exploration is the aforementioned Inception-v3 architecture that is—as GoogleNet—based on the concept of inception modules, but with even more factorized convolutions and more aggressive regularization methods.

All of the models were trained following exactly the same training procedure: each model was trained on 3 different datasets (50%, 75% and 90%) with 10 different learning rates, sampled randomly from $[10^{-6}, 10^{-1}]$ range. Note that we sampled learning rates only once, meaning after getting 10 random values, we stored them in a list which was re-used each time a new model was trained. In every case the learning rate was passed to an Adam optimizer ($\beta_1 = 0.9, \beta_2 = 0.999$, no decay). Individual training lasted 10 epochs, taking approximately 2–3 h to converge. This resulted in 30 different training runs for each CNN. Since all models have 2 outputs (one for benign class and one for malignant), we used the categorical cross-entropy loss function. Two neurons at the end (categorical case) were used instead of a single neuron (binary case) to allow for detailed class activations in future research—in this case it is necessary to have a separate output for each class.

The 256×256 RGB patches were fed into our models using training and validation generators. Batch size of 128 was used for AlexNet, GoogleNet and SqueezeNet, while batch size of 32 was used for Inception-v3 and DenseNet121. The number of examples in a single batch was chosen so that the full GPU memory has been utilized[3]. Before sending data to the model, each of the generators performed image standardization (so that input data has zero mean and unit variance) based on a sample from training patches (500 benign and 500 malignant). No augmentations were applied nor random cropping used. Transfer learning has not been used as the weights' initialization method. Instead, uniform Xavier approach [9] was chosen (this is `keras`' default initialization method for both fully-connected and convolutional layers). Biases were initialized with zeros.

[3] All of the models have been trained on a workstation with Intel i7-8700K CPU, 64 GB of RAM and 2 NVIDIA GeForce GTX 1080 Ti GPUs, each with 11 GB of memory.

5 Discussion

As mentioned in the previous section, we evaluated each CNN type (AlexNet, GoogleNet, etc.) on three datasets (50%, 75% and 90%). For each CNN type, we chose—from 10 runs with different learning rates—the one with lowest validation loss. As a result, we ended up with best AlexNet run for 50% data, best AlexNet run for 75% data, best AlexNet run for 90%, best GoogleNet run for 50% data etc. For each dataset, we present a figure with receiver operating characteristics (ROC) curves to investigate how they perform (see Fig. 1 for details). Additional metrics (accuracy, F1 score and AUC) from models' evaluation on validation data, together with the learning rate used for that specific run, were introduced for a more thorough performance investigation (see Table 2 for details). Accuracy is defined as a ratio of the number of all correctly identified examples (either benign or malignant) over the number of all evaluated examples (accuracy on validation data is termed *validation accuracy*). F1 score is a harmonic mean of precision (proportion of true malignant cases in all cases identified as malignant) and recall (proportion of true malignant cases in all cases that should be identified as such), while AUC represents the area under the ROC curve (type of curve used to examine how a prediction threshold influences the model's performance).

In 50% dataset case, the supremacy of Inception-v3 over other architectures is clearly visible. This is most probably due to the amount of data available in this dataset (51K patches per class) being sufficient to counterbalance the capacity of the network (23M parameters) so that it does not significantly overfit, reaching

Table 2. Additional metrics calculated on validation dataset.

Dataset	Model	Learning rate	Accuracy	F1-score	AUC
50%	AlexNet	0.000006	0.888459	0.890036	0.953168
	GoogleNet	0.000249	0.875173	0.876643	0.945584
	SqueezeNet	0.000374	0.832668	0.832978	0.915294
	DenseNet121	0.000011	0.829579	0.836621	0.924451
	Inception-v3	0.000664	**0.918579**	**0.918403**	**0.973085**
75%	AlexNet	0.000006	0.865107	0.869676	0.935723
	Inception-v1	0.000249	0.848534	0.844594	0.924711
	SqueezeNet	0.000374	0.841228	0.842295	0.911930
	DenseNet121	0.005234	0.841963	0.835888	0.912582
	Inception-v3	0.002838	**0.881092**	**0.883003**	**0.948290**
90%	AlexNet	0.000006	**0.816777**	**0.820889**	0.888870
	Inception-v1	0.000249	0.802649	0.811313	0.867780
	SqueezeNet	0.000374	0.776600	0.789692	0.837452
	DenseNet121	0.000249	0.781236	0.754399	**0.907895**
	Inception-v3	0.005613	0.786976	0.793847	0.849767

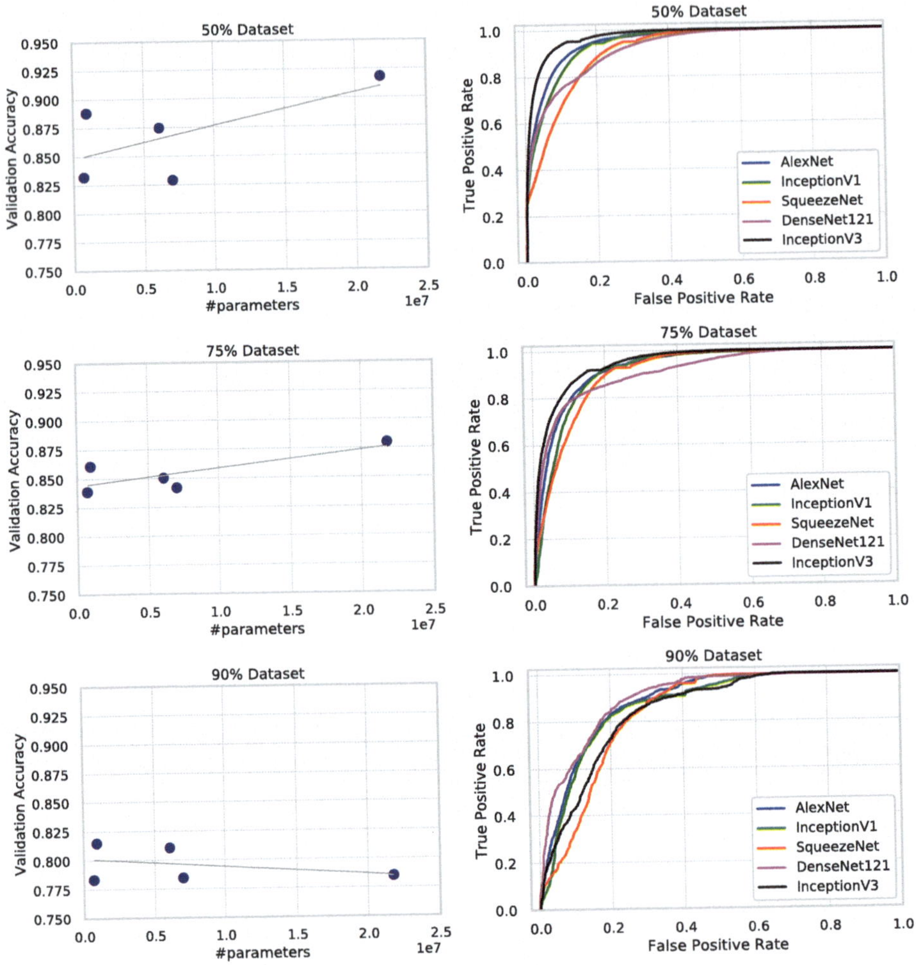

Fig. 1. *Left:* Plots representing how accuracy depends on the number of parameters. *Right:* ROC curves for each dataset.

satisfactory results (validation accuracy over 91%). One would expect that the bigger the network the better the fit if fed with enough data, but it was not the case in this scenario—AlexNet outperformed the rest of remaining architectures with just 1M parameters. The 75% dataset did not bring any surprises—still relatively large amount of training data (24K patches per class) coupled well with Inception-v3's capacity. It outperformed the rest of the models with significant difference. AlexNet, however, again got very close, scoring single percentage points below Inception-v3's results. The 90% dataset changed the situation—little over 2K patches per class was not enough for Inception-v3, which scored almost the worst from all architectures. DenseNet, however, managed to extract the amount of structure enough to get 0.9 AUC. We think that this is the result

Table 3. Comparison of our models with statistics from [3].

Dataset	Model	Absolute sensitivity	Specificity	Positive predicted value	False negative rate	False positive rate
-	FNAC [3]	0.831	0.84	0.983	-	0.014
50%	AlexNet	0.902799	0.874120	0.877630	0.097201	0.125880
	GoogleNet	0.887088	0.863259	0.866441	0.112912	0.136741
	SqueezeNet	0.834525	0.830812	0.831438	0.165475	0.169188
	DenseNet	0.872679	0.786478	0.803423	0.127321	0.213522
	Inception-v3	**0.916431**	**0.920726**	**0.920384**	**0.083569**	**0.079274**
75%	AlexNet	**0.900167**	0.830048	0.841184	**0.099833**	0.169952
	GoogleNet	0.823183	0.873884	0.867149	0.176817	0.126116
	SqueezeNet	0.847995	0.834461	0.836672	0.152005	0.165539
	DenseNet	0.804943	**0.878984**	**0.869307**	0.195057	**0.121016**
	Inception-v3	0.897421	0.864764	0.869041	0.102579	0.135236
90%	AlexNet	0.839735	0.793819	0.802870	0.160265	0.206181
	GoogleNet	**0.848565**	0.756733	0.777194	**0.151435**	0.243267
	SqueezeNet	0.838852	0.714349	0.745976	0.161148	0.285651
	DenseNet	0.671965	**0.890508**	**0.859887**	0.328035	**0.109492**
	Inception-v3	0.820309	0.753642	0.769040	0.179691	0.246358

of DenseNet being able to extensively reuse learned features, which indirectly seems to help it learn from less data. However, AlexNet achieved best validation accuracy and F-score, again shining a new light on the potential of this lightweight architecture. The second small architecture—SqueezeNet—did not perform that well, potentially due to the absence of batch normalization layer (when compared to AlexNet). Overall, it is clearly visible that the performance of our models drops when gradually reducing the amount of data available for training.

Despite not working on the same data, as part of the evaluation process we compared our results with that presented in [3] for ultrasound-guided fine needle aspiration cytology (FNAC) to get the estimate at least in terms of the order of magnitude of scores. The analyses from [3] were performed by experienced pathologists. We selected the following metrics that are comparable to our scenario: absolute sensitivity (in our case understood as pure sensitivity - a measure of the proportion of true malignant examples that are correctly identified as such), specificity (proportion of true benign cases that are correctly identified as such), positive predicted value (precision) for malignant cases, false negative rate and false positive rate. Detailed results were summarized in Table 3. Please note that the comparison is just to get the sense whether metrics' values we achieved were to any extent similar. It is worth mentioning that our metrics have been calculated for all patches and averaged, while the ones from [3] were calculated per case. Nonetheless, absolute sensitivity and specificity of the CNNs are on a satisfactory level. Inception-v3 for the 50% dataset reached a positive predicted value of 0.92, getting relatively close to the examination's value of 0.983. A false negative rate value was not provided in [3] hence it was impossible to compare.

Inception-v3 reached a 7.92% false positive rate, which was reassuring to observe. Overall, the analyses is very encouraging for future research.

Our initial hypothesis was that models with fewer parameters should outperform larger ones when trained and evaluated on limited data (75% and 90% datasets). This could happen because of the learning capacity of architectures like Inception-v3 and DenseNet—they are able to "memorize" the data if it is scarce enough. To verify this hypothesis, plots were created for each dataset to see how accuracy depends on the number of parameters of the model (see Fig. 1). When fitting the line to data points one can observe that, on average, increasing model size with larger dataset (like 50% one) improves the performance. The curve is less steep for 75%, with the performance decreased for 90%.

6 Conclusions

The aim of this work was to explore how modern CNNs could be used to analyze breast cancer FNB specimen. After evaluating five different architectures—AlexNet, GoogleNet, SqueezeNet, DenseNet and Inception-v3—we achieved very promising results. We see Inception-v3 as the best model, although it did not achieve best results in all cases. This is, however, a real-life scenario - it is extremely hard to find a single model that could simply work for all patients, all tissues and all datasets. Getting the results described in this work, especially when compared to manual analyses [3], we are really encouraged to further explore the potential of deep networks for the analysis of cytological specimens.

Acknowledgments. T. Fevens, A. Krzyzak and B. Miselis were supported by the Natural Sciences and Engineering Research Council of Canada under Grants RGPIN-04929-2014 and RGPIN-2015-06412. M. Kowal and R. Monczak were supported by the National Science Centre, Poland (grant no. 2015/17/B/ST7/03704).

References

1. Brancato, B., Crocetti, E., Bianchi, S., Catarzi, S., Risso, G.G., Bulgaresi, P., Piscioli, F., Scialpi, M., Ciatto, S., Houssami, N.: Accuracy of needle biopsy of breast lesions visible on ultrasound: audit of fine needle versus core needle biopsy in 3233 consecutive samplings with ascertained outcomes. Breast **21**(4), 449–454 (2012)
2. Bray, F., Ferlay, J., Soerjomataram, I., Siegel, R.L., Torre, L.A., Jemal, A.: Global cancer statistics 2018: globocan estimates of incidence and mortality worldwide for 36 cancers in 185 countries. CA Cancer J. Clin. **68**(6), 394–424 (2018)
3. Britton, P.: Fine needle aspiration or core biopsy. Breast **8**(1), 1–4 (1999)
4. Deng, J., Dong, W., Socher, R., Li, L.J., Li, K., Fei-Fei, L.: ImageNet: a large-scale hierarchical image database. In: 2009 IEEE Conference on Computer Vision and Pattern Recognition (CVPR), pp. 248–255 (2009)
5. Esteva, A., Kuprel, B., Novoa, R.A., Ko, J., Swetter, S.M., Blau, H.M., Thrun, S.: Dermatologist-level classification of skin cancer with deep neural networks. Nature **542**, 115 (2017)

6. Filipczuk, P., Kowal, M., Obuchowicz, A.: Automatic breast cancer diagnosis based on k-means clustering and adaptive thresholding hybrid segmentation. In: Image Processing and Communications Challenges, vol. 3, pp. 295–302. Springer, Heidelberg (2011)
7. Filipczuk, P., Fevens, T., Krzyżak, A., Monczak, R.: Computer-aided breast cancer diagnosis based on the analysis of cytological images of fine needle biopsies. IEEE Trans. Med. Imaging **32**(12), 2169–2178 (2013)
8. George, Y.M., Bagoury, B.M., Zayed, H.H., Roushdy, M.I.: Automated cell nuclei segmentation for breast fine needle aspiration cytology. Signal Process. **93**(10), 2804–2816 (2013)
9. Glorot, X., Bengio, Y.: Understanding the difficulty of training deep feedforward neural networks. In: AISTATS (2010)
10. Howlader, N., Noone, A., Krapcho, M., Garshell, J., Miller, D., Altekruse, S., Kosary, C., Yu, M., Ruhl, J., Tatalovich, Z., et al.: Seer cancer statistics review, 1975–2012. National Cancer Institute, Bethesda, MD (2015)
11. Huang, G., Liu, Z., Van Der Maaten, L., Weinberger, K.Q.: Densely connected convolutional networks. In: Proceedings of the IEEE Conference on Computer Vision and Pattern Recognition (CVPR), pp. 4700–4708 (2017)
12. Iandola, F.N., Han, S., Moskewicz, M.W., Ashraf, K., Dally, W.J., Keutzer, K.: SqueezeNet: AlexNet-level accuracy with 50x fewer parameters and <0.5 MB model size. arXiv preprint: arXiv:1602.07360 (2016)
13. Ioffe, S., Szegedy, C.: Batch normalization: accelerating deep network training by reducing internal covariate shift arXiv preprint: arXiv:1502.03167 (2015)
14. Kowal, M., Filipczuk, P.: Nuclei segmentation for computer-aided diagnosis of breast cancer. Int. J. Appl. Math. Comput. Sci. **24**(1), 19–31 (2014)
15. Kowal, M., Filipczuk, P., Obuchowicz, A., Korbicz, J., Monczak, R.: Computer-aided diagnosis of breast cancer based on fine needle biopsy microscopic images. Comput. Biol. Med. **43**(10), 1563–1572 (2013)
16. Krizhevsky, A., Sutskever, I., Hinton, G.E.: ImageNet classification with deep convolutional neural networks. In: Advances in Neural Information Processing Systems (NIPS 2012), pp. 1097–1105 (2012)
17. LeCun, Y., Bengio, Y., Hinton, G.: Deep learning. Nature **521**(7553), 436 (2015)
18. Liang, H., Tsui, B.Y., Ni, H., Valentim, C.C., Baxter, S.L., Liu, G., Cai, W., Kermany, D.S., Sun, X., Chen, J., et al.: Evaluation and accurate diagnoses of pediatric diseases using artificial intelligence. Nat. Med. **25**(3), 433–438 (2019)
19. Lin, T.Y., Goyal, P., Girshick, R., He, K., Dollár, P.: Focal loss for dense object detection. In: Proceedings of the IEEE International Conference on Computer Vision (ICCV), pp. 2980–2988 (2017)
20. Parmar, D., Sawke, N., Sawke, G., et al.: Diagnostic application of computerised nuclear morphometric image analysis in fine needle aspirates of breast lesions. Saudi J. Health Sci. **4**(1), 51 (2015)
21. Simonyan, K., Zisserman, A.: Very deep convolutional networks for large-scale image recognition arXiv preprint: arXiv:1409.1556 (2014)
22. Szegedy, C., Ioffe, S., Vanhoucke, V., Alemi, A.A.: Inception-v4, Inception-ResNet and the impact of residual connections on learning. In: Thirty-First AAAI Conference on Artificial Intelligence (2017)
23. Szegedy, C., Liu, W., Jia, Y., Sermanet, P., Reed, S., Anguelov, D., Erhan, D., Vanhoucke, V., Rabinovich, A.: Going deeper with convolutions. In: Proceedings of the IEEE Conference on Computer Vision and Pattern Recognition (CVPR), pp. 1–9 (2015)

24. Underwood, J.C.: Introduction to Biopsy Interpretation and Surgical Pathology. Springer, London (2012)
25. Vandenberghe, M.E., Scott, M.L., Scorer, P.W., Söderberg, M., Balcerzak, D., Barker, C.: Relevance of deep learning to facilitate the diagnosis of HER2 status in breast cancer. Sci. Rep. **7**, 45938 (2017)
26. Weydert, J.A., Cohen, M.B.: Fine needle aspiration: current practice and recent developments. Lab. Med. **34**(12), 851–854 (2003)
27. Zoph, B., Vasudevan, V., Shlens, J., Le, Q.V.: Learning transferable architectures for scalable image recognition. In: Proceedings of the IEEE Conference on Computer Vision and Pattern Recognition (CVPR), pp. 8697–8710 (2018)
28. Żejmo, M., Kowal, M., Korbicz, J., Monczak, R.: Classification of breast cancer cytological specimen using convolutional neural network. J. Phys. Conf. Ser. **783**, 012060 (2017)

Fibroblast Segmentation in Microscopic Brightfield Images with Convolutional Neural Network

Lukasz Roszkowiak[1(✉)], Jakub Zak[1], Krzysztof Siemion[1,2], Joanna Kinasiewicz[1], and Anna Korzynska[1]

[1] Nalecz Institute of Biocybernetics and Biomedical Engineering Polish Academy of Sciences, Ks. Trojdena 4 Str., 02-109 Warsaw, Poland
lroszkowiak@ibib.waw.pl

[2] Central Clinical Hospital of the Ministry of the Interior and Administration, Warsaw, Poland

Abstract. The aim of this study is the comparison of the various deep convolutional neural networks for segmentation of fibroblast in bright-field microscopic images. This investigation compares two main architectures: Unet and Linknet. Every main architecture is equipped with various 'backbone' network creating specific bundle. The experimental dataset consisting of 16 sequences of images of monitored cells' culture have been split into training and validation set. Then it was analysed and used for validation of the networks to establish the best bundle (net architecture and 'backbone'). This study proved that trained deep convolutional neural networks could be used as a segmentation tool in this task.

Keywords: Biomedical engineering · Image processing · Image segmentation · Deep learning · CNN

1 Introduction

Every genetic modification has influence on the cell. Even transduction of the gene responsible for fluorescent protein that allows observation with fluorescent microscopy could be influential on cell behavior. Cell behavior after genetic modification is still a valid problem we are trying to work with. In our previous experiments we analyzed the movement of the human skin fibroblast observed in bright field microscopy [1]. Analyzis of the movement can be used as a quantifiable indicator of cellular response to extracellular cues, physiological stimuli and overall impact of the environment [2]. The main setback was estimation of cell's outline. Previously tested methods did not yield expected results. Therefore, a new approach was needed. In this study we present novel technique for assessment of cell's boundary with deep learning method in the form of convolutional neural network (CNN).

J. Korbicz et al. (Eds.): PCBEE 2019, AISC 1033, pp. 143–151, 2020.
https://doi.org/10.1007/978-3-030-29885-2_13

CNNs seems to be able to cope with different types of images. Among those, medical images present one of the more difficult issues. In our dataset, microscopic images include a limited number of objects with similar textures. They are usually very difficult to process and analyze.

Fully convolutional networks (FCNs) are neural networks where the classification is performed pixel-by-pixel rather than on the whole image. This way segmentation and semantic segmentation can be performed on the images.

The aim of this study is to confirm if the CNN is able to produce reliable results of fibroblast's boundary detection. To facilitate this we tested various models of CNN and trained from scratch as well as fine-tuned to pre-trained models.

2 Materials and Methods

2.1 Image Data

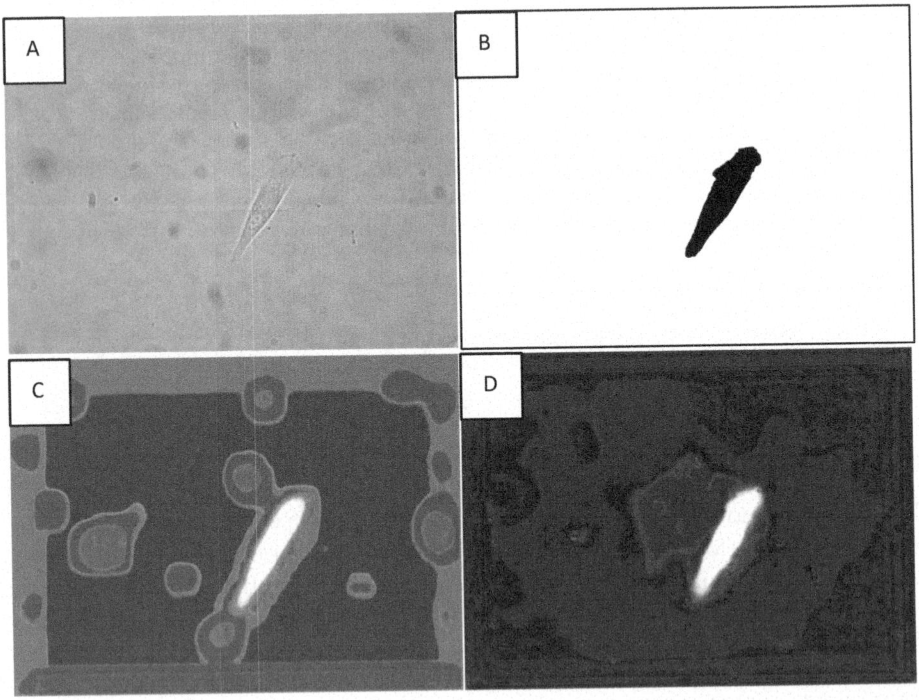

Fig. 1. Example of input image data (A) with related ground truth (B) and results of (C) finetuned Linknet ('resnet50') and (D) Unet ('vgg19') trained from scratch.

Data Collection. The dataset used in this study was created with images of human skin fibroblasts. The experiments procedure was described in detail in previous study [3,4]. The monochromatic image acquisition was performed with camera attached to the microscope (Leica LAS AF DMI 6000B) using 40X objective. The sequences showing cells' movements were acquired automatically with set timestep. As a result we obtained typical brightfield images with artifacts, noise, and low contrast between object and background, as presented in Fig. 1.

Ground Truth Annotation. Previously we used modified level-set method proposed by Li [5] to evaluate the object boundary. We used the previous results to develop the ground truth. As these results of boundary outline were in some cases inaccurate, they were manually modified to better fit the real object boundary. Then, formed boundary was dilated and filled to create the binary mask of full image.

Dataset. The whole dataset consisted of 16 series of images with 1936 images overall. We constructed the training set with 15 of 16 series leaving one out for validation. Furthermore basic image augmentation by random flipping of images horizontally and vertically with 0.5 probability was applied to training set. The validation set was used without augmentation to achieve comparable results.

Backbones. As the tested architectures, namely Unet [6] and Linknet [7], are based on the encoder-decoder representation. It is possible to insert the network with reliable capabilities as the encoder and it's flipped version as decoder. Hence we can name the architecture of a decoder/encoder as a 'backbone' for higher level architecture as depicted in Fig. 2. The 'backbones' we were able to test in this experiment with our hardware: 'inceptionv3', 'resnet18', 'resnet34', 'resnet50', 'resnext50', 'vgg16', 'vgg19'. The main limitation to the complexity of the network was memory capacity of the hardware.

2.2 CNN Training and Validation

We tested 4 types of CNNs with numerous backbones. In this study we focused on Unet [6], Linknet [7], FPNnet and PSPNet [8]. (Unfortunately, the models with FPNnet and PSPNet architecture could not perform in this task, hence their results will not be included in this manuscript.) These architectures have many applications in object recognition and semantic segmentation. Especially Unet which was developed for biomedical image segmentation.

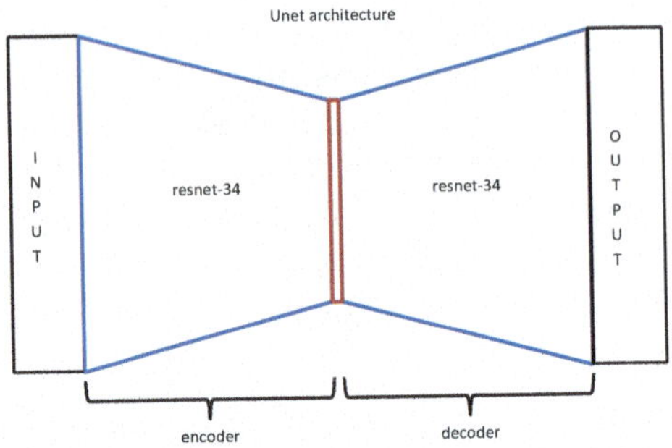

Fig. 2. Encoder-decoder architecture (as in Unet and Linknet) with resnet-34 as a 'backbone'.

Every network was trained two times: first, trained from scratch and second, finetuned. The pretrained weights were obtained from networks trained on 2012 ILSVRC ImageNet dataset [9]. Which was used mainly for classification with localization task. It consisted typical photographs, instead of biomedical data, with 1000 labeled categories.

All experiments were performed in Python by training for 4000 iterations using one of 16 series as a validation set and rest of images from the shuffled dataset as training set. We used Python library segmentation-models and classification-models with default parameters model building. We used Adam optimizer with learning rate and decay set to $1e^{-6}$, and $1e^{-8}$ respectively. Every model was trained with the same optimizer to secure comparable results. All models were trained on Nvidia GPUs and were tested to determine their fitness for estimating boundary of the fibroblast.

The loss function (or optimization score function) is used to compare the true labels (ground truth) with predictions made by network. In this study the binary crossentropy was used. The validation was performed after every epoch so it is possible to monitor the values over process of training. A metric function of binary accuracy was also calculated during training and validation for further assessment. A metric function is similar to a loss function, except that the results from evaluating a metric are not used when training the model.

3 Results

The parameters used for training resulted in steady convergence of loss values for Unet and Linknet as presented in Fig. 3. Both models with various 'backbone' architecture performed with relatively high accuracy.

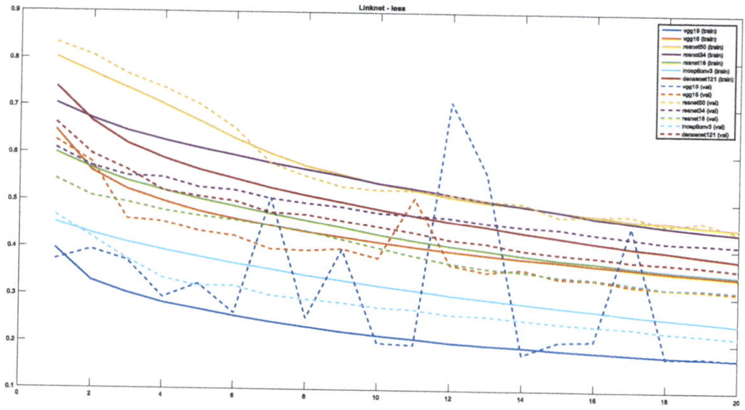

Fig. 3. Steady convergence of loss value during training of Linknet network architecture with various 'backbones'. Presented for training set (solid line) and validation set (dotted line).

We compared the binary accuracy on training and validation set, as well as loss metric. Additionally, we compared the mean time of one epoch for each network. All the results are presented in Table 1.

Table 1. Comparison of accuracy and loss metrics for all the tested networks based on training and validation set. Best performing networks are marked with bold for each set.

	Accuracy		Loss		Time[s]/Epoch
	Train	Val	Train	Val	
'Linknet'					
'finetune'					
'densenet121'	0,875	0,870	0,594	0,576	158
'inceptionv3'	0,968	0,977	0,334	0,324	154
'resnet18'	0,975	0,980	0,242	0,225	88
'resnet34'	0,848	0,825	0,538	0,556	107
'resnet50'	0,993	**0,994**	0,138	0,132	176
'vgg16'	0,967	0,979	0,373	0,403	211
'vgg19'	0,971	0,982	0,278	0,286	236

(contniued)

Table 1. (*continued*)

	Accuracy		Loss		Time[s]/Epoch
	Train	Val	Train	Val	
'scratch'					
'densenet121'	0,974	0,986	0,380	0,360	150
'inceptionv3'	0,985	**0,987**	0,242	0,216	147
'resnet18'	0,967	0,976	0,346	0,314	80
'resnet34'	0,965	0,980	0,437	0,411	98
'resnet50'	0,951	0,975	0,447	0,440	172
'vgg16'	0,993	0,482	0,261	0,881	252
'vgg19'	0,993	0,292	0,170	0,703	233
'Unet'					
'finetune'					
'densenet121'	0,993	0,995	0,306	0,296	169
'inceptionv3'	0,994	**0,997**	0,342	0,345	162
'resnet18'	0,968	0,971	0,475	0,468	103
'resnet34'	0,993	0,995	0,238	0,234	122
'resnet50'	0,994	0,995	0,203	0,199	180
'resnext50'	0,994	0,995	0,217	0,216	265
'vgg16'	0,952	0,952	0,448	0,477	221
'vgg19'	0,966	0,977	0,397	0,422	249
'scratch'					
'densenet121'	0,992	**0,993**	0,210	0,201	162
'inceptionv3'	0,952	0,976	0,441	0,362	155
'resnet18'	0,970	0,982	0,417	0,375	52
'resnet34'	0,994	0,902	0,165	0,878	113
'resnet50'	0,923	0,974	0,536	0,454	171
'resnext50'	0,976	0,982	0,359	0,302	257
'vgg16'	0,989	0,988	0,287	0,295	213
'vgg19'	0,966	0,959	0,447	0,536	243

4 Discussion

The systematic decreasing of loss value and increase of accuracy during training indicates that networks with Unet and Linknet architectures are capable of segmentation fibroblast in microscopic images.

The average accuracy achieved on validation set was better for Unet (0, 977) than Linknet (0, 878). We observed the general tendency that in both cases the metrics were favorable for finetuning than learning from scratch (Unet 0, 985 *vs* 0, 970 and Linknet 0, 944 *vs* 0, 811, respectively). This seems as reasonable since networks that were already trained has established image feature detection. Even thou the pretrained weights were obtained on the training set containing natural images instead of microscopic the networks seem to be able to generalize the feature detection and perform decently. It probably can be related to possibility of generalizing to line and shape detection. Such properties could be easily transferred to different type of images.

Based on the comparison, presented in Table 1, we concluded that, while the final value of accuracy was similar after the training, the convergence for Unet was faster, as presented in Fig. 4. Thus suggesting that the Unet architecture is more robust.

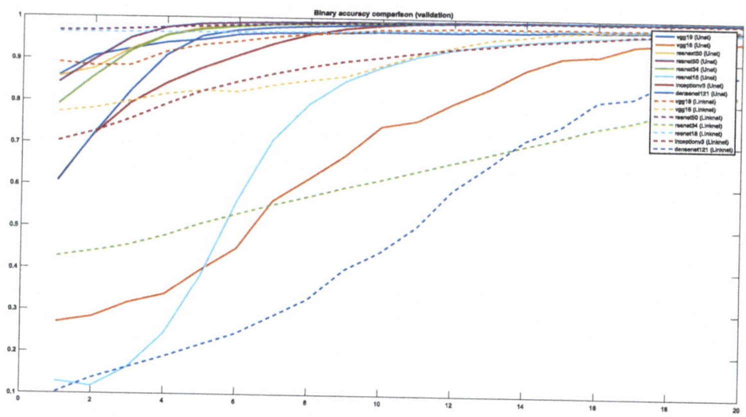

Fig. 4. Comparison of binary accuracy for Unet (solid line) and Linknet (dotted line) with various backbones.

Unet architecture seemed to be more independent from the various 'backbone'. Accuracy of validation set for finetuned Unet was around 0, 995 for majority of 'backbones' while for Linknet architecture the range of accuracy values is wider (as in Table 1).

Also, while training from scratch simple networks like vgg16 and 19 did not achieve good results with Linknet architecture in contrast to Unet. This suggests that Unet architecture overcomes the one of Linknet in this type of task.

The best performing 'backbones' are presented in Table 1 for comparison. The previously established overall tendency of Unet's superiority is also confirmed by this. Furthermore, we noticed the common tendency that more complex architectures (like inception and densenet) yield better results.

Moreover, the time of processing increased with complexity of the 'backbone' architecture. For simpler networks like resnet18 we observed lower times of computation while old network with simple connections like vgg16 ad 19 was usually the slowest.

Additionally to evaluation based on quality coefficients, the visual inspection of the resulting images was performed. We noticed that among finetuned networks those with 'inceptionv3' backbone are visually appealing as the shape estimation is quite definite with sharp edges (easy to establish threshold value on the network's predictions). Apart from that the finetuned Linknet with 'resnet50' backbone has also good results but are less clear. Visually Unet trained from scratch with 'resnet34', 'resnet50' and 'vgg19' surpass other networks results.

One of the main difference observed in Unet and Linknet capabilities was shown by 'resnext50' backbone. While it was possible to train the Unet architecture with this backbone the Linknet caused out of memory error. This proves that Unet has lower needs and better handling in terms of memory.

Generally, it is agreed that CNNs analyze images similarly to human visual processing. Firsly, simple shapes are recognized and with the depth of layers the complexity of shapes increases. At the early stages, CNNs search for oriented line patterns and geometrical shapes. Further down the road, features like edges can be analyzed. As the depth increases, the network discrens specific forms of objects or features very specific to only the type of object presented during training. Therefore, the larger the network, the more subtle differences it should be able to discern. This directly impacts CNN generalization and future finetuning possibilities. Weights in "smaller" networks could sometimes be insufficient in number to incorporate more complex features of the image. "Larger" networks contain many more layers, which largely increases the number of weights involved in processing a feature. Therefore, it can be concluded that the larger the network, the larger the dataset, which the network can generalize. Also, larger networks open more possibilities in finetuning, because of the larger flexibility in changing the weights.

5 Conclusion

We tested the possibility of fibroblast boundary estimation in microscopic images with CNN. This study proved that trained CNN could be used as a segmentation tool in this task. Variety of CNN models was validated on set of microscopic images from experiment. The boundary achieved using proposed method could be further used for cell's movement estimation.

To sum up, we established that Unet performs superior to Linknet in classification and segmentation tasks as it is more independent of used various 'backbone' and performs more robustly. Best performance was achieved while finetuning Unet with 'inceptionv3' backbone.

Acknowledgments. We would like to thank Comtegra S.A. for renting us the two Tesla v100 cards that we used for computing. They had no role in study design, data collection and analysis, decision to publish, or preparation of the manuscript.

This study was supported by statutory funds of Nalecz Institute of Biocybernetics and Biomedical Engineering Polish Academy of Sciences.

References

1. Korzynska, A., Roszkowiak, L., Siemion, K., Zak, J., Zakrzewska, K., Samluk, A., Wencel, A., Pluta, K., Pijanowska, D.: The analysis of the shape of the genetically modified human skin fibroblasts in culture. In: Recent Developments and Achievements in Biocybernetics and Biomedical Engineering, pp. 98–109. Springer, Cham (2017)
2. Pang, F., Liu, Z.: Analyzing temporal dynamics of cell deformation and intracellular movement with video feature aggregation. Biomed. Eng. OnLine **18**(1), 20 (2019)
3. Samluk, A., Zakrzewska, K.E., Pluta, K.D.: Generation of fluorescently labeled cell lines, C3A hepatoma cells, and human adult skin fibroblasts to study coculture models. Artif. Organs **37**(7), E123–E130 (2013)
4. Korzynska, A., Siemion, K., Zak, J., Roszkowiak, L., Wencel, A., Pluta, K., Pijanowska, D.: The analysis of the movement of the genetically modified human skin fibroblasts in culture. In: 2018 IEEE International Conference on Imaging Systems and Techniques (IST). IEEE, October 2018
5. Li, C., Xu, C., Gui, C., Fox, M.: Level set evolution without re-initialization: a new variational formulation. In: 2005 IEEE Computer Society Conference on Computer Vision and Pattern Recognition (CVPR 2005). IEEE (2005)
6. Ronneberger, O., Fischer, P., Brox, T.: U-Net: convolutional networks for biomedical image segmentation. In: Navab, N., Hornegger, J., Wells, W.M, Frangi, A.F. (eds.) Medical Image Computing and Computer-Assisted Intervention – MICCAI 2015, pp. 234– 241. Springer, Cham (2015). ISBN 978-3-319-24574-4
7. Chaurasia, A., Culurciello, E.: LinkNet: exploiting encoder representations for efficient semantic segmentation. In: 2017 IEEE Visual Communications and Image Processing (VCIP), pp. 1–4, December 2017. https://doi.org/10.1109/VCIP.2017.8305148
8. Zhao, H., Shi, J., Qi, X., Wang, X., Jia, J.: Pyramid scene parsing network. In: 2017 IEEE Conference on Computer Vision and Pattern Recognition (CVPR), pp. 6230–6239, July 2017. https://doi.org/10.1109/CVPR.2017.660. ISSN 1063-6919
9. Yakubovskiy, P.: Segmentation models repository. https://github.com/qubvel/segmentation_models/. Accessed 19 Mar 2019

BioTest - Remote Platform for Hypothesis Testing and Analysis of Biomedical Data

Krzysztof Psiuk-Maksymowicz[1,2(✉)], Roman Jaksik[1,2], Aleksander Placzek[3,4], Aleksandra Gruca[4], Sebastian Student[1,2], Damian Borys[1,2], Dariusz Mrozek[4], Krzysztof Fujarewicz[1,2], and Andrzej Swierniak[1,2]

[1] Institute of Automatic Control, Silesian University of Technology, ul. Akademicka 16, 44-100 Gliwice, Poland
krzysztof.psiuk-maksymowicz@polsl.pl

[2] Biotechnology Centre, Silesian University of Technology, ul. Krzywoustego 8, 44-100 Gliwice, Poland

[3] Research and Development Department, WASKO S.A., ul. Berbeckiego 6, 44-100 Gliwice, Poland

[4] Institute of Informatics, Silesian University of Technology, ul. Akademicka 16, 44-100 Gliwice, Poland

Abstract. Modern multi-omics studies introduce a major challenge for biomedical data storage, processing and integration. The ability to utilize multiple different measurement techniques in order to provide a comprehensive view on the studied processes is becoming a standard in molecular biology and medicine, increasing the need for the development of new strategies used for handling various types of data in a single environment. To address this need we developed a remotely accessible BioTest platform which allows to process multi-omics data, obtained using various measurement techniques, including next-generation sequencing (NGS), oligonucleotide microarrays, and mass spectrometry. This allows to process various biomedical data from genomic level, trough transcriptomics, up to the proteomic level on a scalable environment that can handle even the largest datasets. Integrated version control for library files and intermediate results as well as the possibility of designing workflows comprised of various algorithms, ensures that all of the results are highly reproducible. Incorporated support for MapReduce jobs and Big Data processing on a high performance computer cluster allows to test multiple various hypotheses at the same time.

BioTest is the answer to the information explosion observable in life sciences, especially in a cancer research. The ability to handle large amounts of information and possibility to integrate data from various experiments is essential for high quality research, being one of the most important steps towards the development of personalized medicine.

Keywords: Bioinformatics · Biomedical data analysis · Data integration

J. Korbicz et al. (Eds.): PCBEE 2019, AISC 1033, pp. 152–165, 2020.
https://doi.org/10.1007/978-3-030-29885-2_14

1 Introduction

Modern bioinformatics in large extent is concentrated on the analysis of high-throughput sequencing technologies. Massive amounts of data produced by Next Generation Sequencing (NGS), oligonucleotide microarrays, and mass spectrometry presents a challenge for both data storage and analysis requiring advanced systems and tools for the successful application of this technology. There are already many systems for biomedical data management and analysis [3,14,19], some of which are in particular dedicated to NGS data analysis [1,18] or post-processing of such data by means of *e.g.* Gene Ontology tools [8].

The BioTest is an answer for the emerging problems with processing of the biomedical, especially multi-omics, data, namely with a problem of large number of data, often stored in various formats; necessity to utilize statistical tests in order to assess the significance of the findings; need to include additional data in the study (often time/version) dependent; need to combine various analysis methods with incompatible input/outputs.

The paper is structured as follows. Section 2 describes briefly the methodology of the data analysis incorporated in the platform. Section 3 gives the description about data storage and higher level of the data analysis. In Sect. 4 the reader can find an example of processing the genomic data. Finally, Sect. 5 presents our conclusions.

2 Materials and Methods

The BioTest platform constitutes a comprehensive set of tools for processing the biomedical data of different origin. It includes proteomic, genomic, transcriptomic data, which in turn can be integrated with demographic and clinical data and data acquired from various medical databases (e.g. GODB, KEGG, UniProt, etc.). An additional aspect of data processing is their variability over time (also reference data); this aspect of the system is discussed in the subsequent section.

The overall structure of the platform consists of different subsystems including: core web-application for data management, storage and flow; Galaxy server connected with HPC Ziemowit cluster settled for data analysis; Neo4j graph database and Apache Cassandra NoSQL database management system for a higher level of analysis and data visualisation. The individual elements of the system are virtualized and are scalable (see e.g. [18]).

For the needs of the BioTest system, the well-known Galaxy server was employed [1]. Galaxy was originally written for genomic data analysis, particularly data from the Next Generation Sequencing (e.g. tools for mapping the reads against reference sequences). For our needs, we have significantly extended the functionality of the Galaxy. Among the others, one can distinguish machine learning methods, and further methods connected with the functional interpretation of genes or proteins.

2.1 Supervised and Unsupervised Machine Learning Methods and Tools

The supervised and unsupervised large scale data analysis functionality is implemented in BioTest platform using SPICY system (SPecialized Classification sYStem) [6,17,21] created in the R language. Thanks to the modular construction of the system, it provides a high degree of flexibility in creating various structures of classifiers and analysis workflows. Workflows concept is easier to configure than traditional scripting. We can record the whole workflow and therefore, make it possible to repeat the whole analysis pipelines. The Galaxy server provides a possibility to change the workflow by a visual tool (see example workflow for supervised data analysis in Fig. 1). We can choose between the unsupervised and supervised pipelines independently. Using the unsupervised data analysis part, user can use Principal Component Analysis (PCA) to find the directions of most variation and subjectively discover the differences between complex samples. As an analysis result, the user can download the data result tables and, in addition, the visual representation. The analysis contains sample cluster analysis and feature clusterization, which is important to find out the source of data variability. The example of such visualization is shown in Fig. 2. The user can change many analysis parameters to achieve the intended analysis result. The supervised data analysis part contains complex data classification system. The user can choose between different feature selection/extraction methods, for example Partial Least Squares (PLS), t-test, ANOVA, different classification methods, for example Support Vector Machines (SVM), Linear Discriminant Analysis (LDA), Random Forests (RF) and different model validation methods: Bootstrap, k-fold cross-validation, Leave-One-Out, etc. It is easy to change the part of the analysis sequence by changing the selected module or selecting other method parameters. In each analysis, we can use a framework for accessing and executing remote Web Services and run all computation on a remote server or computer cluster. It is particularly important in a large-scale classification analysis, where we are dealing with complex and big datasets.

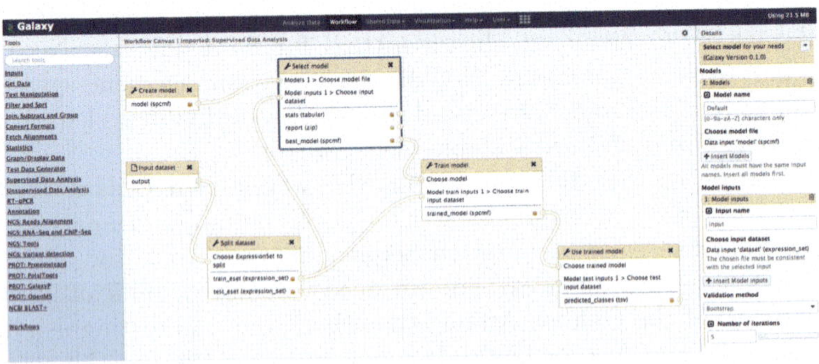

Fig. 1. Visual representation of the predefined workflow for supervised data analysis.

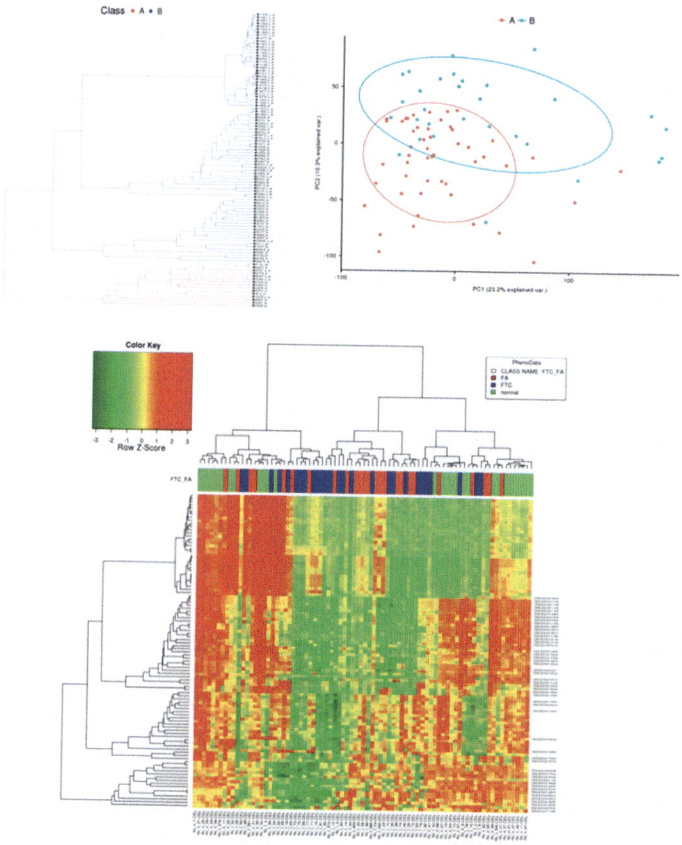

Fig. 2. Visualization of unsupervised data analysis result. The PCA analysis (upper right panel) is supported by hierarchical clustering in feature and sample space and is showed on the heatmap figure (bottom panel). Additionally, it is possible to group the feature sets (upper left panel).

2.2 A Tool for Aggregation of Gene Ontology Terms

The last step of analysis of gene or protein groups obtained as the result of supervised or unsupervised analysis is their functional interpretation. The main purpose of such analysis is to provide information about functions, pathways and biological processes that are common for analyzed set of genes or proteins. Such knowledge can support biological hypotheses related to the experimental design and help biologists or medical doctors to understand results and draw biological conclusions from the performed laboratory experiment.

Gene Ontology Enrichment Analysis. The most typical approach to automated functional interpretation of gene groups is the so-called Gene Ontology enrichment analysis. The Gene Ontology (GO) database provides controlled vocabulary in the form of a directed acyclic graph. The nodes in the GO graph are called Gene Ontology terms and can be used to functionally annotate genes and gene products in terms of their associated biological processes, cellular components and molecular functions [2].

In order to perform GO enrichment analysis, genes from the group of interest and from the reference group are annotated with their corresponding GO terms. Then, based on GO term occurrences in both groups, a statistical test (e.g. chi-square or hypergeometric test) is applied to discover enriched GO terms in the analyzed group of genes. The result of such analysis is provided in the form of a list of statistically significant GO terms which is interpreted as a functional description of an investigated gene group.

A number of tools designed to perform GO enrichment analysis have been developed. Some examples include web application DAVID [10], R library GOstats [5] or stand-alone Cytoscape application BiNGO [15]. Another set of methods that use Gene Ontology terms for functional analysis provide a list of co-appearing combinations of GO terms in the form of so-called logical rules [4,9]. Here, instead of providing a list of single GO terms, a list of statistically significant GO term combinations is returned as a functional description.

Biotest Gene Ontology Clustering And summarization Tool. The main drawback of GO enrichment analysis presented in the previous paragraph is the fact that usually a list of statistically significant GO terms consists of several hundreds of elements. Therefore, it is very difficult to analyze and interpret the obtained functional description.

In order to integrate the information included in a result list of GO terms, the BioTest platform provides a tool that uses a hierarchical clustering algorithm to aggregate GO terms based on their similarities [8]. Groups of GO terms obtained as a result of such analysis generalize functional information and therefore provide a more concise and comprehensive description.

To create a similarity matrix used by the clustering algorithm, the similarities among all statistically significant GO terms have to be computed. As the Gene Ontology database is represented in the form of a hierarchical graph, the most suitable similarity measures are semantic similarity measures. In the BioTest platform, to obtain similarity matrix, the user can choose among the following, state-of-the-art GO similarity measures: Resnik [20], Jiang and Conrath [12], Lin [13], and Wang [22]. Based on the similarity matrix, the hierarchical clustering algorithm is used to compute clusters. By providing the number of clusters, the user can decide how specific should be functional information included in the output clusters.

Clustering results can be processed by another tool implemented on BioTest platform, which allows summarizing information on GO clusters. By mapping GO terms back to their annotating genes and applying BioTest cluster summarization tool, the user can calculate statistics for individual clusters of GO terms. The statistic can be used either to compare the number of up- and down-regulated genes, used in the overrepresentation tests for GO terms or to compare the number of differentially expressed genes between two sample pairs. The algorithm also provides a graphical representation of the results table with proportions of genes specific for each group.

3 Data Storage and High-Level Data Analysis

Methods presented earlier in this article concentrate on raw data integration, data preprocessing and feature selection. As a result of their usage, we achieve the fundamentals to the next level exploration, called a *high-level analysis*. The analysis requires data from different, independent sources, but when integrated together drive the analysis to a discovery of new knowledge. In the BioTest project, these sources covered clinical data from the hospital information system, genomic, proteomic, and/or transcriptomics data collected during medical examination and processed by tools implemented in the Galaxy Server platform, and data gathered at many publicly hosted biological databases. Their integration into one database schema would allow for multidimensional analyses. However, such integration was not possible due to many different use cases related to direct or indirect goals set by researchers, which were not known at the moment of system implementation.

For this reason, we decided to implement the concept of *data lake*, which assumes storing data its native format, both structured and unstructured, until is it needed. Our analyses were focused on genomic data, which gathering has become more and more easier in recent years, however, it consumes a lot of storage resources. For the storage space, we chose the Cassandra database system, which is a flexible and scalable solution. The main features of that platform are:

1. Decentralization which means that there are no single point of failure and no networks bottlenecks, which is very important when processing many biological data with time-consuming algorithms.
2. Replication, which means that data are automatically replicated to other nodes in the cluster for fault tolerance.
3. Wide column storage where data are partitioned, rows are organized into tables and clustered.
4. Incorporated support for MapReduce jobs and Big Data processing.

Another reason for choosing the storage space without a dedicated data structure was the fact that during the analysis of functions of genes and proteins, it is necessary to use publicly hosted, referential databases. These databases provide additional information about molecules identified at the *low-level analysis*. They are periodically updated and vary in time, especially with hierarchical structures. The BioTest system should allow for the analysis of results in the context of different versions of the data schema. To ensure that historical data is matched to the purpose of the higher-level analysis, we used the Apache Spark to process MapReduce jobs [16]. Implemented mechanisms were described in more details in [19]. It is worth emphasizing that the defined Map and Reduce tasks can follow each other repeatedly, where the set of output data from one task becomes a set of input data for another operation. The Apache Spark has been integrated with an external computing cluster that is launched when additional calculations are needed for building ad-hoc queries or creating data aggregations taking into account different types of the data hierarchy.

All these integrated solutions allowed us to create a dynamic data storage layer on the basis of the concept of *schema on read* and enabled to search hidden relations between independent data. Contrary to schema on write, schema on read creates the schema only when reading the data and allows to store the unstructured data. Relations can be described by direction and additional attributes. Graph databases simplify the presentation of such important information since they allow not only to pay attention on the data itself, but also on the relations between the data. In the BioTest project, we integrated the Neo4J, one of the most popular graph databases, which provides a lot of drilling features and an easy to understand query language. We also use Cypher, the elastic query language, for visualization of complex dependencies in an accessible way. The results can be presented in graphical or tabular form. In the case of a graphical form, this is not just a static image, but it is possible to move selected elements, to click to view more information about the object (e.g., additional information about the gene) or to open the object to drill down and search for further dependencies (see Fig. 3).

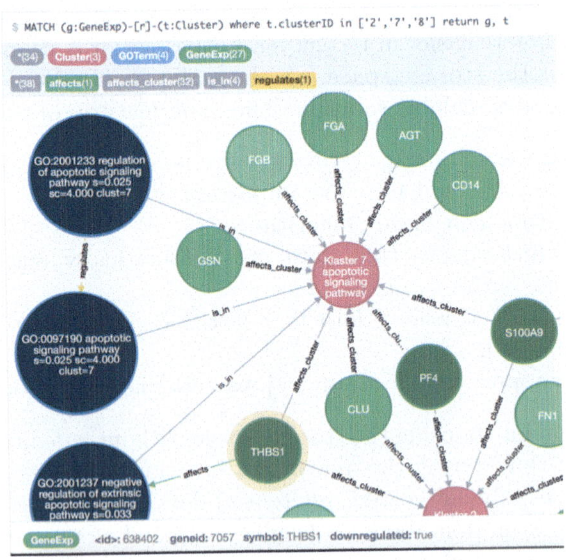

Fig. 3. Relation between clusters, genes and other terms.

Monitoring of the course of the research and the data collection, processing, and aggregation for further analytical purposes is handled by the BioTest application being the backbone of the BioTest platform. It transfers data to lower-level calculations and analyses, where most likely set of key features is identified. The application implements functionalities that allow searching of external biological databases, mapping them to data structures on the platform. It finally takes care of the proper transfer of subsets of data to the graph database, which is the researcher's multidimensional and multi-version analysis tool.

The results of clustering described in the previous chapters can be combined in a simple way with data from biological databases or from systems that record scientific articles. For example, starting from genes belonging to cluster No. 7, the user can go to find points shared with another cluster, he can drill down further to find articles describing connections or genes also located in the cluster and unknown to the researcher. Clinical characteristics of real patients can be discovered, which in the medical examination did not seem crucial. This example is presented in Fig. 3.

4 Example Results

As an example, we show a way of processing a set of microarray data, which originate from a biopsy taken from patients at the Maria Sklodowska-Curie Institute - Oncology Centre, branch in Gliwice. The dataset includes 15 samples that represent two types of thyroid cancer FTC, MTC and control - healthy cells. After copying the data to the BioTest system, the user can run a predefined workflow in Galaxy. Here a workflow of supervised data analysis is performed. It uses SPICY tools and contains e.g. normalization step or supervised data analysis step with a division of the input data into the training and testing data. A successfully launched workflow in Galaxy server is presented in Fig. 4.

This analysis provides a table of standardized expression intensities, information on which genes show significant differences in expression levels between the

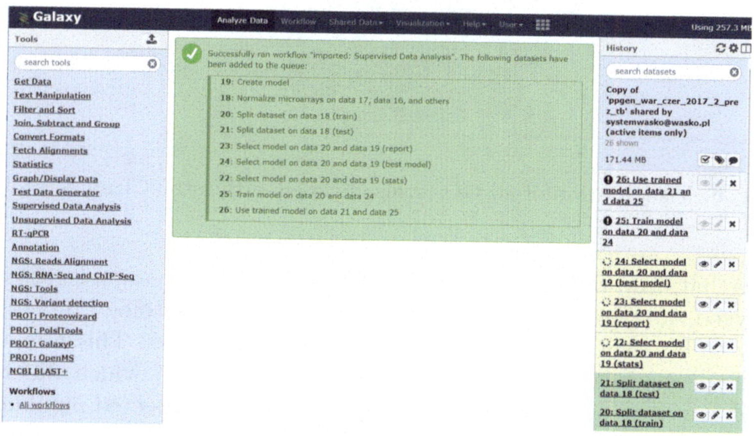

Fig. 4. A process of launching a predefined workflow of the supervised analysis on microarray data.

individual sample types (FTC, MTC and normal) and results of the supervised data analysis which provides information on how well can we predict the sample type given the expression level data alone.

Fig. 5. Visualization of generated 5 clusters of GO terms in Neo4j graph database.

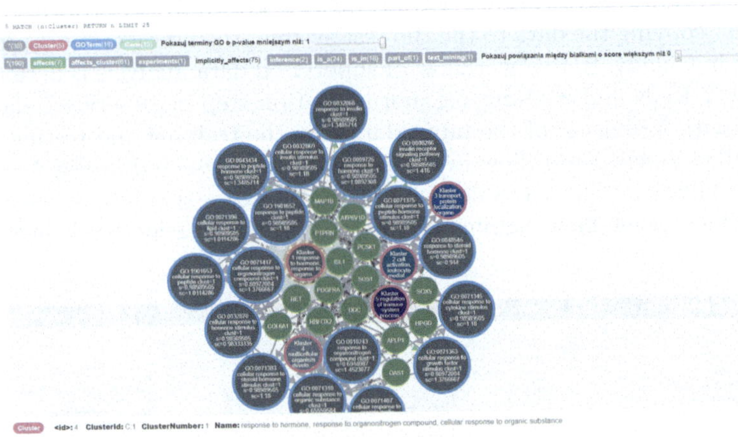

Fig. 6. Visualization of all GO terms enclosed within the Clusters No 1.

Genes that show the highest differences, in terms of expression levels, between the analyzed samples were then used to identify Gene Ontology processes that are likely to be associated with the intracellular differences. This was achieved using a hypergeometric test which identifies processes in which differentially expressed genes are overrepresented, compared to their expected numbers based on random selection.

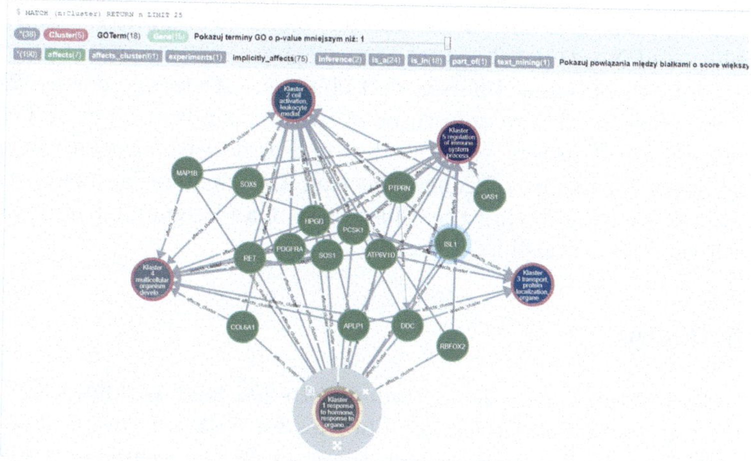

Fig. 7. Visualization of relations between clusters by means of genes effecting the clusters.

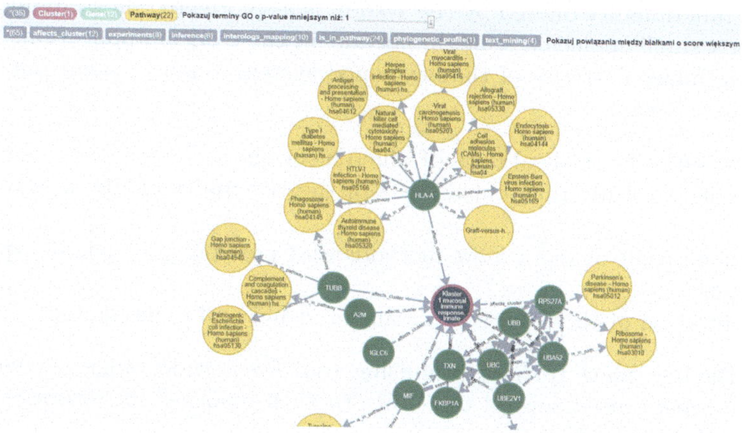

Fig. 8. Visualization of genes effecting the Cluster No 1 and signalling pathways connected with those genes.

Such analysis usually results in a large number of GO processes, that are difficult to interpret. In order to aid this process we use another step in which GO terms are aggregated into clusters (see details in Sect. 2.2). As a result, we achieve 5 clusters with GO terms, their visual representation is presented in Fig. 5 already in a Neo4j framework.

By means of linkage of data from different biomedical databases and dynamical capabilities of Neo4j database we gain an enormous amount of possibilities of finding relations between different GO clusters, GO terms, genes, signalling pathways, etc. We reveal here only three of them. Visualization of all GO terms enclosed within the Clusters No 1 visible in Fig. 6. Visualization of relations between clusters by means of genes affecting the clusters is shown in Fig. 7. Visualization of genes affecting the Cluster No 1 and signalling pathways connected with those genes visible in Fig. 8.

5 Conclusion

The BioTest platform is capable of running pre-designed workflows with a constant set of parameters and reference libraries for each sample provided, additionally monitoring any custom changes that can affect the final outcomes. This feature is very important for clinical applications as it allows to control the reproducibility of the studies. The system supports methods commonly used in basic sciences to identify regulatory interactions and genomic alterations of the malignant cells by utilizing high throughput measurement methods. However, the functionality offered by the system is also oriented on studying various effects on a single molecule scale, important in the clinical applications.

Among many novelties of the developed system, one can point out at least those most valuable:

- integration of biomedical data of different origin;
- possibility to study different aspects of the variability of the data (or methods);
- implementation of many new methods and possibility of making own data workflows;
- interpretation of the results by means of visual and dynamic tool.

The BioTest platform is a promising tool for a wide range of analysis. It has been already used several times in practical analysis, for example in identification of factors that affect reproducibility of mutation calling methods in data originating from the next-generation sequencing [11]; or in thyroid cancer proteome research [7].

Compared to the data presentation methods used by Gene Ontology most popular website, the possibilities are significantly broader when it comes to conducting research. Many limitations have been lifted, like restrictions for searching information about one GO term, a graph is no more a static image, but a dynamic one, the user can drill GO terms in the graph to show further dependencies. In the graph view, all levels above/below a GO term are displayed - depending on the user's preferences. Users can read a list of genes and show the relationship between genes and terms. Finally, one can perform queries to generate tabular statements or present several GO terms on one graph.

Acknowledgement. This work was partially supported by The National Centre for Research and Development grant No. PBS3/B3/32/2015 and Strategmed2/267398/4/NCBR/2015. The work was partially carried out within the statutory research project of the Institute of Informatics, BK-204/RAU2/2019, and the Institute of Automatic Control, 02/010/BK_18/0102. The presented system was developed and installed on the infrastructure of the Ziemowit computer cluster (www.ziemowit.hpc.polsl.pl) in the Laboratory of Bioinformatics and Computational Biology, The Biotechnology, Bioengineering and Bioinformatics Centre Silesian BIO-FARMA, created in the POIG.02.01.00-00-166/08 and expanded in the POIG.02.03.01-00-040/13 projects.

References

1. Afgan, E., Baker, D., van den Beek, M., Blankenberg, D., Bouvier, D., Čech, M., Chilton, J., Clements, D., Coraor, N., Eberhard, C., Gruning, B., Guerler, A., Hillman-Jackson, J., Von Kuster, G., Rasche, E., Soranzo, N., Turaga, N., Taylor, J., Nekrutenko, A., Goecks, J.: The galaxy platform for accessible, reproducible and collaborative biomedical analyses: 2016 update. Nucleic Acids Res. **44**(W1), W3–W10 (2016). https://doi.org/10.1093/nar/gkw343
2. Ashburner, M., Ball, C.A., Blake, J.A., Botstein, D., Butler, H., Cherry, J.M., Davis, A.P., Dolinski, K., Dwight, S.S., Eppig, J.T., Harris, M.A., Hill, D.P., Issel-Tarver, L., Kasarskis, A., Lewis, S., Matese, J.C., Richardson, J.E., Ringwald, M., Rubin, G.M., Sherlock, G.: Gene ontology: tool for the unification of biology. The gene ontology consortium. Nat. Genet. **25**(1), 25–29 (2000). https://doi.org/10.1038/75556
3. Bensz, W., Borys, D., Fujarewicz, K., Herok, K., Jaksik, R., Krasucki, M., Kurczyk, A., Matusik, K., Mrozek, D., Ochab, M., Pacholczyk, M., Pieter, J., Puszynski, K., Psiuk-Maksymowicz, K., Student, S., Swierniak, A., Smieja, J.: Integrated system supporting research on environment related cancers. In: Recent Developments in Intelligent Information and Database Systems. Studies in Computational Intelligence, vol. 642, pp. 399–409. Springer (2016). https://doi.org/10.1007/978-3-319-31277-4_35
4. Carmona-Saez, P., Chagoyen, M., Rodriguez, A., Trelles, O., Carazo, J.M., Pascual-Montano, A.: Integrated analysis of gene expression by association rules discovery. BMC Bioinformatics **7**, 54 (2006). https://doi.org/10.1186/1471-2105-7-54
5. Falcon, S., Gentleman, R.: Using GOstats to test gene lists for GO term association. Bioinformatics **23**(2), 257–258 (2007). https://doi.org/10.1093/bioinformatics/btl567
6. Fujarewicz, K., Student, S., Zielanski, T., Jakubczak, M., Pieter, J., Pojda, K., Swierniak, A.: Large-scale data classification system based on galaxy server and protected from information leak. In: Intelligent Information and Database Systems. Lecture Notes in Computer Science, vol. 10192, pp. 765–773. Springer (2017). https://doi.org/10.1007/978-3-319-54430-4_73
7. Gawin, M., Wojakowska, A., Pietrowska, M., Marczak, L., Chekan, M., Jelonek, K., Lange, D., Jaksik, R., Gruca, A., Widlak, P.: Proteome profiles of different types of thyroid cancers. Mol. Cell. Endocrinol. **472**, 68–79 (2018). https://doi.org/10.1016/j.mce.2017.11.020

8. Gruca, A., Jaksik, R., Psiuk-Maksymowicz, K.: Functional interpretation of gene sets: Semantic-based clustering of gene ontology terms on the biotest platform. In: Man-Machine Interactions 5. Advances in Intelligent Systems and Computing, vol. 659, pp. 125–136. Springer (2018). https://doi.org/10.1007/978-3-319-67792-7_13
9. Gruca, A., Sikora, M., Polanski, A.: RuleGO: a logical rules-based tool for description of gene groups by means of gene ontology. Nucleic Acids Res. **39**(Web Server issue), W293–W301 (2011). https://doi.org/10.1093/nar/gkr507
10. Huang, D.W., Sherman, B.T., Tan, Q., Kir, J., Liu, D., Bryant, D., Guo, Y., Stephens, R., Baseler, M.W., Lane, H.C., Lempicki, R.A.: DAVID bioinformatics resources: expanded annotation database and novel algorithms to better extract biology from large gene lists. Nucleic Acids Res. **35**(Web Server issue), W169–W175 (2007). https://doi.org/10.1093/nar/gkm415
11. Jaksik, R., Psiuk-Maksymowicz, K., Swierniak, A.: Identification of factors that affect reproducibility of mutation calling methods in data originating from the next-generation sequencing. In: Computer and Information Sciences. Communications in Computer and Information Science, vol. 935, pp. 264–271. Springer (2018). https://doi.org/10.1007/978-3-030-00840-6_29
12. Jiang, J.J., Conrath, D.W.: Semantic similarity based on corpus statistics and lexical taxonomy. In: The Proceedings of ROCLING X, Taiwan, pp. 19–33, September 1997
13. Lin, D.: An information-theoretic definition of similarity. In: Proceedings of the Fifteenth International Conference on Machine Learning, ICML 1998, pp. 296–304. Morgan Kaufmann Publishers Inc., San Francisco (1998). http://dl.acm.org/citation.cfm?id=645527.657297
14. Luo, J., Wu, M., Gopukumar, D., Zhao, Y.: Big data application in biomedical research and health care: a literature review. Biomed. Inform. Insights **8**, 1–10 (2016). https://doi.org/10.4137/BII.S31559
15. Maere, S., Heymans, K., Kuiper, M.: BiNGO: a cytoscape plugin to assess over-representation of gene ontology categories in biological networks. Bioinformatics **21**(16), 3448–3449 (2005). https://doi.org/10.1093/bioinformatics/bti551
16. Malysiak-Mrozek, B., Baron, T., Mrozek, D.: Spark-IDPP: high-throughput and scalable prediction of intrinsically disordered protein regions with spark clusters on the cloud. Cluster Comput. **22**(2), 487–508 (2018). https://doi.org/10.1007/s10586-018-2857-9
17. Pojda, K., Jakubczak, M., Student, S., Swierniak, A., Fujarewicz, K.: Comparing different data fusion strategies for cancer classification. In: Proceedings of the International Conference on Information Technology & Systems (ICITS 2018). Advances in Intelligent Systems and Computing, vol. 721, pp. 417–426. Springer (2018). https://doi.org/10.1007/978-3-319-73450-7_40
18. Psiuk-Maksymowicz, K., Mrozek, D., Jaksik, R., Borys, D., Fujarewicz, K., Swierniak, A.: Scalability of a genomic data analysis in the biotest platform. In: Intelligent Information and Database Systems. Lecture Notes in Computer Science, vol. 10192, pp. 741–752. Springer (2017). https://doi.org/10.1007/978-3-319-54430-4_71
19. Psiuk-Maksymowicz, K., Placzek, A., Jaksik, R., Student, S., Borys, D., Mrozek, D., Fujarewicz, K., Swierniak, A.: A holistic approach to testing biomedical hypotheses and analysis of biomedical data. In: Beyond Databases, Architectures and Structures. Advanced Technologies for Data Mining and Knowledge Discovery. Communications in Computer and Information Science, vol. 613, pp. 449–462. Springer (2016). https://doi.org/10.1007/978-3-319-34099-9_34

20. Resnik, P.: Using information content to evaluate semantic similarity in a taxonomy. In: Proceedings of the 14th International Joint Conference on Artificial Intelligence - Volume 1, IJCAI 1995, pp. 448–453. Morgan Kaufmann Publishers Inc., San Francisco (1995). http://dl.acm.org/citation.cfm?id=1625855.1625914
21. Student, S., Pieter, J., Fujarewicz, K.: Multiclass classification problem of large-scale biomedical meta-data. Procedia Technol. **22**, 938–945 (2016). https://doi.org/10.1016/j.protcy.2016.01.093
22. Wang, J.Z., Du, Z., Payattakool, R., Yu, P.S., Chen, C.F.: A new method to measure the semantic similarity of GO terms. Bioinformatics **23**(10), 1274–1281 (2007). https://doi.org/10.1093/bioinformatics/btm087

Detection of Epileptic Seizures via Deep Long Short-Term Memory

Krzysztof Patan(✉) and Grzegorz Rutkowski

Institute of Control and Computation Engineering, University of Zielona Góra, ul. Szafrana 2, 65-516 Zielona Góra, Poland
k.patan@issi.uz.zgora.pl

Abstract. The paper deals with the designing and implementation of a computer-aided system capable to detect seizures by classification of EEG records. The system is based on deep learning method using a recurrent long short-term memory neural network. The main purpose of the system is to help neurologists in detecting seizures fast and reliably. The research was carried out using real EEG recordings of epileptic patients as well as healthy subjects prepared with the cooperation of the medical staff of the Clinical Ward of Neurology of the University Hospital of Zielona Góra, Poland.

Keywords: EEG signals · Seizure detection · Deep learning · Long short-term memory · Classification

1 Introduction

Electroencephalography (EEG) is one of the most popular methods used to measure the activity of neurons on the surface of the cerebral cortex [9]. Analysis of EEG records plays a very important role in the diagnosis of serious neurological disorders such as epilepsy [3,4]. Approximately 1–2% of world's population has epilepsy, about 5% may have at least one seizure during their lifetime, and about 25% of epileptic patients cannot be treated sufficiently by any available therapy [3]. Epileptic seizures can be divided into partial, generalized, unilateral and unclassified [16]. In turn EEG recordings of epileptic patients can be classified into two categories: *inter-ictal* – abnormal activity recorded between epileptic seizures, which has the form of occasional short-term transient waves, and *ictal* – activity recorded during the seizure in the form of long-term polymorphic waves [16,18].

The most popular method of analysis of EEG recordings used by neurologists is a visual inspection. Unfortunately, this method is very time-consuming method and it can be very inefficient. Moreover, there is a very high probability of discrepancies in diagnoses posed by different experts which may result in the assessment of the seizure detection accuracy. What is more, even diagnoses

J. Korbicz et al. (Eds.): PCBEE 2019, AISC 1033, pp. 166–178, 2020.
https://doi.org/10.1007/978-3-030-29885-2_15

carried out by the same expert but in different periods of time may differ significantly. Therefore, the computer-aided system for automatic seizure detection is of great worth.

So far, many different seizure detection systems have been proposed [2,4, 6,11]. Systems working in the time-domain can investigate statistic properties of EEG signals as basic or regularity statistics [18], approximate entropy [4], pattern match regularity statistics [13], and others. Unfortunately, a majority of papers investigated long-term seizures. On the other hand a large number of publications is devoted to the analysis of EEG recordings in the time-frequency domain. Very popular methods used in this scope are discrete wavelet transform (DWT) [7,14,17], matching pursuit (MP) [2,18] and recently a Stockwell transform (ST) [10–12]. In the previous works of the authors time-frequency approaches were successfully applied to short-term seizures detection obtaining even 99.2% of recognition accuracy [11]. However, time-frequency methods are time-consuming. Taking into account that, a method has to transform data recorded at 16 channels the processing time is very long. For example, analysis of entire EEG record lasting 15 min can last longer than the visual inspection. That in the reason that in this work a method operating in the time-domain is considered. A method is based on deep learning using a specific kind of recurrent neural networks called a long short-term memory (LSTM). This structure was proposed in order to cope with the vanishing gradient problem and it is well suited to model dependencies between samples in time sequences. In the paper a deep LSTM network is used to classify EEG sequences into two classes labelled as "normal" and "epileptic".

2 Seizure Database

In cooperation with the medical staff of the Clinical Ward of Neurology of University Hospital of Zielona Góra, Poland, a database of neurological disorders was prepared. EEG data were recorded with 16-channel device using the 10/20 setting. The data was acquired from both epileptic patients and healthy subjects. With the help of neurologists, 588 seizures from 104 patients (both females and males, children and adults) were recorded and analyzed. Simultaneously, 588 sequences from 71 healthy subjects were derived. Eventually, the database consisted of 1176 EEG recordings. An expert in clinical analysis of EEG signals (a neurologist) inspected every record visually to score epileptic and normal sequences. Each record consists of 16 signals. In Fig. 1 it is presented a section of a EEG recording representing a seizure occurred at 835 s. We can see a significant change of signal characteristics, especially in channels P3_O1, P4_O2, T4_T6, T5_O1 and T6_O2. Unfortunately, not only epileptic seizures cause a visible change in the EEG course. A body activity and/or influence of the equipment can significantly complicate the visual inspection. Every additional activity that is not of cerebral origin is called artifact, which is highly undesirable as it can impede the final diagnosis of epilepsy.

Our objective is to detect the inter-ictal seizures pending less than 1 s. Due to the fact that considered seizures are characterized by a short-time occurrence,

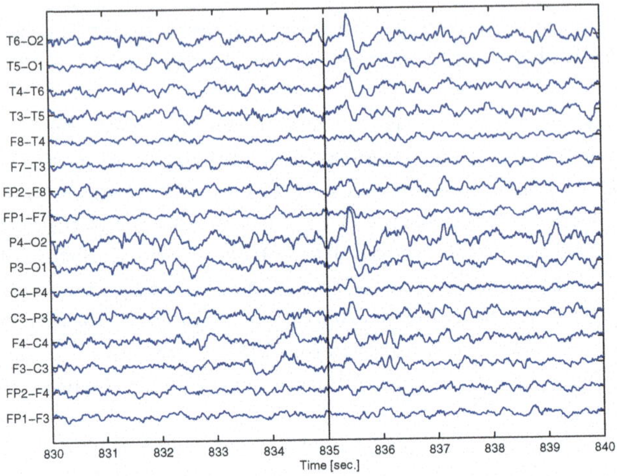

Fig. 1. EEG recording with the beginning of the epileptic seizure marked with the line

the problem of seizure detection becomes non trivial problem to solve. Moreover, after consultations with specialists in the field of neurology we found out that the application should be developed is such a way as to minimize the time spent by a specialist to analyze the record. In fact, in a hospital the EEG signals are recorded by a technician first and then are analyzed by a neurologist.

For each EEG record a neurologist provided the information about classification of the record to one of two classes: "epileptic patient" or "healthy subject". Additionally, for epileptic patients a neurologist marked the beginning time of seizures occurrence. However, it should be pointed out that the duration time of each seizure is not provided. As we deal with short-time seizures, with consultation with experts we assumed that the duration time of short-time seizures is less than one second. Then, in order to prepare data for classifier training, it was required to label the output data. Therefore, we assumed that each seizure marked by a neurologist lasts one second.

3 Statistical Analysis

One of possible methods for seizure detection in the time domain is the analysis of the statistical parameters of the EEG signal. A popular method of EEG analysis is checking the standard deviation of the recorded signal. The basic idea relies on the calculation of the standard deviation of the signal using a moving window. As our objective is to detect short-time seizures a window of the length of one second was used. Using the sample frequency of 100 Hz it gives sequences containing 100 samples. The standard deviation σ of the achieved in this way sequence x is calculated as follows:

$$\sigma = \sqrt{\frac{1}{100}\sum_{i=1}^{100}(x_i - \bar{x})^2}, \tag{1}$$

where $\bar{x}$ is the mean value of x. The standard deviation is calculated for each channel of the EEG record. As a result, a matrix $S \in \mathbb{R}^{16\times n}$ is obtained, where n is the number of time sequences possible to cut off using the moving window. In order to detect abrupt changes in a standard deviation course a simple thresholding was applied. Let s_i be the i-th row of the matrix $S_{i,j}$. A threshold for the i-th channel can be calculated as:

$$T_i = t_\alpha \sigma_i + \bar{s}_i \tag{2}$$

where T_i is the threshold imposed on the i-th channel, σ_i is the standard deviation of the signal s_i, $\bar{s}_i$ stands for the mean value of s_i, and t_α represents a random value of the normal distribution assigned to the significance level α according to:

$$\alpha = P\left(\left|\frac{S_i - \bar{s}_i}{\sigma_i}\right| > t_\alpha\right), \tag{3}$$

where P is the probability operator. A significance level α qualifies a probability that the signal s_i exceeds the value of t_α. Thresholds determined in this way are applied to each sequence $s_i, i = 1, \ldots, 16$.

In order to evaluate the seizure detection system a number of quality indexes are applied. A sensitivity index or a true positive rate measures the proportion of correctly detected seizures:

$$tpr = \frac{n_{tp}}{n_p}100\%, \tag{4}$$

where n_{tp} is the number of detected seizures, n_p is the total number of seizures pointed out by the doctor. In the ideal case $tpr = 100\%$. A specifity index or a true negative rate measures the proportion of correctly identified healthy sequences and it is defined as:

$$tnr = \frac{n_{tn}}{n_n}100\%, \tag{5}$$

where n_{tn} is the number of sequences marked as normal ones, and n_n it the total number of healthy sequences. In the ideal case $tnr = 100\%$ An accuracy index represents the overall accuracy of the system defined in the following way:

$$acc = \frac{n_{tp} + n_{tn}}{n_p + n_n}100\%. \tag{6}$$

The index takes into account both the properly classified seizures and normal conditions. In the ideal case $acc = 100\%$.

4 Long Short-Term Memory

A long short-term memory (LSTM) is a recurrent neural network proposed in 1997 by Hochreiter and Schmidhuber [5] as a solution for the problem of the blowing-up or vanishing gradient that was frequently observed when training the traditional recurrent models. The name long short-term relates to the fact that LSTM is a model for the short-term memory that can last for a long time. A common structure of LSTM consists of a memory cell, an input gate, an output gate and a forget gate. The memory cell is responsible for storing data for arbitrary time periods. In turn, gates control data flow through connections of the whole model. The basic representation of the single LSTM unit is given by the set of equations. The input gate is represented as:

$$\boldsymbol{i}(k) = \sigma_s(\boldsymbol{W}_i\boldsymbol{x}(k) + \boldsymbol{V}_i\boldsymbol{h}(k-1) + \boldsymbol{b}_i), \tag{7}$$

where $\boldsymbol{W}_i$, $\boldsymbol{V}_i$ are the weight matrices and $\boldsymbol{b}_i$ is the bias vector of the input gate, $\boldsymbol{x}(k)$ and $\boldsymbol{h}(k)$ are the input and the output of the LSTM unit, and σ_s is the activation function. In a similar manner, the output and forget gates

$$\boldsymbol{o}(k) = \sigma_s(\boldsymbol{W}_o\boldsymbol{x}(k) + \boldsymbol{V}_o\boldsymbol{h}(k-1) + \boldsymbol{b}_o), \tag{8}$$

$$\boldsymbol{f}(k) = \sigma_s(\boldsymbol{W}_f\boldsymbol{x}(k) + \boldsymbol{V}_f\boldsymbol{h}(k-1) + \boldsymbol{b}_f) \tag{9}$$

are described, where $\boldsymbol{W}_o$, $\boldsymbol{V}_o$ and $\boldsymbol{b}_o$ are parameters of the output gate and $\boldsymbol{W}_f$, $\boldsymbol{V}_f$ and $\boldsymbol{b}_f$ are parameters of the forget gate. Using (7), (8) and (9), the memory cell is given by

$$\boldsymbol{c}(k) = \boldsymbol{f} \circ \boldsymbol{c}(k-1) + \boldsymbol{i}(k) \circ \sigma_c(\boldsymbol{W}_c\boldsymbol{x}(k) + \boldsymbol{V}_c\boldsymbol{h}(k-1) + \boldsymbol{b}_c) \tag{10}$$

and the cell output:

$$\boldsymbol{h}(k) = \boldsymbol{o}(k) \circ \sigma_h(\boldsymbol{c}(k)), \tag{11}$$

where $\boldsymbol{W}_c$, $\boldsymbol{V}_c$ and $\boldsymbol{b}_c$ are weight matrices and bias vector of the memory cell, σ_c and σ_h stand for the activation functions, $\circ$ is the Hadamard product (element-wise product), and the initial values are $\boldsymbol{c}(0) = \boldsymbol{0}$ and $\boldsymbol{h}(0) = \boldsymbol{0}$.

Either a memory cell or gate units can convey useful information about the current state of the network. The memory cell stores the network state, for either a long or a short period. This is achieved by means of the identity activation function. The input gate controls the extent, to which a new value (or a state) flows into the memory cell. The forget gate controls the extent, to which a value remains in the memory cell; and finally the output gate controls the extent, to which the value in the memory cell is used to compute the output of the memory cell. A single LSTM unit is shown in Fig. 3. One can design larger networks using the LSTM units. The topology of the entire network consists of one input layer, one hidden layer and one output layer. The fully self-connected hidden layer includes a number of memory cells and corresponding gate units. For structural details the reader is referred to the work [5] (Fig. 2).

A block scheme of the proposed detection system is presented in Fig. 3. The input sequences are processed to the LSTM layer, where LSTM can be used

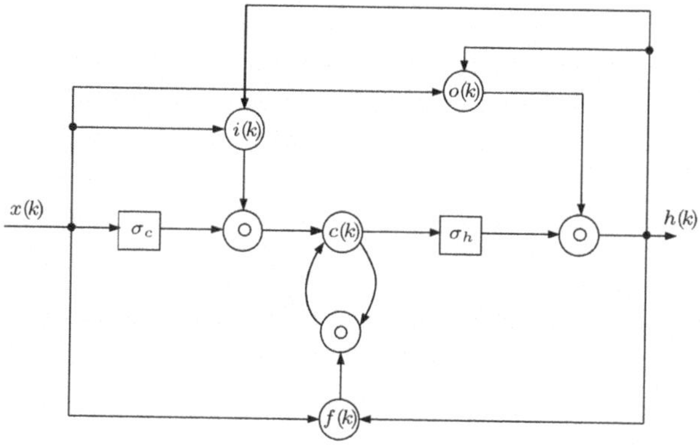

Fig. 2. A simplified structure of an LSTM node; $\circ$ stands for Hadamard product

to remember long-short term dependencies between samples in the input time-sequence. This knowledge flows through a fully connected layer to a softmax layer. The softmax layer is used to assign decimal probabilities to each class of the considered problem. Then the layer has the same number of units as the output layer. Furthermore, these decimal probabilities have to add up to 1. The operation performed by this layer can be represented as follows:

$$P(y = j|x) = \frac{e^{x_j}}{\sum_{k=1}^{K} e^{x_k}}, \quad j = 1, \ldots, K \tag{12}$$

where j is the class index, K is the number of classes, x is the softmax layer input. Finally, a classification layer takes the values from the softmax layer and assigns each input to one of the K mutually exclusive classes using the cross-entropy function:

$$loss = -\sum_{i=1}^{N}\sum_{j=1}^{K} t_{ij} \ln(y_{ij}), \tag{13}$$

where N stands for the number of samples, t_{ij} is the indicator that the i-th sample belongs to the j-th class, and y_{ij} is the output of the softmax unit. In order to increase the classification capabilities of the system instead of one LSTM layer we can use more layers. However, this operation significantly increases the number of adaptable parameters of the entire network and directly influences the computational burden of the training.

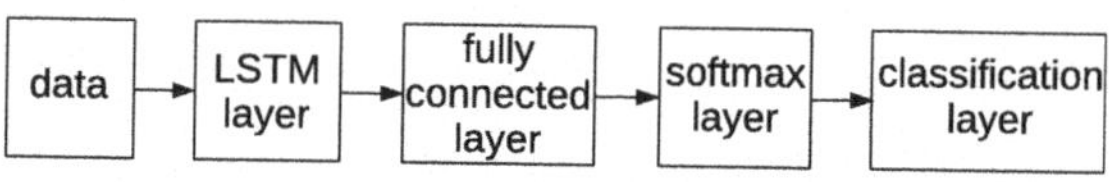

Fig. 3. A block scheme of the classification system

Currently, LSTM networks are widely used in the field of deep learning. Deep learning is based on learning data representation contrary to the task specific methods. In deep learning, each layer of a neural network is trained in order to transform input data into a slightly more abstract and composite representation. The very efficient algorithm for training deep neural networks is Adam (Adaptive momentum estimation). Adam is based on a stochastic gradient descent method and computes individual adaptive learning rates for different parameters using first and second moments of the gradient. In fact, Adam is a fusion of two methods: AdaGrad [1] and RMSProp [15]. The comprehensive description of the algorithm is provided in [8]. The important issue is that the user should provide the initial learning rate η, the gradient decay factor β_1, the squared gradient decay factor β_2 and a mini-batch size.

5 Experiment

5.1 Statistical Analysis

Table 1 includes results of statistical analysis of EEG records carried out using the algorithm presented in Sect. 3. Analyzing the true positive rate by (4) it is obvious that in case of 24 subjects (which is about 23% of a total number of patients) the method did not detect any epileptic seizure. In case of 49% of patients the method detected at least a half of epileptic seizures pointed out by the expert. All seizures were detected in case of 7 patients only. On the other hand reviewing the results listed in Table 1 it is clear that true negative rate (5) achieves high values (from 62.9% to 90.4%) for each considered case. However, the ideal value of *tnr* is 100%. It means that there is a quite relevant number of false seizures detection for each considered EEG record. The last examined index is the overall classification quality represented by the accuracy index (6). As one can see in Table 1 the *acc* index takes the similar values to those of *tnr*. The easy explanation of this phenomenon is a relative small number of time sequences representing seizure to the number of sequences representing the normal work of the brain. For example, for an EEG record lasting 15 min

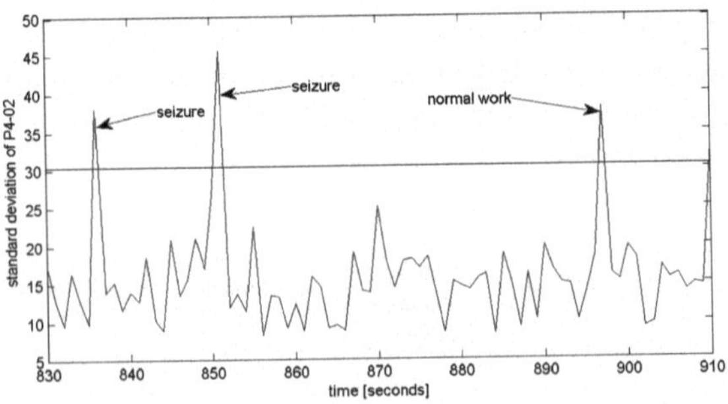

Fig. 4. Example of seizure detection using standard deviation statistic.

Table 1. Seizure detection indexes

id	*tpr* (%)	*tnr* (%)	*acc* (%)	id	*tpr* (%)	*tnr* (%)	*acc* (%)	id	*tpr* (%)	*tnr* (%)	*acc* (%)	*id*	*tpr* (%)	*tnr* (%)	*acc* (%)
1	50	88.8	88.5	27	0	75.6	75.3	53	33.3	82.6	82.4	79	0	82	81.6
2	57	75.3	75.1	28	71.4	77.2	77.2	54	66.6	74.3	74.2	80	33.3	68.8	68.6
3	100	69.3	69.3	29	20	78.8	78.4	55	80	77.7	77.7	81	55.6	64.9	64.8
4	100	70	70.2	30	70	77.5	77.4	56	80	80.2	80.2	82	100	71.3	71.4
5	66.7	83.7	83.6	31	80	85.7	85.7	57	0	88.7	88.2	83	100	82.7	82.7
6	66.7	76.1	76.1	32	88.9	65.5	65.8	58	25	82.6	82.1	84	0	80.2	80.1
7	25	84.7	84.5	33	0	82.6	82.1	59	85.8	83.7	83.7	85	0	81.8	81
8	66.7	81.4	81.4	34	16.7	77.1	76.8	60	50	81.1	80.9	86	88.9	78.8	78.9
9	50	78.4	78.4	35	80	86.2	86.2	61	0	77.6	77.1	87	0	73.3	72.9
10	25	81.9	81.7	36	50	76.1	75.9	62	71.4	75.7	75.6	88	20	81.6	81.2
11	50	75.5	75.3	37	60	75.7	75.6	63	66.7	69	68.9	89	100	78	78.1
12	0	82.2	81.7	38	20	77.9	77.7	64	66.7	79.9	79.8	90	40	76	75.9
13	0	75	74.8	39	0	81.9	81.6	65	0	62.9	62.7	91	80	75.7	75.7
14	25	76.9	76.7	40	50	75.1	75	66	33.3	84.4	84.3	92	20	79.3	79
15	14.3	84.6	84.1	41	25	74	73.7	67	57.1	86.2	86	93	75	82	81.9
16	80	84.6	84.5	42	0	74	73.7	68	66.7	79.6	79.5	94	0	86.4	85.8
17	50	83.6	83.5	43	60	77.3	77.2	69	100	80.5	80.5	95	83.3	70	70.1
18	37.5	75.7	75.4	44	0	78.9	78.4	70	62.5	80.7	80.5	96	25	71.9	71.4
19	87.5	71.7	71.9	45	0	88.6	88	71	33.3	73.8	73.7	97	80	70.2	70.3
20	100	70.7	70.9	46	66.7	80.1	80	72	0	81.7	81.1	98	66.7	84.6	84.6
21	80	74.3	74.4	47	50	79.1	79.1	73	70.6	83.7	83.4	99	20	82.6	81.9
22	75	86.5	86.4	48	33.3	80.8	80.4	74	33.3	69.3	69.2	100	66.7	78	78
23	16.7	78.8	78.3	49	0	73.7	72.9	75	0	90.4	90.1	101	75	76.9	76.8
24	0	85.9	85.7	50	0	74.6	74.1	76	50	83.9	83.7	102	0	85.6	84.9
25	60	74.9	74.8	51	0	84.9	84.3	77	40	82.8	82.6	103	0	73.7	73.3
26	20	81.4	81.1	52	40	82.1	81.9	78	36.3	81.1	80.7	104	50	76.9	76.9

there are 900 time sequences of the length of 1 s. Among them the neurologist indicated 7 representing the epileptic seizure. Then $n_{tn} \gg n_{tp}$ as well as $n_n \gg n_p$. Consequently, in such cases $acc \approx tnr$.

Figure 4 illustrates detection of seizures in one of channels using the standard deviation. For the considered case, a neurologist diagnosed seizures at 835-th and 853-th second. These two seizures were reliably detected. However, we can observe a sudden change of standard deviation around 897-th second which was not treated by a neurologist as a seizure. Summarizing, any artifact will be signalled as a seizure obscuring proper work of the detection system.

Table 2. Investigated network structures

Network structure	Number of LSTM nodes			Activation function		
	1st layer	2nd layer	3rd layer	σ_c	σ_s	σ_h
net1	40	–	–	hyperbolic tangent	sigmoid	linear
net2	60	–	–	hyperbolic tangent	sigmoid	linear
net3	100	–	–	hyperbolic tangent	sigmoid	linear
net4	400	–	–	hyperbolic tangent	sigmoid	linear
net5	60	50	–	hyperbolic tangent	sigmoid	linear
net6	100	70	–	hyperbolic tangent	sigmoid	linear
net7	50	30	10	hyperbolic tangent	sigmoid	linear
net8	100	70	40	hyperbolic tangent	sigmoid	linear

5.2 Deep LSTM Classification

Data Preparation. In order to properly train LSTM network entire dataset was divided into two separable sets: training and testing ones using the splitting ratio equal to 0.5. It means that both sets includes the same number of sequences. The input space (called also a feature space) contained time-sequences recorded at each channel of the EEG equipment. In result, the number of inputs was 16. Taking into account that the seizure time is less than one second we used training sequences of the length of 3 s. Each sample in the sequence needs to be classified to one of the two classes: "normal" or "seizure".

Network Structure Selection. The important problem when dealing with neural systems is a proper selection of the topology in such a way as to guarantee that the neural network is able to solve the problem. We used neural model represented in Fig. 3 starting with one LSTM layer trying different number of nodes. After that we investigated deep models with two and three LSTM layers and with different number of nodes. The specification of investigated neural models is presented in Table 2. The results of seizure detection using proposed neural structures are presented in Table 3 and commented later on in this section in the paragraph *Result evaluation.*

Network Training. The value of the initial learning rate is of a crucial importance. Too large value of η causes large problems with convergence of the training. Different values of η from the interval $(0.0001 - 0.5)$ were tested and we found out that $\eta = 0.02$ guarantees the stable and fast training of the neural classifier. Also different sizes of the mini-batch were tested starting from 10 finishing on 128. The best training results were obtained using the mini-batch size equal to 35. For decay parameters we used default values namely $\beta_1 = 0.9$ and $\beta_2 = 0.999$.

Table 3. Seizure detection results for LSTM

Network topology	Recognized (%)		Unrecognized		*tnr* (%)		*tpr* (%)		*acc* (%)	
	train	test	train	test	train	test	train	test	train	test
net1	51	52	19.9	28.8	81	81	49	49	70	70
net2	55	55	37.5	54.4	83	83	51	52	72	73
net3	54	55	33.2	43.6	83	83	52	54	73	73
net4	60	56	32.5	50.7	83	84	56	54	74	74
net5	73	74	13.5	18.5	81	82	66	67	76	77
net6	80	81	10.5	16.5	89	89	75	74	84	84
net7	76	75	9.2	13.8	77	77	68	68	74	74
net8	92	91	1.5	1.5	90	90	85	85	88	88

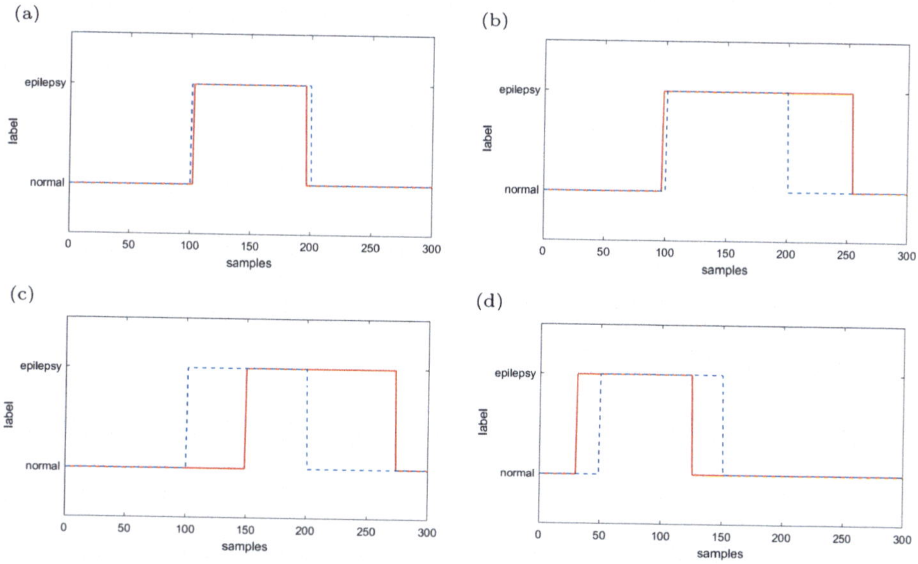

Fig. 5. Examples of seizure detection using LSTM.

Seizure Detection Results Evaluation. Table 3 includes results of EEG recordings classification using different network structures described in the paragraph *Network structure selection*. The seizure is treated as surely recognized when *tpr* is greater or equal to 50%. On the other hand the seizure is assumed to be unrecognized when *tpr* is equal to zero. Each network was trained 10 times and the network with the lowest value of the loss function was selected for further evaluation. The first observation is that using one LSTM layer and increasing the number of hidden units an increasing recognition of seizures can be achieved. Unfortunately, the recognition accuracy is not so high (60% for the training set).

Table 4. Epilepsy diagnosis vs patients

id	Seizures number	*tpr* (%)	id	Seizures number	*tpr* (%)	id	Seizures number	*tpr* (%)	id	Seizures number	*tpr* (%)
1	6	100	27	3	100	53	3	100	79	6	83.3
2	7	87.7	28	7	100	54	6	100	80	6	100
3	2	100	29	5	100	55	5	100	81	9	100
4	5	100	30	10	80	56	5	100	82	5	80
5	6	83.3	31	5	80	57	5	100	83	2	100
6	3	100	32	9	88.9	58	8	100	84	2	100
7	4	100	33	5	100	59	7	85.7	85	8	87.5
8	3	100	34	6	100	60	6	83.3	86	9	77.8
9	2	100	35	5	100	61	5	80	87	6	100
10	4	100	36	6	100	62	7	100	88	5	100
11	6	83.3	37	5	80	63	6	83.3	89	2	100
12	4	75	38	5	80	64	9	88.9	90	5	100
13	3	100	39	4	50	65	5	100	91	10	100
14	4	75	40	6	83.3	66	3	100	92	5	100
15	7	85.7	41	4	100	67	7	71.4	93	8	100
16	10	90	42	5	80	68	3	100	94	8	100
17	4	75	43	5	100	69	1	100	95	6	100
18	8	87.5	44	5	80	70	8	87.5	96	8	87.5
19	8	100	45	6	83.3	71	3	100	97	5	80
20	6	100	46	9	88.9	72	5	100	98	3	66.7
21	6	100	47	2	100	73	17	100	99	10	90
22	5	75	48	6	100	74	3	100	100	3	66.7
23	4	83.3	49	9	100	75	3	100	101	4	100
24	7	71.4	50	5	100	76	8	75	102	7	85.7
25	5	80	51	6	100	77	5	100	103	4	100
26	5	80	52	10	100	78	11	81.8	104	2	100

Other quality indexes: *tpr*, *tnr*, and *acc* are not so good as well. But the main inconvenience is that in each case we observed a large number of unrecognized seizures. Increasing the number of LSTM layers may improve the recognition score. Indeed, the results achieved for network structures net5 and net6 clearly show that values of all quality indexes were improved significantly. However, only after using three LSTM layers with a number of hidden units large enough we can obtain acceptable results of unrecognized seizures. For net8 there are only 1 or 2 unrecognized sequences which can be treated as a pretty good result. Seizure detection examples are illustrated in Fig. 5. It is assumed that each seizure lasts one second. The desired decision is marked with the dashed blue line.

The predicted classification provided by the LSTM network is marked with the solid red line. Figure 5a presents the typical example of the properly detected seizure. Both the predicted beginning and ending time of the seizure are very close to those presupposed as true. However, not all detected seizures are represented in that way. In Fig. 5b it is portrayed a case when the LSTM network pointed out that the seizure lasts 1.5 s which is longer that assumed seizure duration. This is a very interesting case showing that the system is able to provide the information not only about the beginning time of the seizure but also about a seizure duration. In fact in real-life short-time seizures are not exactly a second long. Yet another interesting examples comprise cases where the classification system made a decision about a seizure later that neurologist decision (Fig. 5c) or even earlier (Fig. 5d). This can be another valuable source of information about a true seizure beginning time. A neurologist pointed out the approximated time of seizure occurrence only. However, we do not know how uncertain such a decision is.

Epilepsy Diagnosis. As presented in the previous paragraph the seizure is considered as detected when *tpr* value is greater or equal to 50%. However, for each patient the neurologist pointed out a particular number of seizures as presented in Table 4. In this work we assumed that the epilepsy is diagnosed when at least a half of pointed out regions are detected as the epileptic seizures. The results of epilepsy diagnosis vs patients are presented in Table 4. It is evident that in all patients the epilepsy was diagnosed surely. Moreover, for 60 patients all seizures were detected. Comparing the achieved results with those obtained for the standard deviation method (see Table 1) it is clear that LSTM network performs pretty well and it is able to give much more accurate diagnosis.

6 Concluding Remarks

The paper presents the application of deep LSTM network to detect short-time epileptic seizures. Analysing the achieved results we can conclude that the proposed approach detects epileptic seizures with a pretty high quality comparing to other methods defined in the time-domain. Moreover, the LSTM neural networks can provide a fast on-line decision making and can be perceived as an alternative solution to time-frequency methods. Finally, the proposed system may be a valuable source of information for a neurologist about the exact time of seizure occurrence as well as the seizure duration. Our future research works will be focused on the optimal structure selection of LSTM network as well as on the analysis of the entire EEG records acquired not only from epileptic patients but also from healthy subjects.

References

1. Duchi, J., Hazan, E., Singer, Y.: Adaptive subgradient methods for online learning and stochastic optimization. J. Mach. Learn. Res. **12**, 2121–2159 (2011)

2. Durka, P.: Adaptive time-frequency parametrization of epileptic spikes. Phys. Rev. E **69**, 051914-1–051914-5 (2004)
3. Engel, J.: Seizure and Epilepsy. FA, Davis, Philadelphia (1989)
4. Guo, L., Rivero, D., Pazos, A.: Epileptic seizure detection using multiwavelet transform based approximate entropy and artificial neural networks. J. Neurosci. Methods **193**, 156–163 (2010)
5. Hochreiter, S., Schmidhuber, J.: Long short-term memory. Neural Comput. **9**, 1735–1780 (1997)
6. Hopfengartner, R., Kerling, F., Bauer, V., Stefan, H.: An efficient, robust and fast method for the offline detection of epileptic seizures in long-term scalp EEG recordings. Clin. Neurophysiol. **118**, 2332–2343 (2007)
7. Kahn, Y., Gotman, J.: Wavelet based automatic seizure detection in intracerebral electroencephalogram. Clin. Neurophysiol. **114**, 898–908 (2003)
8. Kingma, D., Ba, J.: Adam: a method for stochastic optimization. arXiv preprint arXiv:1412.6980 (2014)
9. Kong, W., Jiang, B., Fan, Q., Zhu, L., Wei, X.: Personal identification based on brain networks of EEG signals. Int. J. Appl. Math. Comput. Sci. **28**, 745–757 (2018)
10. Rutkowski, G., Patan, K.: Extraction of attributes of the EEG signals based on the Stockwell transform. Pomiary Automatyka Kontrola **59**(3), 208–211 (2013). (in Polish)
11. Rutkowski, G., Patan, K., Leśniak, P.: Comparison of time-frequency feature extraction methods for EEG signals classification. Lect. Notes Artif. Intell. **7895**, 187–194 (2014)
12. Rutkowski, G., Patan, K., Leśniak, P.: Computer aided on-line seizure detection using Stockwell transform. Adv. Intell. Syst. Comput. **230**, 279–289 (2014)
13. Shiau, D.S., Halford, J.J., Kelly, K.M., Kern, R.T., Inman, M., Chien, J.H., Pardalos, P.M., Yang, M.C.K., Sackellares, J.C.: Signal regularity-based automated seizure detection system for scalp EEG monitoring. Cybern. Syst. Anal. **46**, 922–935 (2010)
14. Subasi, A.: EEG signal classification using wavelet feature extraction and a mixture of expert model. Expert Syst. Appl. **32**, 1084–1093 (2007)
15. Tieleman, T., Hinton, G.: RMSProp. Technical report (2012). Lecture 6.5, COURSERA: Neural Networks for Machine Learning
16. Tzallas, A.T., Tsipouras, M.G., Tsalikakis, D.G., Karvounis, E.C., Astrakas, L., Konitsiotis, S., Tzaphlidou, M.: Automated epileptic seizure detection methods: a review study. In: Stevanovic, D. (ed.) Epilepsy - Histological, Electroencephalographic and Psychological Aspects. InTech (2012). ISBN 978-953-51-0082-9
17. Yuan, Q., Zhou, W., Zhang, L., Zhang, F., Xu, F., Leng, Y., Wei, D., Chen, M.: Epileptic seizure detection based on imbalanced classification and wavelet packet transform. Seizure **50**, 99–108 (2017)
18. Z-Flores, E., Trujillo, L., Sotelo, A., Legrand, P., Coria, L.N.: Regularity and Matching Pursuit feature extraction for the detection of epileptic seizures. Journal of Neuroscience Methods **266**, 107–125 (2016)

Breast Cancer Computer-Aided Diagnosis System Using k-NN Algorithm Based on Hausdorff Distance

Marcin Skobel(✉), Marek Kowal, and Józef Korbicz

Institute of Control and Computation Engineering, Faculty of Computer, Electrical and Control Engineering, University of Zielona Góra, Zielona Góra, Poland
M.Skobel@issi.uz.zgora.pl

Abstract. World statistics indicate that the breast cancer is the most common worldwide type of cancer among women. The development of computer-aided diagnosis techniques may contribute to a more effective therapy against this type of cancer. In this work, we present preliminary research regarding cell nuclei classification based on the Hausdorff distance. The obtained results indicate that using only Hausdorff distance to the classification of individual cell nuclei allows us to achieve 75% accuracy. Moreover, the speed of calculations and the possibility of using additional features describing cell nuclei open new paths to computer-aided diagnosis support systems development.

Keywords: Breast cancer · k-Nearest neighbour · Classification · Hausdorff distance · Computer-aided diagnosis

1 Introduction

The world statistics shows that the breast cancer has been the most common type of cancer among women in 2018 [1]. Estimated statistics shows that more than 2 million new cases of breast cancer were detected in 2018 [1]. Moreover, the number of newly detected breast cancer cases increases each year. Therefore, apart from training new medical staff, it is also worth focusing on the development of computer-aided diagnostic methods.

Diagnosis of breast cancer very often is carried out in three stages. In the beginning, a palpation examination is performed. If the tumor is suspected, then the patient is referred for a mammography or ultrasound examination. When the presence of the tumor is confirmed, it is necessary to verify the type of cancer. To this end, a tumor sample is taken. Depending on the method of acquisition of biological material, tissue fragments are examined (histopathology) or samples of free cells are examined (cytopathology). This paper focuses on the issue of automatic analysis of cytological samples, which were collected by fine-needle biopsy. The cytological examination is less invasive thus more comfortable for the patient but unfortunately, it is also less accurate.

J. Korbicz et al. (Eds.): PCBEE 2019, AISC 1033, pp. 179–188, 2020.
https://doi.org/10.1007/978-3-030-29885-2_16

The material taken from the tumor is placed on the glass slide and then fixed and stained. Then the slide is scanned and stored in a computer database in the form of an image. Pathologists can analyze the sample under a microscope or on a computer screen after connecting via a computer network to an image repository. The analysis of the cytological sample is very time-consuming and tedious because pathologists have to inspect the whole sample, which can contain tens of thousands of cells. In the beginning, pathologists look at the sample at a small magnification and then some cell clusters analyze at higher magnification so that they can see the precise shape, color, and texture of individual nuclei. Finally, they decide whether the cells have the characteristics of a malignant or benign tumor based on their features. Unfortunately, to properly discriminate the type of cancer, pathologists have to practice very long to gain experience by watching thousands of different preparations. We can help pathologists by providing tools for automatic analysis of cellular structures. The first stage of such system is detecting the cell nuclei in the image and then segment them. A lot of different approaches have been already proposed for this problem [3,4,7,9,11,12,15,18–21]. After nuclei detection/segmentation we are able to extract their features and classify them [9]. We can generate an enormous number of different features to describe cell nuclei. However, most of them will not contain the information useful to determine if they represent benign or malignant tumor. Therefore, we need to select the ones that are most informative [16]. Noninformative features have to be rejected because they can interfere with the classification process. The authors of scientific reports concerning the classification of cytological of histopathological samples reports various classification accuracies, which ranges from 75% [6] to 84,0% [14].

In this work, we are addressing the problem of breast cancer classification, but without extracting the features from segmented cell nuclei. Instead, we have used a Hausdorff measure to evaluate cell nuclei similarity based directly on their silhouettes [8,10,13]. A number of cell nuclei were manually segmented and classified either as benign or malignant by the pathologists from the University Hospital of Zielona Góra, Poland. Next, these nuclei were used as a training dataset to create the k-nearest neighbor (kNN) classifier. To classify a new cell nucleus, it has to be segmented and then we need to find the k nearest cell nuclei in the training dataset using Hausdorff distance. Thanks to the fact that cell nuclei in the training dataset are of known type, we can finally determine the class of the new nucleus.

Because in this work we do not deal with the problem of cell nuclei segmentation then we have used for studies cell nuclei that were segmented manually. The remainder of this paper is organized as follows. The materials used in the experiments are described in Sect. 2. The details of the method are presented in Sect. 3. Section 4 describes the experiment and results. Concluding remarks are given in Sect. 5.

2 Materials

The experiment was carried out using cytological samples, obtained as a result of cooperation with the Department of Pathomorphology of the University Hospital in Zielona Góra. Cytological images were based on samples collected from 50 patients (25 malignant and 25 benign cases). Then each of the images was manually segmented. As a result of the manual segmentation, we get 6444 cell nuclei. In order to reject wrongly segmented cell nuclei an outlier detection procedure was deployed. The procedure consist of two steps. In the first step, we rejected all cell nuclei with area greater than 9,000 pixels. During the second step, we rejected nuclei whose *Roundness* was less than 0.4 (this value was chosen arbitrary based on the distribution of roundness in the training dataset). *Roundness* is given by the following formulea:

$$Roundness = \frac{4A}{\pi M^2}, \tag{1}$$

where A - area, M - major axis length.
In the first step, 1 cell nucleus was rejected, which was significantly exceeding the fixed threshold. In the second step, 119 cell nuclei with highly irregular shapes were eliminated. Finally, there are 6324 nuclei left in the dataset (4366 benign and 1958 malignant).

3 Methods

The manual segmentation of cell nuclei was carried out with the help of ImageJ software. In practice, the task of manual segmentation boils down to precise tracing of the edge of the nuclei objects. As a result, we get ROI selection, which can be saved in the .roi file (multiple ROIs were saved in the .zip file). In the next step, the obtained ROIs are transformed into silhouettes (binary masks). Every nucleus is presented as a binary mask of the size 150 by 150 pixels (see Fig. 1). Before further processing, every nucleus was translated to the center of the image. The set of cell nuclei masks was randomly divided four times into a training and test dataset. The principle of selecting sets assumes

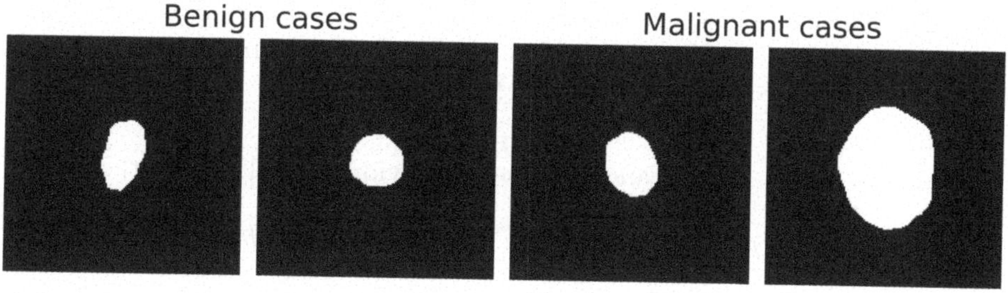

Fig. 1. Examples of binary masks of nuclei

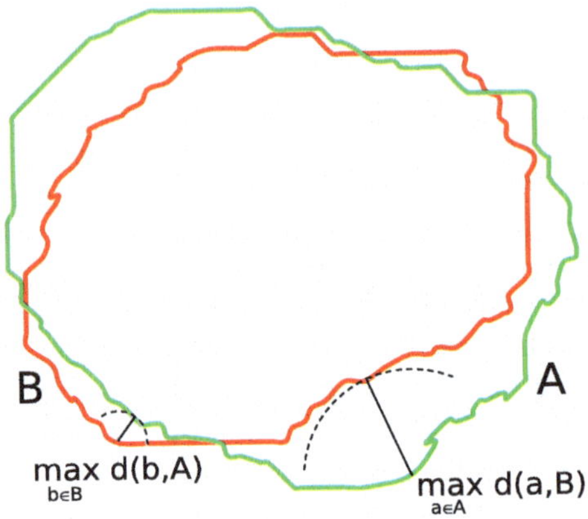

Fig. 2. Computing Hausdorff distance

that images from a single patient go exclusively to the training or test set. This approach increases the credibility of the experiment. Given 4 different training and test data collections, we have used a cross-validation technique in order to increase the statistical significance of the experiment carried out. Moreover, we also carried out the experiment using the reclassification method.

The prepared training sets were subjected to the clustering process. The Hausdorff metric is used in the clustering of various objects [2] therefore, it can also be used in the case of cell nuclei. In our approach, clustering was necessary for several reasons. Firstly, we had to significantly reduce the number of cell nuclei in the training data set in order to reduce the computational burden of the proposed method. Secondly, we observed that the training data set contains a lot of cell nuclei that are very similar so there was no reason to keep them all in the atlas. So, we used agglomerative clustering to reduce the number of cell nuclei in the training dataset and for balancing the benign and malignant classes. Hierarchical clustering was realized with the linkage defined with the help of Hausdorff distance (HD) [17]. After clustering the training dataset was reduced to 1000 cell nuclei where 500 prototype cells represent benign cases and 500 prototypes represent malignant cases).

A single cell nucleus is a set of pixels: $A = \{a_1, a_2, a_3, ..., a_m\}$ in the image space. The Euclidean distance between the individual elements of the set A and the set $B = \{b_1, b_2, b_3, ..., b_n\}$ in two-dimensional Euclidean plane, we can define as:

$$d_E(a, b) = \sqrt{(x_b - x_a)^2 + (y_b - y_a)^2} \quad (2)$$

Then we can define the distances between individual pixels to another set of pixels:

$$d(a, B) = \min_{b \in B} d_E(a, b); \qquad d(b, A) = \min_{a \in A} d_E(b, a) \quad (3)$$

Finally, we can get Hausdorff distance as (see Fig. 2):

$$d_H(A, B) = \max\{\max_{a \in A} d(a, B), \max_{b \in B} d(b, A)\} \quad (4)$$

However, Hausdorff distance during the binary nuclei masks processing is reduced to examining the edges between masks. By using this approach, the calculation of Hausdorff distance can be significantly accelerated.

The next step, after constructing the clusters, is to find their centers. Cluster centers are objects whose sum of distances to the remaining objects in the cluster is the smallest. The obtained cluster centers become a test set. Finally, a classification is performed based on the obtained test set consisting of 1000 cell nuclei. The kNN algorithm was used to perform the classification (Fig. 3).

4 Results

The Table 1 shows the results of the accuracy classification for individual validation methods. In the majority of prepared test sets, the highest classification result was obtained for k = 17. Depending on the selected test and training set, the classification accuracy result ranges from 0,62 to 0,82. The average value of classification accuracy for 4 data sets was 0.67. The results indicate that the selection of training and test sets has the greatest impact on the classification

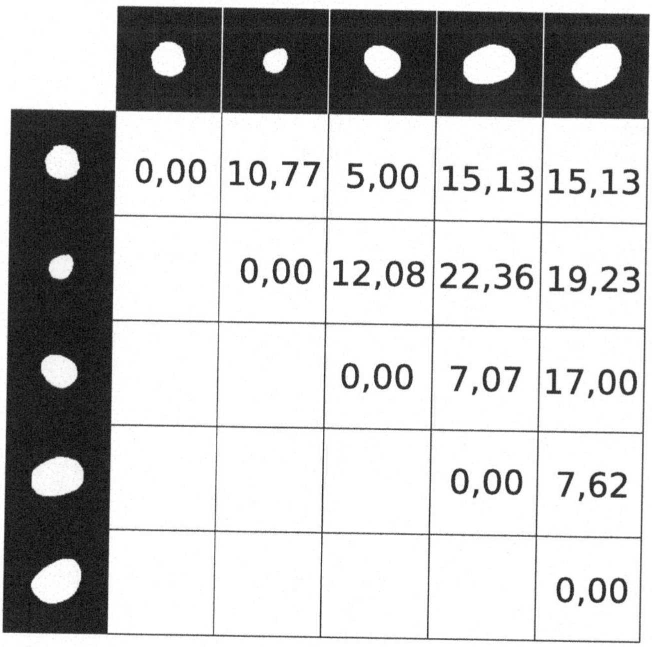

Fig. 3. Hausdorff distance results for binary masks of nuclei

Table 1. The results of accuracy of validation methods depending on k parameter

k	1	3	5	7	9	11	13	15	17	19	21
Cross-validation											
1st set	0,7058	0,7676	0,7942	0,8004	0,8112	0,8115	0,8169	0,7942	**0,8209**	0,8201	0,8169
2nd set	0,5959	0,6233	0,6227	0,6318	0,6343	**0,6377**	0,6222	0,6227	0,6312	0,6273	0,6258
3rd set	0,5856	0,6037	0,6127	0,6156	0,6150	0,6178	0,6212	0,6127	**0,6241**	0,6193	0,6193
4rd set	0,5822	0,6014	0,6105	0,6139	0,6133	0,6178	0,6207	0,6105	**0,6224**	0,6176	0,6181
Avg.	0,6174	0,6490	0,6601	0,6654	0,6684	0,6712	0,6703	0,6601	**0,6746**	0,6711	0,6700
Reclassification											
All	**0,7508**	0,7301	0,7343	0,7373	0,7440	0,7472	0,7478	0,7343	0,7498	0,7473	0,7503

accuracy. At the bottom of the table, the results of validation of the classification using the reclassification method are included. The reclassification method consists in creating one set which is used for the training and testing of the classifier. For this reason, the value of the classifier's accuracy increased to 0.75.

As a result of the processing of the generated test sets, 50 independent results of the percentage of malignant cell nuclei in the images were obtained (see Fig. 4). The results indicate that the average number of cell nuclei classified as malignant

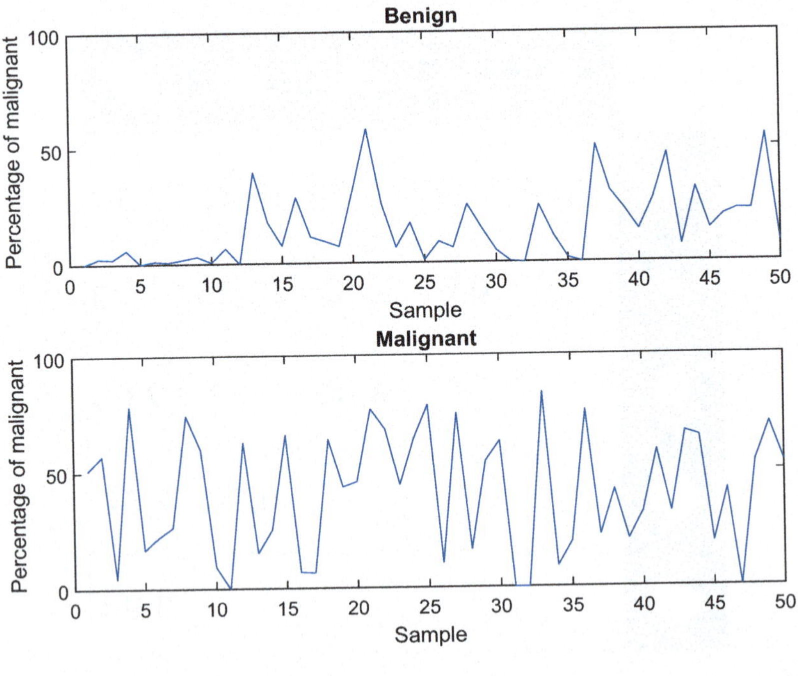

Fig. 4. Percentage of malignant cell nuclei in the single image - Cross-validation

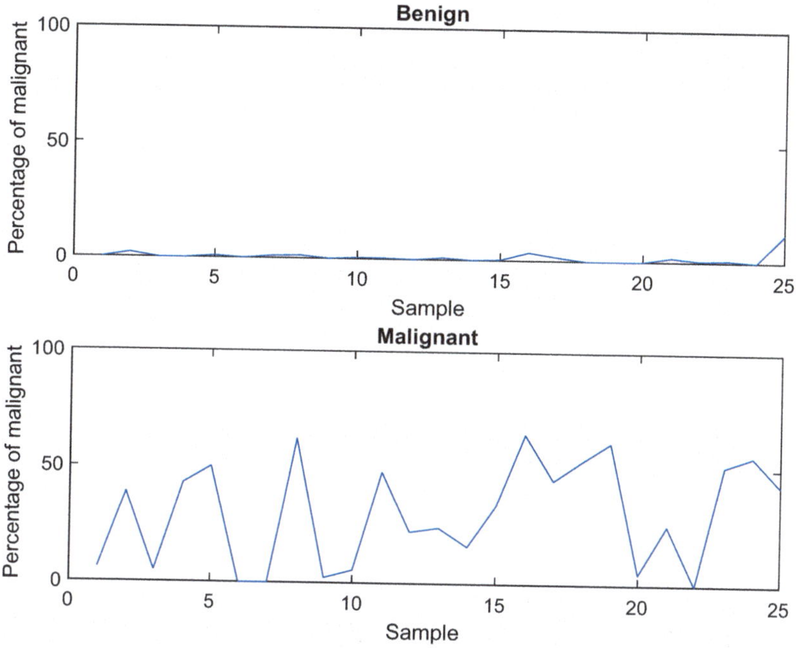

Fig. 5. Percentage of malignant cell nuclei in the single image - Reclassification

is lower for benign images. Unfortunately, among malignant images there are those cases in which the percentage of detected single malignant cell nuclei is 0%. There are also many cases of benign images with a high percentage of cell nuclei classified as malignant.

Obviously, the results obtained for validation using reclassification are much better (see Fig. 5). The majority of results for benign images oscillate around 0% of nuclei classified as malignant. Unfortunately, also among malignant images there are several cases in which the number of cell nuclei classified as malignant was 0%.

The last figure (Fig. 6) shows the accuracy of classification depending on the percentage of malignant nuclei. The maximum classification accuracy value for simple cross-validation is 0.75 for 32% of objects classified as malignant nuclei in the image. The results for classification after validation using reclassification, shows that the maximum accuracy is 0.9. In addition, such a high result is stays for low levels of malignant nuclei in the image. Unfortunately, even for validation using reclassification, malignant images with zero content of malignant classified nuclei occur.

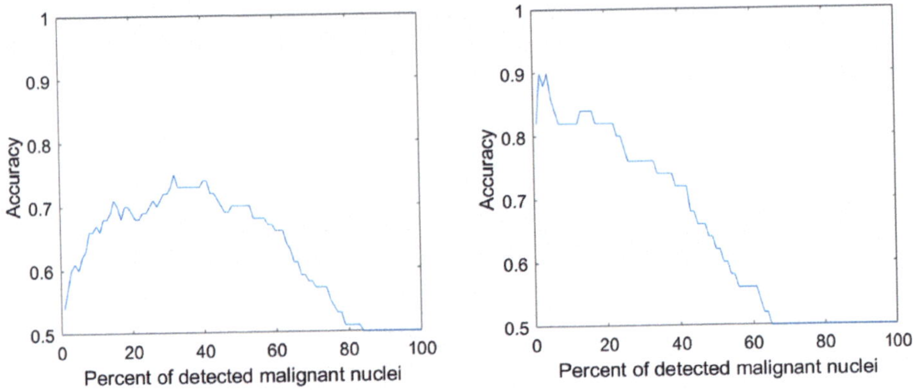

Fig. 6. Accuracy depending on the percentage of malignant nuclei in the image, left - cross-validation, right - reclassification

5 Conclusions

The shape and size of malignant and benign cell nuclei are an important diagnostic indicator. Unfortunately, the differences between individual malignant and benign nuclei are extremely small. Therefore, unexplored factors determine the medical diagnosis. Finding these factors may affect the acceleration and improvement of the diagnostic process based on the cytological images.

In the presented experiment it was possible to prove that a group of morphometric features (e.g.: area, perimeter, roundness etc.) can be supplemented by one metric in classification process. Further work should focus on finding the parameters which can replace a set of colorimetric and set of textural features. These parameters should support the Hausdorff distance on classification process. The weakness of Hausdorff distance is the fact that it is not an invariant metric. In other words, the cell nucleus rotated by a few degrees can get a high value of the Hausdorff distance relative to the same, not rotated nucleus. However, in literature are attempts to resolve the weakness of Hausdorff's metric [5].

The presented experiment was performed on manually segmented nuclei. In future work, the experiment should be repeated on the automatically segmented nuclei.

Acknowledgement. The research was supported by National Science Centre, Poland (2015/17/B/ST7/03704).

References

1. Bray, F., Ferlay, J., Soerjomataram, I., Siegel, R.L., Torre, L.A., Jemal, A.: Global cancer statistics 2018: globocan estimates of incidence and mortality worldwide for 36 cancers in 185 countries. CA Cancer J. Clin. **68**(6), 394–424 (2018). https://doi.org/10.3322/caac.21492

2. Chavent, M.: A Hausdorff distance between hyper-rectangles for clustering interval data. In: Banks, D., McMorris, F.R., Arabie, P., Gaul, W. (eds.) Classification, Clustering, and Data Mining Applications, pp. 333–339. Springer, Heidelberg (2004). https://doi.org/10.1007/978-3-642-17103-1_32
3. Cheng, F., Hsu, N.: Automated cell nuclei segmentation from microscopic images of cervical smear. In: 2016 International Conference on Applied System Innovation (ICASI), pp. 1–4 (2016). https://doi.org/10.1109/ICASI.2016.7539846
4. Cui, Y., Zhang, G., Liu, Z., Xiong, Z., Hu, J.: A deep learning algorithm for one-step contour aware nuclei segmentation of histopathological images. arXiv preprint: arxiv:1803.02786 (2018)
5. Dubuisson, M., Jain, A.K.: A modified Hausdorff distance for object matching. In: Proceedings of 12th International Conference on Pattern Recognition, vol. 1, pp. 566–568 (1994). https://doi.org/10.1109/ICPR.1994.576361
6. Fondón, I., Sarmiento, A., García, A.I., Silvestre, M., Eloy, C., Polónia, A., Aguiar, P.: Automatic classification of tissue malignancy for breast carcinoma diagnosis. Comput. Biol. Med. **96**, 41–51 (2018). https://doi.org/10.1016/j.compbiomed.2018.03.003
7. Husham, A., Hazim Alkawaz, M., Saba, T., Rehman, A., Saleh Alghamdi, J.: Automated nuclei segmentation of malignant using level sets. Microsc. Res. Tech. **79**(10), 993–997 (2016). https://doi.org/10.1002/jemt.22733
8. Irshad, H., Veillard, A., Roux, L., Racoceanu, D.: Methods for nuclei detection, segmentation, and classification in digital histopathology: a review—current status and future potential. IEEE Rev. Biomed. Eng. **7**, 97–114 (2014). https://doi.org/10.1109/RBME.2013.2295804
9. Kowal, M., Skobel, M., Nowicki, N.: The feature selection problem in computer-assisted cytology. Int. J. Appl. Math. Comput. Sci. **28**(4), 759–770 (2018). https://doi.org/10.2478/amcs-2018-0058
10. Kumar, N., Verma, R., Sharma, S., Bhargava, S., Vahadane, A., Sethi, A.: A dataset and a technique for generalized nuclear segmentation for computational pathology. IEEE Trans. Med. Imaging **36**(7), 1550–1560 (2017). https://doi.org/10.1109/TMI.2017.2677499
11. Naylor, P., Laé, M., Reyal, F., Walter, T.: Nuclei segmentation in histopathology images using deep neural networks. In: IEEE 14th International Symposium on Biomedical Imaging (ISBI 2017), pp. 933–936 (2017). https://doi.org/10.1109/ISBI.2017.7950669
12. Paramanandam, M., O'Byrne, M., Ghosh, B., Mammen, J.J., Manipadam, M.T., Thamburaj, R., Pakrashi, V.: Automated segmentation of nuclei in breast cancer histopathology images. PLoS ONE **11**(9), 1–15 (2016). https://doi.org/10.1371/journal.pone.0162053
13. Paramanandam, M., Thamburaj, R., Manipadam, M.T., Nagar, A.K.: Boundary extraction for imperfectly segmented nuclei in breast histopathology images – a convex edge grouping approach. In: Barneva, R.P., Brimkov, V.E., Šlapal, J. (eds.) Combinatorial Image Analysis, pp. 250–261. Springer International Publishing, Cham (2014). https://doi.org/10.1007/978-3-319-07148-0_22
14. Reis, S., Gazinska, P., Hipwell, J.H., Mertzanidou, T., Naidoo, K., Williams, N., Pinder, S., Hawkes, D.J.: Automated classification of breast cancer stroma maturity from histological images. IEEE Trans. Biomed. Eng. **64**(10), 2344–2352 (2017). https://doi.org/10.1109/TBME.2017.2665602
15. Sadanandan, S.K., Ranefall, P., Le Guyader, S., Wahlby, C.: Automated training of deep convolutional neural networks for cell segmentation. Sci. Rep. **7** (2017). https://doi.org/10.1038/s41598-017-07599-6

16. Szemenyei, M., Vajda, F.: Dimension reduction for objects composed of vector sets. Int. J. Appl. Math. Comput. Sci. **27**(1), 169–180 (2017). https://doi.org/10.1515/amcs-2017-0012
17. Tian, K., Yang, X., Kong, Q., Yin, C., He, R., Yau, S.S.-T.: Two dimensional Yau-Hausdorff distance with applications on comparison of DNA and protein sequences. PloS ONE **10**, e0136577 (2015). https://doi.org/10.1371/journal.pone.0136577
18. Veta, M., van Diest, P.J., Kornegoor, R., Huisman, A., Viergever, M.A., Pluim, J.P.W.: Automatic nuclei segmentation in H&E stained breast cancer histopathology images. PLoS ONE **8**(7) (2013). https://doi.org/10.1371/journal.pone.0070221
19. Wang, P., Hu, X., Li, Y., Liu, Q., Zhu, X.: Automatic cell nuclei segmentation and classification of breast cancer histopathology images. Sig. Process. **122**, 1–13 (2016). https://doi.org/10.1016/j.sigpro.2015.11.011
20. Wienert, S., Heim, D., Saeger, K., Stenzinger, A., Beil, M., Hufnagl, P., Dietel, M., Denkert, C., Klauschen, F.: Detection and segmentation of cell nuclei in virtual microscopy images: a minimum-model approach. Sci. Rep. **2**, 503 (2012). https://doi.org/10.1038/srep00503
21. Yang, X., Li, H., Zhou, X.: Nuclei segmentation using marker-controlled watershed, tracking using mean-shift, and Kalman filter in time-lapse microscopy. IEEE Trans. Circuits Syst. I Regul. Pap. **53**(11), 2405–2414 (2006). https://doi.org/10.1109/TCSI.2006.884469

E-health and Telemedicine

Taking Videoplethysmographic Measurements at Alternative Parts of the Body – Pilot Study

Radosław Mędrala and Piotr Augustyniak(✉)

AGH University of Science and Technology, Krakow, Poland
august@agh.edu.pl

Abstract. Videoplethysmography (VPG) has been studied for more than a decade and recently found useful as a tool for touchless measurement of the human pulse. The principle of measurement lies in subtle changes of skin color due to filling of subcutaneous capillary vessels with blood and emptying in the rhythm of pulse. Most authors assume the measurement video footage is taken at the face or forehead due to good perfusion, dense capillary network, most fragile skin and good visibility in most humans. However, a setup requiring the subject facing to the camera is impractical and we found interesting to explore how the accuracy of the method degrades with moving the area of observation to other uncovered parts of the body. In this paper we compared the VPG from three simultaneously captured video tracks and the ECG-based reference beat recorder as sources of the pulse information. The principal finding is that consistency between VPG pulse measurements made at the forehead, cheek and interclavicular region is high (1.28 bpm), despite occasional high absolute error (up to 3.9 bpm). The reliability of the VPG measurement is the highest at the forehead and drops in other regions without affecting the measurement accuracy.

Keywords: Videoplethysmography · Remote patient monitoring · Heart rate detection · Video signal processing

1 Introduction

Among a variety of methods recently available for measurement of the heart rate in human, the videoplethysmography (VPG) is particularly interesting for two reasons: (1) it provides a touchless measurement with accurate results and (2) it uses simple consumer-grade equipment such as worldwide spread webcams. Since a reliable measurement is taken from the image of face and the face is commonly uncovered, most algorithms start with face detection to ensure the displayed result is the value of pulse and nothing else. We found it a serious limitation of the assumed setup (not the VPG method itself) since many scenarios could be proposed where the face is either not available (e.g. measurement in neonates [1] or in sleeping humans) or irregularly illuminated (e.g. in drivers or computer operators). In such cases alternative measurement regions (e.g. in the neck, the arm or the palm) could provide VPG signals of sufficient reliability. In case the ambient light is not sufficient or not appropriate for capturing the sequence, the VPG is supported by an artificial spotlight. Here again,

J. Korbicz et al. (Eds.): PCBEE 2019, AISC 1033, pp. 191–200, 2020.
https://doi.org/10.1007/978-3-030-29885-2_17

illuminating the face is intrusive and alters the subject's behavior, while illuminating other parts of the body is tolerable and can be not even noticed.

2 Related Work

First steps towards contactless optical measurements of the human heart rate are attributed to Verkruysse et al. [2], who applied a consumer-grade digital photo camera (320 × 240 and 640 × 480, 24-bit RGB, 30 fps) to show that blood pulse signal can be extracted from a series of frames with the human face acquired in ambient light. Next milestone has been reached three years later [3, 4] with a method that compensates the subject's movements using a blind source separation. The authors applied an automated face detection algorithm, what improved their results but implied limitations challenged in the present paper. Finally, various color spaces have been studied for optimal video taking: from a single green channel [2], through an ICA-derived combination of RGB [3], infrared detection (in a range of 700 nm to 1000 nm) [5] to the extended color space (ROGCB: red-orange-green-cyan-blue) and its subspaces with additional orange and cyan sensors. The studies of various video recording equipment are summarized in [6]. The research for optimal lighting also corresponds to the work by Krolak [7], who applied the VPG to subjects with various skin tones. Recently Wang et al. [8] found that extracting a continuous heart-rate measurement from three-dimensional RGB signal may not eliminate more than two interference sources with a linear combination. They proposed to mathematically increase the dimensionality of the data and showed reliable measurement results even during fitness exercises.

3 Materials and Methods

3.1 Study Conditions

Two white men (of the age 18 and 27) were videotaped in a still sitting position from a distance of two meters. A uniform green fabric was used as a background, consequently the automatic white balance in the camera was similarly set in all cases. Three lighting sources were used:

- Warm lighting (color temperature of 2700 K) with two incandescent bulbs of 320 lm,
- Neutral lighting (color temperature of 4000 K) with two semiconductor sources (LED's) of 570 lm,
- Cold lighting (color temperature of 6500 K) with natural sunlight of the order of 450 lm.

The video footage was taken with Sony SLT- A58 digital Single-Lens Reflex camera with a resolution of 1440 × 1080 pixels and a speed of 25 frames per second.

In the presented study three body parts were investigated as possible VPG measurement regions (Fig. 1):

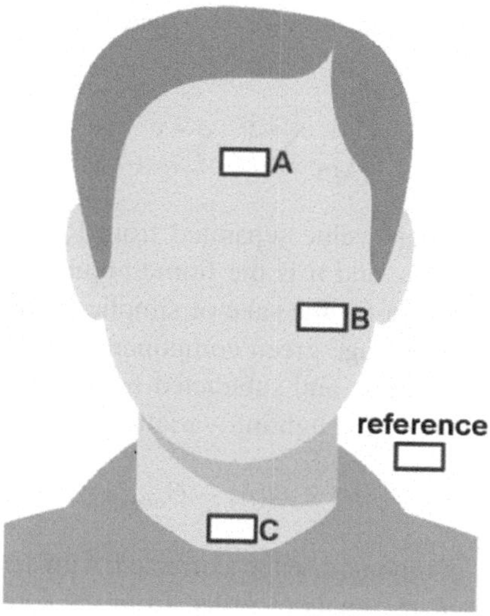

Fig. 1. Three VPG measurement regions considered in the study: (A) the forehead, (B) the cheek and (C) the interclavicular region.

- The forehead (most frequently selected by other authors and thus producing a reference for other VPG signals),
- The cheek,
- The central interclavicular region exposed above the neckline.

As the measurement was dependent on the external lighting, an extra region has been defined in the background as a visual reference area.

Five trials were performed with each volunteer and each lighting source. Four 20 s video footages (i.e. 500 frames each) were simultaneously taken from each body part. In result 120 video strips were processed to get 90 time series containing the pulse information.

The chest strap pulsometer based on two-leads recording of electrical heart activity (Polar H10, ± 1 ms heart beat detection accuracy) was used to provide a ground-truth information about the pulse. This measurement was, however, processed by the firmware and reported as the average value of four consecutive intervals between adjacent R peaks.

3.2 Image and Signal Processing

Following the guidelines in [2] and considering findings reported in [7] we decided to use a single green component of the RGB image series. The light intensity captured by the lens with automatic aperture was transformed by the 20 megapixel CMOS sensor to a Full HD 4:3 matrix (i.e. 1440 $\times$ 1080). Therefore, pixel values have been normalized to the measurement range 0…255 and the G component is expressed as a discrete dimensionless quantity and together with derived spectra will be reported as such throughout this paper. Depending on the body part, the VPG sensing zone was of different size and thus the values of green component $G_{i,j}(n)$ of all pixels have been averaged:

$$G(n) = \frac{1}{MN}\sum\nolimits_{i=1}^{M}\sum\nolimits_{j=1}^{N} G_{ij}(n) \tag{1}$$

where G is the green component value separated from RGB attributes of each pixel, M and N are region dimensions, and n is the frame number.

We used a rectangular region for the sake of simplicity, but the area shape may be adjusted upon necessity. The average green component in the background $G_{bkg}(n)$ has been calculated for each n-th frame and subtracted from the result of body part under test $G(n)$ in order to compensate for lighting variations:

$$G_C(n) = G(n) - G_{bkg}(n) \tag{2}$$

where $G_C(n)$ is the green component value corrected with respect to the background illumination. Green component signals recorded from the forehead, the cheek and the interclavicular region from Person 1 at cold lighting are presented in Fig. 2.

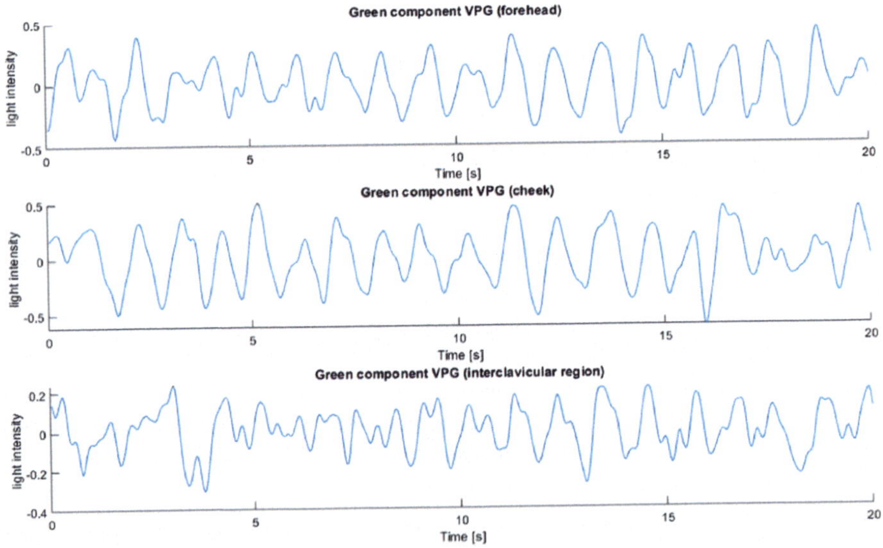

Fig. 2. Example green component signals recorded from the forehead, the cheek and the interclavicular region from Person 1 at cold lighting.

Next processing step aimed at further reduction of influence from the lighting changes and consisted in normalizing the time series $G_N(n)$:

$$G_N(n) = \frac{G_C(n) - avg(G_C)}{std(G_C)} \tag{3}$$

and filtering them with a bandpass Butterworth filter to the spectrum range of 0.7–4 Hz. The *avg*() and *std*() symbols stand for average and standard deviation values respectively.

In next step the Welch approach was applied to calculate the Power Spectrum Density (PSD). To this point the signal was split into a series of overlapping segments. The segments were windowed and fed to the discrete Fourier transform:

$$\hat{G}_k(k) = \sum\nolimits_{n=0}^{N-1} g_n(n) \cdot e^{-\frac{2\pi i}{N}nk}, k = 0, \ldots, N-1 \tag{4}$$

Finally, spectra of the segments were averaged in order to reduce the variance of the PSD estimation. The frequency corresponding to the maximum of the PSD is the pulse value. Spectra calculated for VPG signals recorded from the forehead, the cheek and the interclavicular region from Person 1 at cold lighting are presented in Fig. 3.

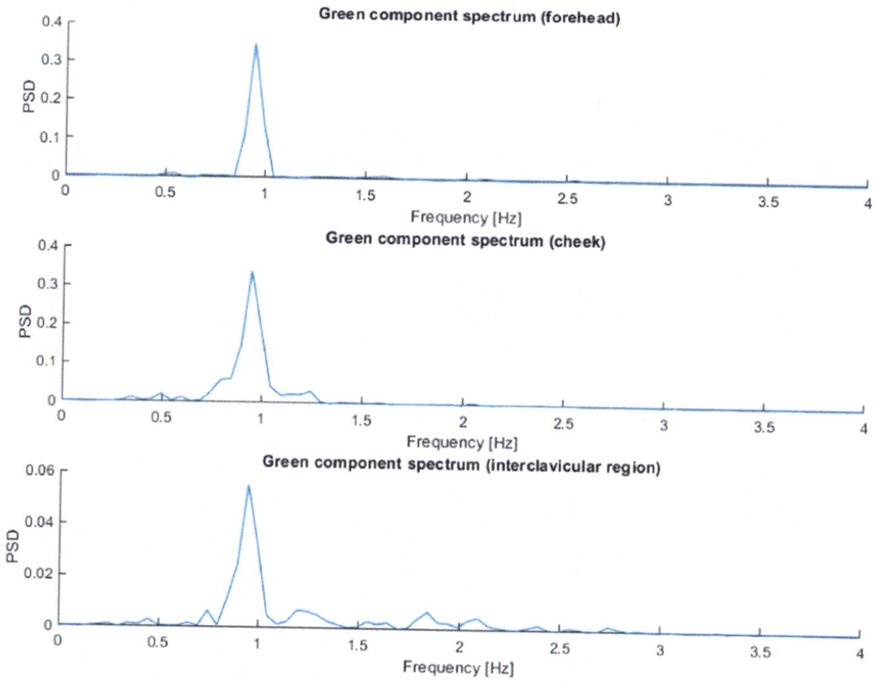

Fig. 3. Spectra calculated for VPG signals recorded from the forehead, the cheek and the interclavicular region from Person 1 at cold lighting (trials corresponding to those in Fig. 2).

The highest peak of the spectrum is expected to represent the frequency related to the blood pulse. Since the blood wave propagates through flexible main arteries, the velocity of wave front propagation is a widely recognized marker of arterial tissue elasticity [9] and thus of the risk of coronary flow blockage. It is believed that the velocity of the blood wave front could be estimated from the phase shift between VPG curves (see Fig. 2) measured in two points of a known distance on the same artery. To this point calculation of the cross-correlation between these curves was calculated and the time shift between two best matching pulse waves was taken as representative for blood propagation.

4 Results

Three questions were addressed during the experiments:

- What is the difference of the pulse value measured at the forehead, cheek and interclavicular region?
- What is the reliability of the pulse measurement at these points (expressed by the maximum value of the PSD)?
- What is the time shift between the pulse wave at these points (expressed by the time shift value of the $G(n)$ sequences cross-correlation peak)?

4.1 Results of Measurement Accuracy

The accuracy of VPG-based pulse measurement was evaluated by comparison to the ground truth value from the pulsometer. The reference pulsometer readout is based on electrical activity of the heart detected by two conductive rubber electrodes embedded in the textile chest strap. The battery-operated H10 sensor (by Polar) detects the heart beat peaks with a millisecond accuracy, averages four time intervals between adjacent validated peaks and sends the value via Bluetooth Low Energy link accordingly to an open communication protocol. Table 1 displays values of simultaneous VPG results and pulsometer readout.

Table 1. Average pulse values [bpm] measured at the forehead, cheek and interclavicular region (IcRe).

Subject	Lighting	Pulsometer	Forehead VPG	Cheek VPG	IcRe VPG
Person 1	Warm	61	61.2	61.2	61.2
	Neutral	67	64.3	64.3	63.1
	Cold	58	56.8	56.8	56.8
Person 2	Warm	69	69.2	69.2	70.4
	Neutral	66	62.5	62.5	62.5
	Cold	65	64.3	64.3	64.3

Maximum error of the pulse value was 3.9 beat per minute i.e. 6.18%. Average error of the pulse value was 1.28 bpm i.e. 2.07%.

4.2 Results of Measurement Reliability

The VPG measurement of the pulse is based on the maximum of PSD function. Thus the reliability of the result is expressed in the amplitude of this maximum. Thanks to normalization of the green component time series, the values of their spectra can be compared. Table 2 displays PSD maximum values of normalized green component time series.

Table 2. Average PSD maxima (dimensionless quantity) measured at the forehead, cheek and interclavicular region (IcRe).

Subject	Lighting	Forehead VPG	Cheek VPG	IcRe VPG
Person 1	Warm	0.400	0.221	0.121
	Neutral	0.386	0.018	0.169
	Cold	0.449	0.232	0.097
Person 2	Warm	0.119	0.037	0.017
	Neutral	0.288	0.047	0.077
	Cold	0.146	0.100	0.046

Average maximum of the VPG spectra at the forehead was 0.298 (taken as the reference value), at the cheek was 0.109 (i.e. 36.6% of the reference value) and at the interclavicular region was 0.0878 (i.e. 29.5% of the reference value).

4.3 Results of Pulse Wave Velocity

As the video frames were simultaneously taken at different parts of the body, the time shift between best correlated green component time series represents the pulse wave velocity (see Sect. 3.2). Table 3 displays results for the time shift of the pulse wave at the cheek and at the interclavicular region related to the pulse wave at the forehand.

Table 3. Average time shift [ms] of the pulse wave at the cheek and interclavicular region (IcRe) related to the pulse wave at the forehead.

Subject	Lighting	Cheek VPG/Forehead VPG	IcRe VPG/Forehead VPG
Person 1	Warm	−1920	1032
	Neutral	112	−640
	Cold	8	−24
Person 2	Warm	24	48
	Neutral	24	144
	Cold	56	80

The results calculated for person 1 with warm lighting and in case of interclavicular region also with neutral lighting cannot be considered as reliable. The values are comparable to a double, single or a half of the heart cycle. In other cases the measurements yield believable results, however the accuracy was not verified. On the other hand, measurement of pulse wave velocity shows considerable granulation due to low frequency of video frames.

5 Discussion

The experimental results show high degree of consistency between VPG pulse measurements made at the forehead, cheek and interclavicular region (Fig. 1, Table 1). However, the absolute accuracy of the method is only satisfactory with the average result difference of 1.28 beats per minute (or 3.9 pbm in worst case) from the pulsometer readout. Maximum difference between VPG and the pulsometer values occur in neutral light (LED sources) regardless the region of measurement.

The reliability of the VPG measurement, expressed by the maximal PSD peak is the highest at the forehead. In other regions it drops, but the measurement accuracy was only marginally affected. This was expected due to the properties of the skin and increasing depth of subcutaneous capillary vessels. In case of hairy skin on the chest or calf in men further drop of the measurement reliability is expected, although these locations are usually covered with clothing and were not considered in this study. Nevertheless, properly illuminated the indicated alternative locations provide acceptable quality of pulse information and extend the range of applicability of the VPG method. Having simultaneous measurements at multiple body parts allows for constant evaluation of lighting and dynamic selection of the optimal source in case of lighting variation (e.g. in case of ambient light or illumination by a computer display).

Having two simultaneous measurements of the pulse is also necessary to estimate the pulse wave velocity, which is a useful parameter for further examination of the vascular system. In this case, the measurement of phase delay requires two distant VPG signal sources which in turn cannot guarantee similar reliability of the signal. It is worth to be noticed that the measurement quality in each studied body part depends on color temperature of the lighting. The reliable measurement of pulse wave velocity has been found affected by poor synchronization of two waves. In deficient lighting conditions the dominance of pulse-related spectrum peak is questionable and the best correlation between the pulse waves is found one or two heart cycles apart. Moreover, even in case of correct synchronization, the accuracy of the reported time delay is limited by camera frame rate, which is usually 25 or 30 fps. This finding suggests that either high speed cameras or advanced mathematical tools (such as signal or spectrum modeling) should be used to detect the subtle differences of pulse wave phase in the video sequences.

A serious limitation of our study was the number of participants. For this reason we call it a pilot study. Results from just two people are not supporting general conclusions based on the statistics. However, main achievement of the reported research is that VPG pulse measurements made at the forehead, cheek and interclavicular region yield highly consistent results. This statement challenges the common belief about the face as the unique VPG measurement point and encourages further research on other areas such as palms or feet not always clothed.

This work paves the way for a more systematic study on the decrease of result quality in various regions making the readout reliability predictable for each of these regions. Such study, however, requires a large population of volunteers, perfect control of the illumination and various degree of physical load applied. This is a point of departure for inventing an automatic zone selecting algorithm that tracks several VPG measurements in the same individual and optimizes the readout quality based on the features of spectra. Such solution is also believed to meet expectances of real-time surveillance, where the pulse rate from several people sharing a common area need to be simultaneously and continuously captured by several video cameras of adjacent visual fields [10].

The other way of possible further development is related to detailed study of optimal illumination for each VPG recording zone. Most of currently performed VPG recordings benefit from the ambient light thus the measurement is imperceptible. Nevertheless, illuminating the skin with small spots of controlled spectral characteristics makes the measurement noticeable, but still far from being invasive. Such approach assume first selecting the measurement zones accordingly to the particular medical purpose (e.g. on the same artery for distant measurement of pulse wave velocity). Then optimal illumination intensity and color temperature is set based on the dominance of the pulse-related peak of the spectrum. Finally, the motion of the monitored individual is tracked and the positions of illuminating spots are adjusted accordingly.

Acknowledgment. Research supported by the AGH University of Science and Technology in year 2019 from the subvention granted by the Polish Ministry of Science and Higher Education.

References

1. Mestha, L.K., Kyal, S., Xu, B., Lewis, L.E., Kumar, V.: Towards continuous monitoring of pulse rate in neonatal intensive care unit with a webcam. In: 36th Annual International Conference of the IEEE Engineering in Medicine and Biology Society (EMBC), pp. 3817–3820 (2014)
2. Verkruysse, W., Svaasand, L.O., Nelson, J.S.: Remote plethysmographic imaging using ambient light. Opt. Express **16**(26), 21434–21445 (2008)
3. Poh, M.Z., McDuff, D.J., Picard, R.W.: Non-contact, automated cardiac pulse measurements using video imaging and blind source separation. Opt. Express **18**(10), 10762–10774 (2010)
4. Poh, M.-Z., McDuff, D.J., Picard, R.W.: Advancements in noncontact, multiparameter physiological measurements using a webcam. IEEE Trans. Biomed. Eng. **58**(1), 7–11 (2011)
5. McDuff, D., Gontarek, S., Picard, R.W.: Remote detection of photoplethysmographic systolic and diastolic peaks using a digital camera. IEEE Trans. Biomed. Eng. **61**(12), 2948–2954 (2014)
6. Przybylo, J., Kantoch, E., Jablonski, M., Augustyniak, P.: Distant measurement of plethysmographic signal in various lighting conditions using configurable frame-rate camera. Metrol. Meas. Syst. **23**(4), 579–592 (2016)

7. Krolak, A.: Influence of skin tone on efficiency of vision-based heart rate estimation. In: Augustyniak, P., Maniewski, R., Tadeusiewicz, R. (eds.) Recent Developments and Achievements in Biocybernetics and Biomedical Engineering, Proceedings of the 20-th Polish Conference on Biocybernetics and Biomedical Engineering. Advances in Intelligent Systems Computing, pp. 44–56. Springer International Publishing AG, Cham (2017)
8. Wang, W., den Brinker, A.C., Stuijk, S., de Haan, G.: Robust heart rate from fitness videos. Physiol. Meas. **38**(6), 1023–1044 (2017)
9. Trawinski, Z.: Two-point method for arterial local pulse wave velocity measurement by means of ultrasonic RF signal processing. Arch. Acoust. **35**(1). https://doi.org/10.2478/v10168-010-0001-9
10. Przybylo, J., Jablonski, M., Kantoch, E., Augustyniak, P.: Distant pulse measurement system for real-time surveillance applications. In: Computing in Cardiology 2017, vol. 44 (2017). https://doi.org/10.22489/cinc.2017.073-212

Multimodal Measurement Systems for Health and Behavior Analysis in Living Environment

Piotr Augustyniak(✉)

AGH University of Science and Technology, Krakow, Poland
august@agh.edu.pl

Abstract. This review reveals and briefly discusses problems of capturing, archiving and analyzing the human behavior in a natural dwelling environment. Ambient assisted living systems are defined and justified with current social needs. Sensing paradigms and various examples of sensors are presented with a focus to their imperceptibility and data safety. Two paradigms of behavioral data storage are presented and examples of behavior predictive methods conclude the paper.

Keywords: Ambient Assisted Living · Body sensor network · Human behavior · Physiological measurements

1 Introduction

The continuous aging of developed societies and life-long health education of the citizens called more attention to wide range of senior-related problems from social insurance policies to assistive services in smart environments. The last category is the area where tools for technical support of medicine meet surveillance techniques and enter to everyday living of the growing population of the elderly. The assistive technologies are designed to support the human assistance with the pervasive and ubiquitous service, but as independent living became one of the most appreciated values, some seniors prefer a discrete digital assistant to a usually strained social worker.

Ambient Assisted Living (AAL) systems are similar to medical long-term diagnostics (e.g. ECG Holters) because they are based on physiological recordings such as electrocardiography, blood pulse, motion and posture. One of the possible outcomes of AAL systems are alerts to social or medical services or automated delivery of pharmaceuticals. Another similarity is the longitudinal behavioral record which usually inherits the structure from a patient health record. The AAL systems are also digital surveillance systems, mainly due to the use of visual input and common image processing and pattern recognition techniques. Although, while the goal of the surveillance is to protect a given area or object and record their infringement by any alien agent, the AAL is focused on a living subject and captures any activity including internal physiological processes as well as interactions with the surrounding.

The necessity of AAL solutions is supported by three principal observations:

- Majority of accidents take place in households and concern senior adults so substantial social and economic costs can be saved due to prevention.

J. Korbicz et al. (Eds.): PCBEE 2019, AISC 1033, pp. 201–210, 2020.
https://doi.org/10.1007/978-3-030-29885-2_18

- Each appliance in the household is engineered with specific assumptions about user abilities and technical performance; both factors decrease with time narrowing the safety margin.
- Functional impairment immediately follows subtle pathologic changes at the cell level; performance drop precedes other measurable markers of several diseases.

Three elements (sometimes referred to as layers [1]) are commonly distinguished in AAL systems:

- Sensor layer, physical wearable and infrastructure (IoT) sensors, the data collection and management system, secure data transport.
- Behavioral record layer with preprocessing, storage and retrieval sub-systems,
- Inference and prediction machine, interface to alerting and intelligent home (environment) subsystems.

These layers will be further discussed in this paper with examples from the literature and author's experience.

2 Sensors and Sensor Networks

2.1 Sensor Types and Characteristics

Sensors are the input of the surveillance system of any kind. While usually the accuracy and stability are two key parameters of sensors, in case of behavioral studies the unobtrusive or even imperceptible operation is the most welcome feature. In case of continuous measurements in human, the unobtrusiveness corresponds to the lack of influence of the act of measurement to the process of interest what is a fundamental rule of metrology.

Main role of sensors is to collect the information, however capturing and collecting functions rarely can be combined in a wearable and imperceptible device. What is more important, the information collected from several sensors of various types cooperating within a system is expected to be available without delay for a supervising device. For this reason, sensors are usually equipped with a communication interface which secures the acquired data transmission and receives management data.

In most cases the collected information has an electrical (and currently: digital) form. Except for cases of voltage measurement, a transducer is an obligatory component of the sensing device. Even if electrophysiological signals are concerned, the voltages of few millivolts cannot be directly stored or transmitted and need a kind of processing to get their final (e.g. digital) form. The on-site processing of the information is usually adapted to the purpose and working conditions of the sensor and hard-wired into sensor architecture. In some sensors, however, it can be programmed via a bidirectional communication interface from the supervising device or self-adaptive with regard to the previously collected data. This last category of measurement devices we refer to as intelligent sensors.

Several physical principles of sensing were successfully applied in behavioral studies including optical, visual (i.e. image-based), inertial, gravitational, pressure-based, tensional, chemical, microwave- and radiation-based, acoustic and ultrasonic, finally: electrical. The choice is basically supported by the physical nature of the recorded phenomena, but usually the physiological action at cells', organs' and organisms' levels manifests itself in several dimensions. The selected measurement modalities may thus be principal (best representative for the phenomenon of interest) and auxiliary (used to increase the reliability of the result).

Two types of sensing paradigms are used depending on the sensor's reference [10]:

- infrastructure or embedded sensors, fixed in a specific location to monitor changes in a given range of the space (environment),
- wearable sensors, fixed to the monitored individual and capturing changes in his or her body or in the environment he or she interacts with.

The infrastructure sensors due to their static position are part of the intelligent environment created for surveillance, active safety or commodity purposes. Their operation area is limited and thus the freely moving individual may not be continuously followed by the system. However, their fixed location is advantageous for unlimited power supply and wired communication infrastructure of high security.

The wearable sensors are a seamless companion of the monitored moving individual [11]. Their results are continuous and unambiguously related to the person of interest. However, in order to preserve the natural behavior, they should be unobtrusive (or even imperceptible). Their designs usually involve powering issues and the use of wireless communication with particular measures of data security.

2.2 Intelligent Sensors

Intelligent sensors are equipped with a sensing part or transducer, a processing unit, and a transmission and/or storage interface. Their operation can either be modified internally as a result of measurement history or externally according to the needs of cooperation within a sensing system.

Two areas are currently explored by researchers of intelligent sensors:

- reducing the power consumption and size of the sensor with maintaining limited programmability, and
- increasing the on-site processing power and agility without the significant growth of size and energy consumption.

The pursuit for the smallest yet smart sensors for electrophysiological signals was marked by milestones as the proposal of ultra-low power front-end amplifier by Tseng et al. [2], who applied 0.18 μm CMOS technology to achieve the SNR value of over 54 dB at a supply voltage of 0.4 V (0.09 μW). The sensor consists of a chopper-stabilized instrumental amplifier with common mode interference cancellation and a programmable gain amplifier (0 to 30 dB in frequency bands 0.5 to 100 or 10 to 400 Hz). The option of miniature multi-purpose biosignal data sensor with 8 GB in-site data storage was proposed by Bailey et al. [3]. The recorder has programmable sensitivity and sampling frequency to adapt to various types of biosignals. Three versions of ASIC-based conditioning modules for 64 brain penetrating array of 15 μm electrodes integrated with a digitizer was proposed by Zoladz et al. [4]. The simultaneous sampling frequency is 14 ks/s with the resolution of 12 bits and the total data stream is transmitted to the PC via USB. In a subsequent paper [5] the same research team demonstrated how to improve the CMRR by more than 30 dB.

The other branch of development is focused on complex systems according to a acquire-transform-archive working paradigm with the programmability added at all of these three stages. In a prototype by Liu et al. [6] a four-channel 16-bit signal, such as the EMG, ECG and outcome of accelerometer transducers, is acquired at 1 kHz. Next it can either be saved in a microSD card or sent via ZigBee wireless interface (a range up to 100 m). The device weights 102 g, requires 430 mW at full power (6 h with two AAA batteries) and is designed for field measurements of human activity. Another complete data logger [7] follows the Intelligent Electrode and Active Cable concept [8] and includes a custom-designed digital controller to alternately support the network management and biosignal measurements. A mixed-signal system-on-chip with analog front-end (45–63 dB, 0.5–1000 Hz), 8-bit ADC and digital core was made in 0.18 μm CMOS technology and consumes 20 μW from a 1.2 V supply. Digital communication with external signal storage equipment is made wirelessly with a separate low-energy Bluetooth chip. Also for high-density recordings of brain signals (up to 400 channels) a configurable matrix system was proposed [9] consisting of an analog front-end, ADC and digital signal processing unit and a wireless transmitter. The prototype was fabricated in 0.65 μm CMOS technology and a single amplifier of 2.5 μV noise and 10 kHz bandwidth consumes 17.2 μW from a 1 V supply. The onboard signal processing routines include EEG-specific feature extraction and data clustering.

2.3 Imperceptible Sensors

A pursuit for imperceptible sensors is driven by the need to capture an unbiased behavioral record and is not intended for pervasive invigilation as some may suspect. Similarly to medical records, the collected behavioral data are highly confidential and as such need responsible processing and storage complying to the legal regulations for sensitive data. Making the sensor imperceptible augments the comfort of daily living, since the subject does not feel to be investigated all the time and does not see additional devices (such as cameras or wristbands) in the living space.

Regular domestic appliances can be fitted with usage sensors for increasing the security of independently living users and for identifying their habits and detection of unexpected activity as a possible health setback [12]. The authors propose to modify an electric kettle, a microwave oven, and a kitchen tap to screen their usage for correctness and safety, but also to analyze the information about regularity of hot food and drink intake. Despite of limited studies, significant inter-personal differences were detected, and high repeatability of each subject's action supports the assumption that activities performed in an unusual way are discriminative markers in case of health setback or operation difficulty.

An intelligent sensor of energy usage was proposed as an imperceptible single-sensor behavior monitoring device [13]. Such devices are already widely applied for profiling of energy consumption and are able to deduct the usage of electrically powered domestic appliances from the detailed analysis of their dynamic electrical characteristics [14]. Assuming that at least some devices in the household are directly operated by the human, these maintenance-free, mains-powered and WiFi-enabled sensors provide information that in combination with the usage rules reveals important facts about the operator's functional health. In this proposal original equipment was used without any modification, thus any appliance available on the marked may be used as behavioral sensor. Unfortunately only electrically powered appliances may be monitored and only directly operated ones provide information on operator's activity.

An imperceptible sensor can also provide information on human blood pulse thanks to videoplethysmographic technique. This method uses subtle rhythmic changes of the skin color due to the blood flow in subcutaneous capillary vessels. It has been recently proven that this measurement offers adequate accuracy in various lighting conditions and can use inexpensive and widespread equipment such as webcams [15]. This technique has been applied for human in motion [16] and is also recently studied for possible detection of emotions in computer gamers.

Imperceptible sensors can also be integrated in smart garments. A working prototype of the smart shirt equipped with a processor, wearable sensors, power supply and telemedical interface was presented in [17]. The smart shirt collects accelerometer, optical pulse, ambient light and body temperature sensors data, extracts behavioral data features with the ARM-family processor and transmits them to the LAN with a Bluetooth Low Energy interface. All elements are sewn on the shirt and arranged into a wired Personal Area Network for the maximum measurement reliability and comfort to the user.

2.4 Sensor Network

A sensor network provides organization, management, and data transmission between the sensors (permanent and optional) and the supervising device, in a scenario requiring immediate availability of the measurement result.

The organization includes sensor identification, authentication and broadcasting of data encryption keys together with initialization of corresponding data collection structures and processing threads. A general design assumes that sensors are connected, initialized and deactivated, disconnected on the fly, what is frequently used in behavioral recording in the case of optional sensors, low quality connection or when the moving individual enters/exits the connection range.

The management of a sensor network includes occasional or time-based verification of sensor status (e.g. battery status), update of the cryptographic keys and broadcasting of the synchronizing time stamp. Management also includes programming commands for intelligent sensors.

The data transmission is necessary for collecting the sensed data (raw or processed in-site). This function is not present in the sensors with local data storage (e.g. an SD card), however many sensors connected to the network of unknown reliability are also accompanied by local buffers allowing for temporary data storage in case of disconnection. The data from the local backup are available to the supervising device with an unpredictable delay and can be synchronized thanks to the time stamps included.

The network usually adopts the architecture of a star or a mesh. In the first case the central node usually has supervisory functions and communicates directly with each sensor node, but no alternative pathways can be routed in case of disconnection. In the latter case data transmission pathways may be composed of multiple sections what slows down the transmission but enables alternative pathways routing and multiple centers to take over management functions.

Taking into account the specificity of the sensor types, wearable sensors are most suitable as nodes of star-shaped network due to the limited connection range and limited power. This network is managed by a wearable server [18, 19], which in turn is an optional node of the infrastructure sensor network. The infrastructure sensors are organized either according to a star or to a mesh topology. The advantage of alternative pathways routing is not particularly important in case of wired network infrastructure, however in case of wireless connections the extension of the surveillance may be easier with mesh-shaped network.

Other sensor network topology used in behavioral studies is a hierarchical star where one of the regular nodes of the master network, besides of its measurement function, also has a supervisory function in the slave sub-network.

2.5 Sensor Information Safety

Information safety is usually referred to in two aspects: immunity and confidentiality of the data. Both of them are of primary importance in case of behavioral recordings since recorded data are expected to support health-related decisions and to cary the information related to the highest level of privacy.

The working example of security measures designed for low-energy transmission within the BSN is based on AES-CCM encryption algorithm. The node pairing starts with distribution of a short term key uniquely generated for this purpose [20]. A short term key for the temporary channel is a string of 6 numeric digits and secures a temporary point-to-point wireless transmission for just few milliseconds. Within this time an exclusive long time key is passed to the sensor node. Since the BSN configuration is a randomly occurring short time process, the man-in-the-middle attacks have little chance to success.

Wireless communication modules such as CC3100MOD Wi-Fi module (Texas Instruments) [21] dedicated for Internet-of-Things applications include an on-board network processor and a power management subsystem. The network processor provides TCP/IP stack and efficient cryptographic procedures (256-Bit AES) among other features. The standby power consumption is 430 μW, increasing to 162 mW during data reception and to 670 mW during the transmission. The transmission duty cycle has thus a paramount impact on the overall power required.

3 Digital Behavioral Record

3.1 Natural Representation of Behavior

The storage of behavioral record usually inherits data structures specific to recorded multimodal data. Proposing an adequate structure is usually not as straightforward as a combination of the component data formats. In case of raw data storage, the difference in sensor-specific data includes:

- phenomena representation scales and units,
- permanent and optional sensors and data streams,
- continuous or event-related data,
- hierarchy of measurements in terms of importance and reliability,
- various data formats: images, signals, isolated values etc.
- delay of data availability.

In most systems the longitudinal record of the human behavior is organized on the time line with all tracks synchronized by time stamps and sensor labels. Each track is usually assigned to a specific sensor with given parameters (such as the amplitude scale, the sampling frequency etc.) regardless of its permanent or optional operation. Some complex data structures (e.g. occasional images) are referenced to by a pointer on the time line. Event-related data containers are also used for archiving temporary (or intermediate) results of the interpretation such as measurements of distance in images or the heart rate in the ECG). In the structural aspect the behavioral record shows high degree of similarity to a personal health record.

Several advantages of raw longitudinal record are commonly recognized: (1) persistence - the self-explanatory data structure can be accessed with any software now and in the future and (2) the scalability – records may be freely extended according to new sensors added upon request to the system. Main drawback of this type of behavioral data storage is the huge size of the record, e.g. collecting data from several video cameras is often not relevant for increasing the accuracy of the measurement or reliability of the final finding.

3.2 Symbolic Representation of Behavior

An alternative way of behavioral data storage is based on behavioral events automatically detected by specialized software. The software usually transforms the physical description of human activity into symbols necessary to represent the action performed at a desired level of abstraction. A representative example is the approach by Crossman et al. [22], who developed a High Level Symbolic Representation (HLSR) for human behavior modeling. The proposal inherits components and processes common to precedent cognitive architectures (e.g. Soar or ACT-R) mimicking human-like levels of knowledge and reasoning. The HLSR is in fact a high-level knowledge-based programming language for defining behavior models and provides compilers for translating these models into executable code for different cognitive architectures.

More recently Wei, Liu and Xing [23] proposed a symbolic representation of human motion by behavior strings (BS). Each motion capture data interval (e.g. a frame) is processed as high-dimensional discrete data point and transformed to clusters based on the local distance between components. Next, each cluster in the frame is represented by a symbol (e.g. a letter) thus in the time line, the continuous action becomes a series of behavior strings (words). By analyzing the BS, the human motion capture data is segmented into distinct behavior segments and the cycles of motion are found. Besides compact representation, the method shows a high rate of repeatability, the ability to finds the cycles of motion and eases extraction of motion clips representing a given behavior from a long original motion sequence.

4 Identifying and Predicting Human Actions

Several approaches were studied for recognizing the activity of the human in his or her residential space. The work by Kelly et al. [24] proposed the hidden Markov Models (HMM) to distinguish daily living activities based on video sequences. Pirsiavash and Ramanan [25] used a temporal pyramids model trained with images from head-mounted camera. Other authors used Neural Networks and Support Vector Machine to classify the behavioral data into a closed-choice set of activity. These early approaches, however, used inference rules based on commonsense knowledge and did not allow for considering of individual habits.

More recently Soran et al. [26] predicted the next probable action by learning the relationship of actions of a specific individual with the HMM. Wang et al. [27] decomposed video sequences into elementary segment units which composes actions while Hassan et al. [28] applied a dynamic Bayesian network to classify daily living activities based on acceleration data recorded with a smartphone. Bang applied a human activity intention model based on recurrent neural network (RNN) trained with the ADL records in a residential space to infer the human activity intentions in real-time [29]. Although the status of RNN's in current time point depends on the history of precedent time points, the inertia is constant and does not adequately reflect behavioral sequences of different length.

Augustyniak and Slusarczyk [30] proposed a graph-based decomposition of compound activity into 7 elementary poses able to represent the unpredicted activity in a very concise record. The representation of any action consists of 7 values of graph nodes (the pose share) and 21 of edge flows (the pose change). Moreover, next probable activity (i.e. intention) can be inferred from the flow in graph edges, and the individual-related values of flow are memorized and used for prediction of future action.

In [29] the authors proposed a behavioral record structure composed of data aggregates called 'percepts'. These containers collect identified actions, poses and objects in the vicinity and predicted intentions. A three step algorithm was proposed to recognition of activity intention. First, the present activity and activity intention ontology are used to recognize the intention of a given percept sequence. Next, the algorithm identifies the activity as complex intentions or as concurrent intentions and decides whether to terminate the previous activity intention in favor of the newly inferred activity intention. Finally, the percept sequence is corrected by an uncertainty handling step which identifies and excludes unreliable intentions.

Acknowledgment. Research supported by the AGH University of Science and Technology in year 2019 from the subvention granted by the Polish Ministry of Science and Higher Education.

References

1. Augustyniak, P.: Layered design of an assisted living system for disabled. In: Piętka, E., Kawa, J. (eds.) Information Technologies in Biomedicine, pp. 498–509. Springer, Heidelberg (2012)

2. Tseng, Y., Ho, Y., Kao, S., Su, C.: A 0.09 μW low power front-end biopotential amplifier for biosignal recording. IEEE Trans. Biomed. Circuits Syst. **6**(5), 508–516 (2012)
3. Bailey, C., Hollier, G., Moulds, A., Freeman, M., Austin, J., Fargus, A., Lampert, T.: Miniature multisensor biosignal data recorder and its evaluation for unsupervised Parkinson's disease data collection. In: Sensordevices 2014, The Fifth International Conference on Sensor Device Technologies and Applications, pp. 84–92 (2014)
4. Zoladz, M., Kmon, P., Rauza, J., Grybos, P., Blasiak, T.: Multichannel neural recording system based on family ASICs processed in submicron technology. Microelectron. J. **45**(9), 1226–1231 (2014)
5. Kmon, P., Gryboś, P., Żołądź, M., Lisicka, A.: Fast and effective method of CMRR enhancement for multichannel integrated circuits dedicated to biomedical measurements. IEEE Electron. Lett. **51**(22), 1736–1738 (2015)
6. Liu, Y.-P., Chen, H.-C., Sung, P.-C.: Wireless logger for biosignals. Int. J. Appl. Sci. Eng. **8** (1), 27–37 (2010)
7. Yang, G.: Hybrid integration of active bio-signal cable with intelligent electrode. Steps toward wearable pervasive-healthcare applications. Doctoral thesis, KTH Information and Communication Technology, Stockholm (2012). http://kth.diva-portal.org/smash/get/diva2:610512/FULLTEXT01.pdf. Accessed 23 Mar 2019
8. Yang, G., Chen, J., Cao, Y., Tenhunen, H., Zheng, L.-R.: A novel wearable ECG monitoring system based on active-cable and intelligent electrodes. In: Proceedings of 10th International Conference on e-health Networking, Applications and Services, HealthCom 2008 (2008)
9. Chandler, R.J.: A system-level analysis of a wireless low-power biosignal recording device. UCLA Electronic Theses and Dissertations (2012). http://escholarship.org/uc/item/1836k3z4. Accessed 23 Mar 2019
10. Augustyniak, P., Smoleń, M., Mikrut, Z., Kańtoch, E.: Seamless tracing of human behavior using complementary wearable and house-embedded sensors. Sensors **14**, 7831–7856 (2014)
11. Wojtowicz, B., Dobrowolski, A., Tomczykiewicz, K.: Fall detector using discrete wavelet decomposition and SVM classifier. Metrol. Meas. Syst. **22**, 304 (2015)
12. Augustyniak, P., Kantoch, E.: Turning domestic appliances into a sensor network for monitoring of activities of daily living. J. Med. Imaging Health Inform. **5**(8), 1662–1667 (2015)
13. Augustyniak, P.: Detection of behavioral data based on recordings from energy usage sensor. In: Rutkowski, L., et al. (eds.) Proceedings of 15th International Conference Artificial Intelligence and Soft Computing, ICAISC 2016, pp. 137–146 (2016)
14. Zoha, A., Gluhak, A., Imran, M.A., Rajasegarar, S.: Non-intrusive load monitoring approaches for disaggregated energy sensing: a survey. Sensors **12**, 16838–16866 (2012)
15. Przybylo, J., Kantoch, E., Jablonski, M., Augustyniak, P.: Distant measurement of plethysmographic signal in various lighting conditions using configurable frame-rate camera. Metrol. Meas. Syst. **23**(4), 579–592 (2016)
16. Wang, W., den Brinker, A.C., Stuijk, S., de Haan, G.: Robust heart rate from fitness videos. Physiol. Meas. **38**(6), 1023–1044 (2017)
17. Kantoch, E.: Recognition of sedentary behavior by machine learning analysis of wearable sensors during activities of daily living for telemedical assessment of cardiovascular risk. Sensors **18**, 3219 (2018). https://doi.org/10.3390/s18103219
18. Kim, H., Kim, S., Van Helleputte, N., Artes, A., Konijnenburg, M., Huisken, J., Van Hoof, C., Yazicioglu, R.F.: A configurable and low-power mixed signal SoC for portable ECG monitoring applications. IEEE Trans. Biomed. Circuits Syst. **8**(2), 257–267 (2014)
19. Augustyniak, P.: Remotely programmable architecture of a multi-purpose physiological recorder. Microprocess. Microsyst. **46**, 55–66 (2016)

20. Padgette, J., Scarfone, K., Chen, L.: Security guide to Bluetooth - Recommendations of the National Institute of Standards and Technology, Special Publication 800-121 Rev. 1 (2012)
21. https://www.ti.com/product/CC3100MOD. Accessed 23 Mar 2019
22. Crossman, J., Wray, R.E., Jones, R.M., Lebiere, C.: A high level symbolic representation for behavior modeling. http://cc.ist.psu.edu/BRIMS/archives/2004/Papers/04-BRIMS-051.pdf. Accessed 23 Mar 2019
23. Wei, R., Liu, W., Xing W.: A symbolic representation of motion capture data for behavioral segmentation. In: Proceedings of 21st International Conference on Distributed Multimedia Systems, DMS 2015, pp. 78–84 (2015)
24. Kelley, R., Tavakkoli, A., King, C., Nicolescu, M., Bebis, G.: Understanding human intentions via hidden Markov models in autonomous mobile robots. In: Proceedings of Conference on Human Robot Interaction, Amsterdam, Netherlands (2008)
25. Pirsiavash, H., Ramanan, D.: Detecting activities of daily living in first-person camera views. In: Proceedings of the IEEE Computer Society Conference on Computer Vision and Pattern Recognition (CVPR), pp. 2847–2854 (2012)
26. Soran, B., Farhadi, A., Shapiro, L.: Generating notifications for missing actions: don't forget to turn the lights off!. In: Proceedings of the IEEE International Conference on Computer Vision, pp. 4669–4677 (2015)
27. Wang, H., Yang, W., Yuan, C., Ling, H., Hu, W.: Human activity prediction using temporally-weighted generalized time warping. Neurocomputing **225**, 139–147 (2017)
28. Hassan, M.M., Uddin, M.Z., Mohamed, A., Almogren, A.: A robust human activity recognition system using smartphone sensors and deep learning. Future Gener. Comput. Syst. **81**, 307–313 (2018)
29. Kim, J.-M., Jeon, M.-J., Park, H.-K., Bae, S.-H., Bang, S.-H., Park, Y.-T.: An approach for recognition of human's daily living patterns using intention ontology and event calculus. Expert Syst. Appl. **132**, 256–270 (2019)
30. Augustyniak, P., Slusarczyk, G.: Graph-based representation of behavior in detection and prediction of daily living activities. Comput. Biol. Med. **95**, 261–270 (2018)

Assessing the Influence of the Teaching Method on Cognitive Aspects in the Process of Mathematical Education Among Blind People

Michał Maćkowski[1], Piotr Brzoza[1], Katarzyna Rojewska[2], and Dominik Spinczyk[3](✉)

[1] Faculty of Automatic Control, Electronic and Computer Science, Silesian University of Technology, 16 Akademicka, 44-100 Gliwice, Poland
[2] Clinical Hospital No. 1 in Zabrze, Department of Pediatrics, School of Medicine with the Division of Dentistry in Zabrze, Medical University of Silesia in Katowice, 15-18 3th Maja, 41-800 Zabrze, Poland
[3] Faculty of Biomedical Engineering, Silesian University of Technology, 40 Roosevelta, 41-800 Zabrze, Poland
dominik.spinczyk@polsl.pl

Abstract. Teaching math to blind people is a significant challenge for modern education, due to the structural information contained in mathematical formulas. Computer methods through the effective use of an alternative description of structural information, presented in sound form, can help the blind person to become more independent in the process of learning mathematics. Evaluation of the influence of the teaching method on cognitive aspects in the process of mathematical education among blind people was conducted based on a survey prepared by psychologists. Three detailed sub-categories of the assessment of the cognitive aspects were developed: knowledge operationalization, ability of self-correction during the learning process and strategy for solving exercises based on gradation. Two different teaching methods were compared: classic method and proposed alternative multimedia method. The proposed alternative method in two of the three defined sub-categories achieved a statistically significant improvement in results: knowledge operationalization, strategy for solving exercises based on gradation.

Keywords: Cognitive aspect of learning · Alternative math presentation · Alternative math teaching and learning · Computer aided math learning

1 Introduction

Teaching math to blind people is a significant challenge for modern education, due to the structural information contained in mathematical formulas. Standard

J. Korbicz et al. (Eds.): PCBEE 2019, AISC 1033, pp. 211–220, 2020.
https://doi.org/10.1007/978-3-030-29885-2_19

mathematical notation is not available to the blind [1,2]. Nowadays, an important area being developed is the ambient assistant living (AAL), and in particular computer support for teaching and accessibility of information by blind people.

Learning the basics of mathematics is a current issue in the literature. In the world strongly based on science and technology, blind people without basic mathematics skills limit their role in a society.

1.1 Methods and Tools Supporting the Access to Information for the Blind

For basic computer use, people use screen reader programs that read information chosen by a blind person from the computer screen using a speech synthesizer. In the case of books, the DAISY standard (Digital Accessible Information System standard ANSI/NISO Z39.86-2005 R2012 Specifications for the Digital Talking Book) was developed [3,4], which facilitates the navigation via the book at various levels of detail: chapters, sub-chapters, paragraphs, sentences, words and finally characters. Mathematical formulas contained in the book are saved using MathML notation and can be read in the form of alternative descriptions prepared by the DAISY book editor [5,6]. There are also known national variants of Braille notation containing mathematical symbols. However, due to the volume of materials translated using it, it is not widespread. Automatic math expression reading system is another solution, called i-Math, operated with screen reader it produces voice output on a computer, which can read math documents aloud [7].

In the literature, many other works can be found regarding the perception and acquisition of information by blind people. A number of studies have focused on the design of audio-based interfaces and evaluated their impact on learning and cognition [8,9]. The results of the work [10] indicate that sound can be a powerful interface to develop and enhance memory and skills in mathematics among blind children. The authors of that work, however, mainly focused on the assessment of cognitive aspects related to short-term memory.

The most commonly used method in the case of learning mathematics by blind persons is the use of an assistant who participates in a math lesson together with a blind person, writing the mathematical formulas used on his behalf.

1.2 Challenges of Teaching and Learning Mathematics by Visually Impaired

The classical method of learning mathematics used by blind people with an assistant has a number of limitations. Blind people are characterized by low motivation and a number of emotional problems related to teaching mathematics and additional difficulties associated with reduced self-esteem [11–13]. Due to the lack of visual perception, learning is much slower in relation to the people without sight dysfunction. Frequent repetition of content is required. Generally, this makes it difficult to create a positive atmosphere for learning and significantly hinders social behaviors that are important in the teaching process.

Moreover, several studies have evaluated different spatial abilities in blind people to understand the role of vision in cognitive system [14–16].

The aim of the work is the assessing the influence of the teaching method on cognitive aspects in the process of teaching and learning mathematics by blind people.

2 Materials and Methods

Because of the many limitations discussed above, there is a public need in the field of computer-aided teaching of mathematics by visually impaired students. Computer methods through the effective use of an alternative description of structural information, presented in sound form, can help the blind person to become more independent in the process of learning mathematics.

2.1 Developed an Alternative Method of Teaching Mathematics

The developed alternative method of learning mathematics takes into account a number of aspects that assume the possibility of learning by blind people [17]. While developing materials from a selected semantic mathematics area, a set of concepts to be mastered was presented. It is presented in the form of a directed graph ordering the assimilation of concepts, so as to minimize the blind students' effort and limit their negative experience in this area. Each concept from the graph has assigned math exercises that may only include concepts that appear as predecessors in the graph. The knowledge of a particular user of the method is represented in the form of a vector of knowledge containing information about the mastery of selected concepts by the user. At the beginning of the learning process, it is assumed that the user has not yet come up with any idea. Introduced preliminary test allows this information to be verified and adjust the difficulty of the proposed exercises to the current level of the student. In the teaching phase, the knowledge vector as part of the student's progress is updated, thus avoiding excessive repetition of the material already developed.

Individual exercises are subject to the process of decomposition into an orderly sequence of elementary steps. In each step there are contextual sub-messages containing the theory necessary to master a given concept (Fig. 1).

In order to present effectively the structural information contained in the math formulas, a set of rules was prepared to create alternative descriptions, containing additional information not only about the symbols appearing in the formula but also about its structure [18,19]. In order to facilitate the preparation of tasks according to the developed method, an extended notation in the TEX standard was prepared for the generation of the structure of the exercise with a division for elementary steps and conditions for transitions between them. The developed set of exercises from a given area is available in the form of an internet portal, which a blind user can use to learn using a computer and an installed screen reader program.

2.2 Cognitive Aspects and Proposed Method of Their Evaluation

Cognitive science is an interdisciplinary field of science dealing with the study of cognitive processes such as perception, consciousness, memory, understanding, reasoning, linguistic abilities, problem solving, intelligence, cognitive representations and others.

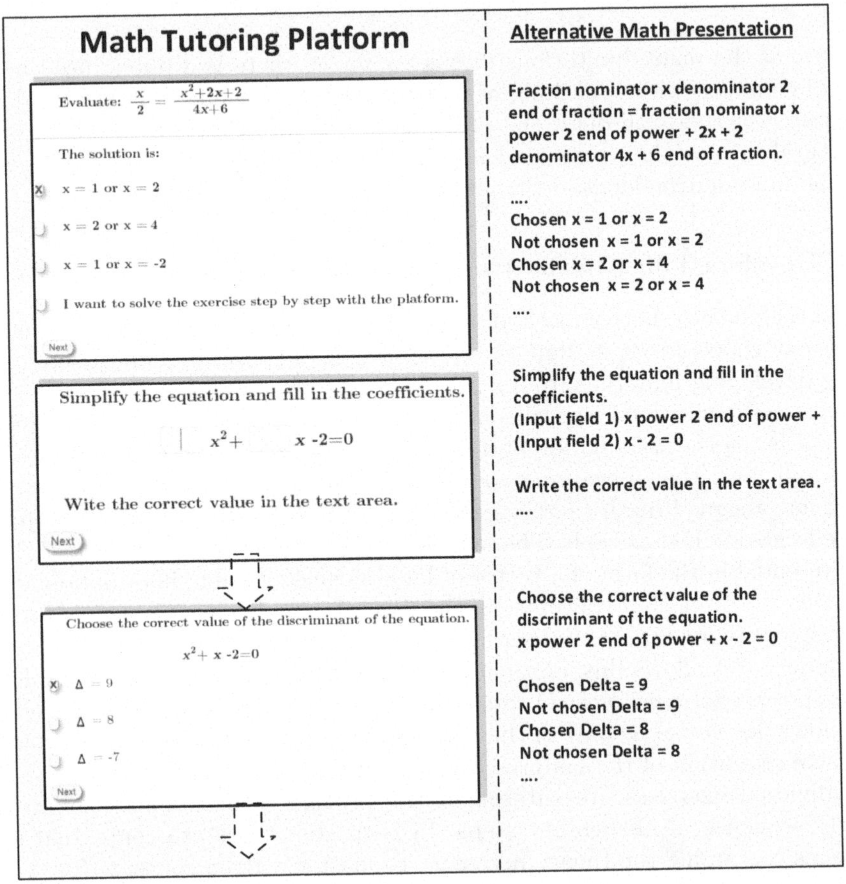

Fig. 1. Example of exercise with alternative math presentation

Cognitive science examines what is cognition, what is it for and how it works. Its research areas include natural and artificial cognitive systems. This science, among others significantly uses clinical neuropsychology, which indicates that the functioning of the brain is shaped by complex factors, forming co-operating consistent systems based on the operation of many mechanisms, including equivalence, complementarity, specialization, neuroplasticity, integration at all levels of brain functioning.

After the literature research, it was decided to distinguish three subcategories:

1. Knowledge operationalization: The ability of an effective operationalization of information reaching through other access channels and information about a given phenomenon already possessed will be essential here. Operationalization can be understood as a process of selecting information that corresponds with the definitions of specific phenomena adopted in the course of conceptualization [20].
2. Ability of self-correction during the learning process: Neurocognitive science emphasizes the important relationship between cognitive processes and the functioning and structure of the brain, whose one of the basic tasks is proper adaptation and auto- correction [21].
3. Strategy for solving exercises based on gradation: The involvement of the nervous system can be observed in the regulation of all human life and functional processes in this learning process as well as in the strategies used in it [22].

In order to assess cognitive aspects of learning mathematics by blind people, brainstorming was conducted among a team of psychologists, during which the following issues were highlighted:

- How do we acquire knowledge?
- How do we represent knowledge in the mind?
- How do we use knowledge in thinking and acting?
- How to improve the acquisition and use of knowledge?

Based on the discussed issues and the analysis of the literature on the subject, a questionnaire was developed Table 1. Answers for all questions were possible in nominate scale with grades form 1 to 5 with step size equals one. Then the sum of points obtained by a particular participant was divided by the maximum number of points possible to obtain and expressed in percent scale.

2.3 Research Group, Control Group and Conducted Experiments

The selected research group consisted of 30 students. Each person was blind or visually impaired. The difficulty of math exercises was at the level of a high school course or an academic beginner. People in the group did not have any other disabilities. All blind or visually impaired persons participating in the study had an acquired defect. The group consisted of 16 women and 14 men, aged 19–24, middle-aged 20 and a standard deviation of 1.52. The following inclusion criteria were adopted in the studies:

1. The group includes persons whose degree of visual impairment or blindness was significant.
2. The vision defect was acquired.
3. The secondary school certificate classes and the first year of technical studies in Silesia have been included in the group.

Table 1. A research questionnaire of cognitive aspect.

Question number	Question	Subcategory
Q 1	I think about different ways to solve a given problem.	Knowledge operationalization
Q 2	I try to combine new things I learn with the things that I've met before	Knowledge operationalization
Q 3	I try to make sure that what I do is correct	Ability of self-correction during the learning process
Q 4	I try to understand my mistake in case I do something wrong	Ability of self-correction during the learning process
Q 5	When I study, I consider only those problems that I have previously done	Ability of self-correction during the learning process
Q 6	When work is difficult, I only solve the easier part	Strategy for solving exercises based on gradation
Q 7	I only do the necessary minimum	Strategy for solving exercises based on gradation

4. Lack of co-occurring emotional disorders.

31 people were found to meet the inclusion criterion, all of which were invited for testing. From among those invited, 30 people took part in the research. According to the sample size calculation, the sample should be 29 - with confidence level 95, fraction size 0.5 and maximum error 0.05.

Research – Stage 1. The group solved exercises without using the developed platform for learning mathematics for the blind - the first experiment (E1).

Research – Stage 2. The group solved exercises using the developed platform for learning mathematics for blind people - the second experiment (E2).

In the first experiment, the order of solved exercises follows the order of the concepts in a directed graph from the easiest to the most difficult ones. The participants did not solve identical exercises in the E1 and E2 experiments, they treated them with the same level of difficulty (for each concept in the graph the authors prepared a set of several exercises). In E1, the participants solved the exercises of self-reliance, while the content of the exercise was read by the teacher.

3 Results

The results obtained as part of the proposed sub-categories in the cognitive aspects are presented in Table 2 and in the graph of Fig. 2.

Table 2. Normalized numerical results (with standard deviation) in individual subcategories.

Cognitive aspects subcategory	Experiment 1	Experiment 2
Knowledge operationalization	75 ± 16	85 ± 13
Ability of self-correction during the learning process	68 ± 9	76 ± 11
Strategy for solving exercises basedon gradation	30 ± 16	50 ± 17

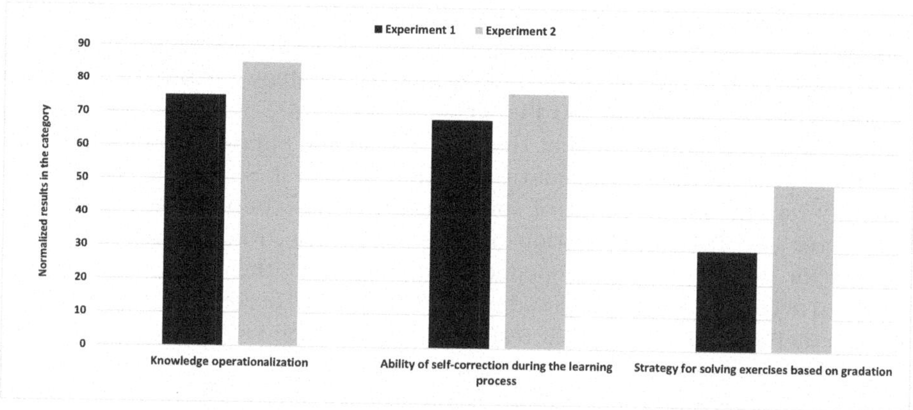

Fig. 2. Normalized results obtained during experiments in individual categories

Statistical analysis (Wilcoxon's test of median equality) was carried out after finding the normality of the distribution (Shapiro-Wilk test). The results obtained are presented in Table 3.

Table 3. Statistical tests on equality of medians in experiments in individual categories of cognitive aspects.

Cognitive aspects subcategory	Calculated p-value between experiments
Knowledge operationalization	False (p-value = 0.0411)
Ability of self-correction during the learning process	True (p-value = 0.1690)
Strategy for solving exercises based on gradation	False (p-value = 0.0087)

4 Discussion

The aim of the work was the assessment of the influence of the teaching method on cognitive aspects in the process of mathematical education among blind people. The conducted research confirmed the importance of choosing the teaching method. Cooperation with psychologists has allowed to define three specific subcategories: knowledge operationalization, ability of self-correction during the learning process and strategy for solving exercises based on gradation.

The obtained results indicate a statistically significant improvement in results in two of the three proposed subcategories: knowledge operationalization and strategy for solving exercises based on gradation (Table 3). The greatest improvement in results was obtained in the subcategory strategy for solving exercises based on gradation (Table 2 and Fig. 1).

The research results show that thank to developing appropriate methods and mechanisms for learning mathematics to blind people, it is possible to influence positively on the improvement of some cognitive aspects of these people and, thus, faster and better acquisition of knowledge. The designed system allows students for more effective independent learning of mathematics areas. It was developed according to the proposed method of exercise decomposition combined with the method of evaluation of results and knowledge vector.

The obtained results have been confirmed by literature reports indicating the significance of the sound descriptions in the process of teaching math's by blind people [8–10]. Through the methods proposed by the authors, the decomposition of exercises and alternative sound descriptions contain additional information about the structure of the mathematical formula, the problem of cognitive overload known from the theory of cognitive load has been limited [23,24].

Based on the experiments carried out and the obtained results, it can be concluded that teaching and learning mathematics by visually impaired students using the proposed method of decomposition of exercises in combination with the method of evaluation of results based on the vector knowledge, contributes to universal education through:

1. Increasing the availability of materials containing structured information (mathematical formulas) for the blind. This has a large impact on the operationalization of knowledge, i.e. the choice of the representation method adapted to the needs of the anticipated operations on the form of knowledge [20]. Operationalization of knowledge for human needs is important especially in interactive systems, including learning systems.
2. The possibility of self-study using the proposed Math Platform. In this case, an improvement in the results in terms of ability of self-correction during the learning process was also observed. However, the difference was not statistically significant, which may indicate that the teacher's participation in the learning process is very important. In addition, the obtained results indicate that it is necessary to continue working on the developed method.

3. Increasing the independence of the student from the teacher. Through the application of the vector of knowledge and the automatic selection by the developed platform of subsequent exercises, the student can increase his knowledge through gradation. The obtained results confirm this thesis.

References

1. Van Garderen, D., Scheuermann, A., Jackson, C.: Developing representational ability in mathematics for students with learning disabilities: a content analysis of grades 6 and 7 textbooks. Learn. Disabil. Q. **35**, 24–38 (2012)
2. Wang, M.T., Fredricks, J.A., Ye, F., Hofkens, T.L., Linn, J.S.: The math and science engagement scales: scale development, validation. and psychometric properties. Learn. Instr. **43**, 16–26 (2016). https://doi.org/10.1016/j.learninstruc.2016.01.008
3. Kerscher, G.: DAISY consortium: information technology for the world's blind and print-disabled population past, present, and into the future. Libr. Hi Tech. **19**, 11–15 (2001). https://doi.org/10.1108/07378830110384520
4. Middleton, H.: Technology, Learning and Working? Blind and Vision-Impaired People's Use of Technology (2007)
5. Brzoza, P., Spinczyk, D.: Multimedia browser for internet online daisy books. In: Presented at the International Conference on Computers Helping People with Special Needs (2006). https://doi.org/10.1007/11788713_158
6. Spinczyk, D., Brzoza, P.: Multimedia system for accessible distant education (2008). https://doi.org/10.1007/978-3-540-68168-7_58
7. Wongkia, W., Naruedomkul, K., Cercone, N.: iMath: automatic math reader for Thai blind and visually impaired students. Comput. Math. Appl. **64**, 2128–2140 (2012). https://doi.org/10.1016/J.CAMWA.2012.04.009
8. Baldis, J.J.: Effects of spatial audio on memory, comprehension, and preference during desktop conferences (2001)
9. Sfinchez, J., Lumbreras, M., Cernuzzi, L.: Interactive virtual acoustic environments for blind children: computing, usability, and cognition. In: CHI 2001 Extended Abstracts on Human Factors in Computing Systems - CHI 2001, p. 65. ACM Press, New York (2001). https://doi.org/10.1145/634067.634109
10. Sánchez, J.H., Flores, H.E.: AudioMath: blind children learning mathematics through audio (2004)
11. Reykowski, J.: From the issues of psychology of motivation. Wydawnictwa Szkolne i Pedagogiczne (1982)
12. Kim, Y.-I.: The effects of assertiveness training on enhancing the social skills of adolescents with visual impairments. J. Vis. Impair. Blind. **97**, 285–297 (2003)
13. Bhuvaneswari, M., Immanuel Selvaraj, C., Selvaraj, B., Srinivasan, T.: Assessment of psychological and psycho-physiological problems among visually impaired adolescents. Iran. J. Psychiatry Behav. Sci. **10**, e3895 (2016). https://doi.org/10.17795/ijpbs-3895
14. Leo, F., Tinti, C., Chiesa, S., Cavaglia, R., Schmidt, S., Cocchi, E., Brayda, L.: Improving spatial working memory in blind and sighted youngsters using programmable tactile displays. SAGE Open Med. **6** (2018). https://doi.org/10.1177/2050312118820028
15. Pasqualotto, A., Proulx, M.J.: The role of visual experience for the neural basis of spatial cognition. Neurosci. Biobehav. Rev. **36**, 1179–1187 (2012). https://doi.org/10.1016/J.NEUBIOREV.2012.01.008

16. Tinti, C., Adenzato, M., Tamietto, M., Cornoldi, C.: Visual experience is not necessary for efficient survey spatial cognition: evidence from blindness. Q. J. Exp. Psychol. **59**, 1306–1328 (2006). https://doi.org/10.1080/17470210500214275
17. Maćkowski, M.S., Brzoza, P.F., Spinczyk, D.R.: Tutoring math platform accessible for visually impaired people. Comput. Biol. Med. **95** (2018). https://doi.org/10.1016/j.compbiomed.2017.06.003
18. Maćkowski, M., Brzoza, P., Żabka, M., Spinczyk, D.: Multimedia platform for mathematics interactive learning accessible to blind people. Multimed. Tools Appl. **77** (2018). https://doi.org/10.1007/s11042-017-4526-z
19. Spinczyk, D., Maćkowski, M., Kempa, W., Rojewska, K.: Factors influencing the process of learning mathematics among visually impaired and blind people. Comput. Biol. Med. **104**, 19 (2019). https://doi.org/10.1016/J.COMPBIOMED.2018.10.025
20. Rubin, A., Babbie, E.R.: Research methods for social work (2017)
21. Johnson, M.H., De Haan, M., Niedźwiecka, A., Pisula, E., Tomalski, P., Fiedorowicz, M., Grupa Wydawnicza Harmonia: Neurokognitywistyka rozwoju? wprowadzenie. Harmonia Universalis - Grupa Wydawnicza Harmonia (2018)
22. Herzyk, A.: Wprowadzenie do neuropsychologii klinicznej. Scholar (2005)
23. Sweller, J.: Cognitive load during problem solving: effects on learning. Cognit. Sci. **12**(2), 257–285 (1988). https://doi.org/10.1207/s15516709cog1202_4
24. Skulmowski, A., Rey, G.D.: Measuring cognitive load in embodied learning settings. Front. Psychol. **8** (2017). https://doi.org/10.3389/fpsyg.2017.01191

Modelling of Physiological Processes

Modeling and Optimization of Radio-Chemotherapy

Andrzej Swierniak[1], Jaroslaw Smieja[1(✉)], Marzena Mura[1,3], and Piotr Bajger[2]

[1] Institute of Automatic Control, Silesian University of Technology, Akademicka 16, 44-100 Gliwice, Poland
{Andrzej.Swierniak,Jaroslaw.Smieja}@polsl.pl

[2] College of Inter-Faculty Individual Studies in Mathematics and Natural Sciences, University of Warsaw, Banacha 2c, 02-089 Warsaw, Poland
p.bajger@uw.edu.pl

[3] Ardigen, ul. Podole 76, 30-394 Cracow, Poland
marzena.mura@gmail.com

Abstract. Pre- or post-operational chemo-radiotherapy have become one of the standard adjuvant therapies in recent years. Though both chemo- and radiotherapy protocols are standardized, the question about the order in which they should be applied, or concurrency, remains an open question. In this work we attempt to answer it with mathematical modeling and optimization of two-dimensional control that represents therapy in a control-theory based approach. In order to address this problem, two issues are discussed. First, two different ways of modeling tumor growth under therapy are compared. For each of them, the necessary conditions for optimal control representing the therapy are presented and discussed. Then, Kaplan-Meier survival curves are compared for standard therapy protocols used in clinics and different approaches to model tumor growth. Finally, a framework for analysis of treatment efficacy is presented, in which optimization and survival analysis are used sequentially.

Keywords: Chemotherapy · Radiotherapy · Modeling · Optimization

1 Introduction

The use of optimal control theory in mathematical modeling of cancer therapies has had more than five decades of history (see, e.g., [1,2] and references therein). Practical application of such theoretical work, however, are scarce. Nevertheless, recent progress in linking various biomolecular markers to the stage of the disease as well as biomedical imaging has provided methods of monitoring cancer growth, studying cancer cell populations and that creates a new opening for

The authors were supported by NCN grant (National Science Centre, Poland) DEC-2016/21/B/ST7/02241. Calculations were partially performed on the Ziemowit computational cluster (http://www.ziemowit.hpc.polsl.pl) created in the POIG.02.01.00-00-166/08 project (BIO-FARMA) and expanded in the POIG.02.03.01-00-040/13 project (Syscancer) and PBS3/B3/32/2015 project (Biotest).

J. Korbicz et al. (Eds.): PCBEE 2019, AISC 1033, pp. 223–233, 2020.
https://doi.org/10.1007/978-3-030-29885-2_20

mathematical modeling in this field and refining control theory methods so that they might gain acknowledgment in clinics.

It should be noted that chemotherapy, as a systemic treatment, combined with radiation, as a local therapy, are one of the most often used forms of combined treatments in clinical practice. Despite that, and despite the fact that the literature devoted to optimization of anticancer treatment is large, in general, such combination has escaped attention of the researchers. There have been only few efforts to model them and to compare concurrent versus sequential chemoradiotherapy (e.g. [3,4]). However, they were focused on testing specific protocols, and no attempt to optimize therapy has been made there. On the other hand, when treatment is dealt with as a control optimization problem, it is not radio- and chemotherapy combined. Moreover, in optimization-oriented publications the models do not relate to survival curves which arguably constitute the only point of interest that could gain the attention of clinicians.

The goal of this paper is to answer two questions, important from the point of view of clinical relevance of modeling and optimization of anticancer treatment: (i) how the necessary conditions derived for a control optimization problem, representing the search for the best therapy protocol, depend on the model structure? and (ii) is the choice of the cancer growth model important for survival curves used to evaluate the ultimate outcome of the therapy?

In that respect, we analyze structural sensitivity of control models involving therapy strategies with respect to different tumor growth terms and different ways of modeling the therapies. Such analysis is not usually performed, as most works dealing with sensitivity analysis of cancer growth models focus on parameter sensitivity (e.g. [5]).

In this work we combine two approaches that usually do not meet in mathematical modeling of anticancer treatment. First, after introducing the model of cancer growth and treatment, optimal control problem is stated. Necessary conditions are provided and analyzed in the context of their clinical relevance. Then, Kaplan-Meier survival curves [6] for patients with cancer are shown for a particular numerical example and the question of their dependence on a particular model form is discussed. Finally, a framework for solving optimal control problems arising in anticancer therapies is proposed.

2 Modeling of Anticancer Treatment

2.1 The Simplest Case

Let us assume that the population of cancers cells, described by its size N, when no treatment is administered, grows according to a Gompertz model, one of the most frequently used in modeling population dynamics:

$$\dot{N} = -\rho_G N ln(N/K), \tag{1}$$

with the initial condition $N(0) = N_0 > 0$,where the parameter K represents the maximum tumor size that could be reached under specific conditions (imposed by some environmental constraints, such as nutrients, space, angiogenic support, etc.).

Let us introduce two-dimensional control vector $[u(t), d(t)]^T$, whose components represent chemotherapy and radiotherapy, accordingly. The chemotherapy usually is assumed to kill the tumor cells according to the log-kill hypothesis [9], hence the kill rate is proportional to chemotherapeutic agent concentration at time t. Radiotherapy effect is usually described in terms of the so called liner-quadratic (LQ) model [10]. Then, the system with control is given by

$$\dot{N} = -\rho_G N ln(N/K) - \gamma u(t)N - \left(\alpha d(t) + \beta d^2(t)\right) N, \tag{2}$$

where parameters γ, α and β correspond to the chemotherapy effect per unit of effective drug concentration and two different radiotherapy effects, associated often with single and double breaks in DNA, respectively.

The goal of a therapy may be mathematically formulated in the following way: for a fixed therapy duration T, find measurable functions u and d that minimize tumor volume at the end of treatment, subject to the following constraints:

$$\begin{array}{l} 0 \le u(t) \le u_{max} \\ 0 \le d(t) \le d_{max} \\ \int_0^T u(t)dt \le U \\ \int_0^T d(t)dt \le D \end{array} \tag{3}$$

for some positive constants u_{max}, v_{max}, U and D.

The optimization problem defined by (2)–(3) is a starting point for further analysis.

If an oversimplified case is assumed, in which a chemotoxic drug instantly affects cancer cells, following its administration (no pharmacokinetics considered) and no DNA repair is taken into account, after a change of variable $x(t) = \frac{\log N(t)}{N_\infty}$ application of the Pontryagin's maximum principle leads to the Hamiltonian in the form

$$H = -p_1\left(\rho_G x + cu + \left(\alpha d + \beta d^2\right)\right) + p_2 u + p_3 d, \tag{4}$$

where p_1, p_2 and p_3 are co-state variables.

Then, the optimal control [7] takes the following form:

1. If $Tu_{max} > U$, then

$$u(t) = \begin{cases} 0 & \text{for } t \in \left[0, T - \frac{U}{u_{max}}\right] \\ u_{max} & \text{for } t \in \left[T - \frac{U}{u_{max}}, T\right] \end{cases} \tag{5}$$

2. If $Tu_{max} < U$, then $u(t) = u_{max}$ for all $t \in [0; T]$
3. If $Td_{max} > D$, then

$$d(t) = \begin{cases} 0 & \text{for } t \in \left[0, T - \frac{D}{d_{max}}\right] \\ d_{max} & \text{for } t \in \left[T - \frac{D}{d_{max}}, T\right] \end{cases} \tag{6}$$

4. If $Td_{max} < D$, then $d(t) = d_{max}$ for all $t \in [0; T]$.

This means that, somehow counterintuitively, both radiation and chemotherapy should be applied after a delay. Moreover, they will always be applied simultaneously over a period $\left(T - \min\left(\frac{D}{d_{max}}, \frac{U}{u_{max}}\right), T\right)$.

However, the model introduced above ignores two treatment-related phenomena: drug metabolism of chemotherapeutic agent and DNA repair of adjacent strand breaks. As they may affect the structure of theoretically optimal treatment protocols, we propose to incorporate their mathematical description into the model.

2.2 Modeling Pharmacokinetics and DNA Repair

The simplest pharmacokinetics (PK) model is of the first order. If $u(t)$ in (2) represents the effective concentration of a drug, it may be described using the following equation:

$$\dot{u} = -\lambda u + c(t) \tag{7}$$

where $c(t)$ denotes the drug dose and the initial condition is given by $u(0) = 0$.

Including PK in the model does not require modification of the Eq. (2) - a change in interpretation of $u(t)$ is sufficient. However, it is not the case when DNA repair should be taken into account. Assuming that the intracellular mechanisms repair double-strand breaks at rate μ, the Eq. (2) should be replaced by

$$\dot{N} = -\rho_G N ln(N/K) - \gamma u(t)N - (\alpha d(t) + \beta d(t) f(t))\, N, \tag{8}$$

where $f(t)$ is the solution of the following equation:

$$\dot{f} = -\mu f + d(t) \tag{9}$$

with $f(0) = 0$.

Similarly as in the case discussed in the preceding section, if $T u_{max} < U$, then $u(t) = u_{max}$ and if $T d_{max} < D$, then $d(t) = d_{max}$, for all $t \in [0; T]$. Otherwise the optimal control is given in the bang-bang form either with one switch u_{max}–0 or two switches 0–u_{max}–0 [7]. If

$$\frac{\log \rho_G - \log \lambda}{\rho_G - \lambda} < T \tag{10}$$

and

$$\frac{\exp\left(\rho_G(U/u_{max} - T)\right) - \exp\left(\lambda(U/u_{max} - T)\right)}{\rho_G - \lambda} < \frac{\exp(\rho_G T) - \exp(\lambda T)}{\rho_G - \lambda}, \tag{11}$$

then

$$u(t) = \begin{cases} 0 & \text{for } t \in (0, t_1) \\ u_{max} & \text{for } t \in (t_1, t_2) \\ 0 & \text{for } t \in (t_2, T) \end{cases} \tag{12}$$

where t_1, t_2 are determined by the relations $p_2(t_1) = p_2(t_2)$ and $t_2 - t_1 = U/u_{max}$.
Otherwise

$$u(t) = \begin{cases} u_{max} \text{ for } t \in \left[0, \frac{U}{u_{max}}\right] \\ 0 \quad\quad \text{for } t \in \left[\frac{U}{u_{max}}, T\right] \end{cases} \tag{13}$$

It has been proved that no singular arcs are part of control trajectories for biologically relevant relations between parameters [7].

Optimal control $d(t)$ is much harder to find but it has been shown in [7] that it contains at most one switch and is given by

$$d(t) = \begin{cases} 0 & \text{for } \; t \in [0, t_3] \\ d_{max} & \text{for } \; t \in [t_3, T] \end{cases} \tag{14}$$

where t_3 can be found numerically.

An important finding here is that while the optimal protocol of radiotherapy has been found to be the same regardless of incorporating or neglecting DNA repair, the form of optimal chemotherapy is affected by including PK in the system description. When PK is neglected, the optimal solution suggests waiting with chemotherapy until the end of the treatment horizon. It should be noted, however, that introducing PK into the problem had a significant effects on the conclusions drawn from optimal control theory. When PK is explicitly included in the model, the optimal solution suggests two possible strategies, where the exact choice depends on model parameters. In one case, the optimal strategy is completely opposite to the solution obtained without PK, i.e. it suggests applying chemotherapy at the beginning of the treatment. In the other case, the optimal strategy involves postponing chemotherapy administration, but not quite to the very end of treatment horizon. This may be attributed to the fact that the treatment effect persists even after drug administration due to PK.

The above finding gives rise to another question, about the implications of assuming only one of possible descriptions of tumor growth.

3 Gompertz vs. Logistic Growth

As shown in the previous section, expanding mathematical description of the system under investigation by incorporating additional processes may lead to qualitatively different results, as far as optimal control is concerned. However, some processes may be described in more than one way. Therefore, it is important to check, if other form of describing the same processes may also lead to a qualitatively different control trajectory. Moreover, even when the optimal control trajectory is found in the bang-bang form, the switching times depend on model parameters. That means that they may be different for individual patients. To check if the choice of growth model affects therapy outcome, instead of deriving necessary conditions for each case separately, in this section the analysis is focused on therapy outcomes for an a priori assumed control trajectory and different models of tumor growth.

In addition to the Gompertz growth model, in this section a logistic one is considered, which, when coupled with therapy modeling leads to the following equation:

$$\dot{N} = \rho_L N(1 - N/K) - \gamma u(t)N - (\alpha d(t) + \beta d^2(t))N, \tag{15}$$

The main difference between the Gompertz and the logistic models is that the first one assumes an exponentially decreasing growth rate, while the second one assumes linearly decreasing growth rate. One of the important implications of this difference is that in the case of Gompertz model, a complete eradication of the tumor is impossible. For a detailed description of the tumor growth models, presented above, and their applications see, e.g., [11].

It should be noted that the parameters ρ_L and ρ_G in the models of logistic and Gompertz growth have slightly different interpretation, though both are related to the growth rate. Therefore, they should be estimated separately for each type of these models [8].

The results from the previous section indicated that a bang-bang solution is optimal for a given fixed terminal time. Moreover, current clinical protocols involve multiple bang-bang controls that may be interpreted as subsequent cycles of therapy, with optimal control applied in each cycle. Therefore, three different multivariable control strategies are checked in this section: sequential chemo- and radiotherapy (CRT), concurrent CRT and radiotherapy only. Realistic clinical protocols ([3] and discussions with cooperating oncology clinicians) were tested, with changes of $c(t)$ and $d(t)$ control variables shown in Fig. 1, where in each case the therapy was started 14 days after tumor detection.

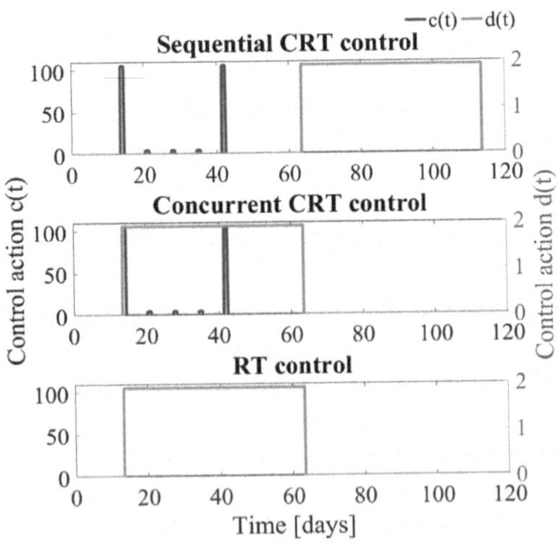

Fig. 1. Control $c(t)$ and $d(t)$ for three analyzed cases.

Recently, parameter values for the presented model (for Gompertz model of tumor growth) were estimated, based on survival curves available in literature [3].

Values of parameters used in this paper are given in Table 1. Additionally, we applied the following assumptions:

- the patient dies when tumor reaches 13 [cm] in diameter (the death condition),
- there is a correlation between the tumor growth rate ρ and the radiosensitivity parameter α (coefficient = 0.87, estimated based on *in vitro* experiments [12]),
- the tumor control probability at the end of treatment is $p = e^{(-N_{cc})}$, where N_{cc} is the number of remaining cancer cells,
- the patient is cured after the treatment if random number generated from uniform distribution [0, 1] is lower than probability p.

Table 1. Values of parameters used in simulations

Parameter	Value
ρ_G, ρ_L	$7.00 \cdot 10^{-5}$ (mean), $7.23 \cdot 10^{-3}$ (std dev.)
α	0.0398 (mean), 0.0168 (std dev.)
γ	0.0093
β	$\alpha/10$

Three treatment strategies were tested on a group of a thousand patients. Each patient was characterized by different ρ and α parameter values, drawn from bivariate normal distribution to reproduce high correlation between parameters. The initial tumor size, $N_0 = 6$ [cm] in diameter, was the same for all patients.

Table 2. Actual clinical therapy protocols analyzed in the paper

Case	Therapy type	Chemotherapy details	Radiotherapy details
(a)	Sequential CRT	Vinblastine 5 mg/m^2, weekly, first 5 weeks cisplatin 100 mg/m^2, days 1 & 29	(starting day 50) 63 Gy/7 weeks
(b)	Concurrent CRT	Vinblastine 5 mg/m^2, weekly, first 5 weeks cisplatin 100 mg/m^2, days 1 & 29	(starting day 1) 63 Gy/7 weeks
(c)	RT only	None	(starting day 1) 63 Gy/7 weeks

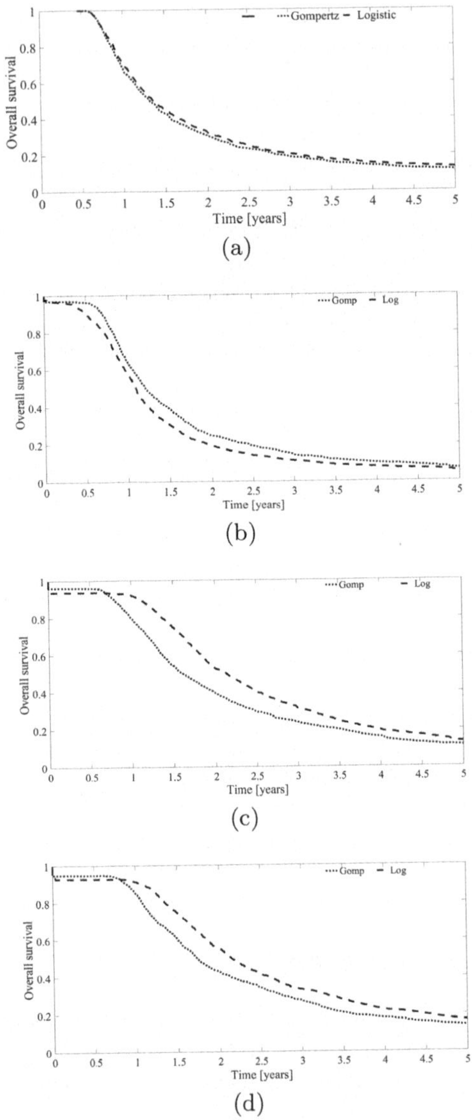

Fig. 2. Kaplan Meier survival curves estimated for thousand patients based on Gompertz and logistic growth models: (a) chemotherapy only; (b) radiotherapy only; (c) concurrent CRT therapy and (d) sequential CRT therapy.

First, we compared results obtained for two growth models for a group of thousand patients grouped by treatment strategy. Results are presented in the form of Kaplan-Meier survival curves (Fig. 2).

Five-year survival was below 20% for both models in all cases. However, except for the chemotherapy as the only treatment, in which case both Gompertz

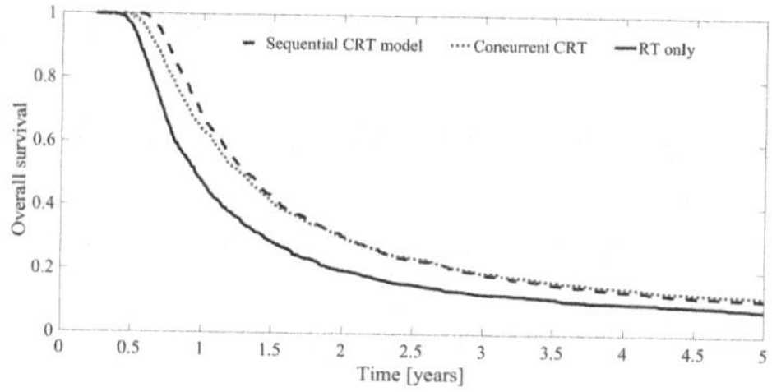

(a) Gompertz model

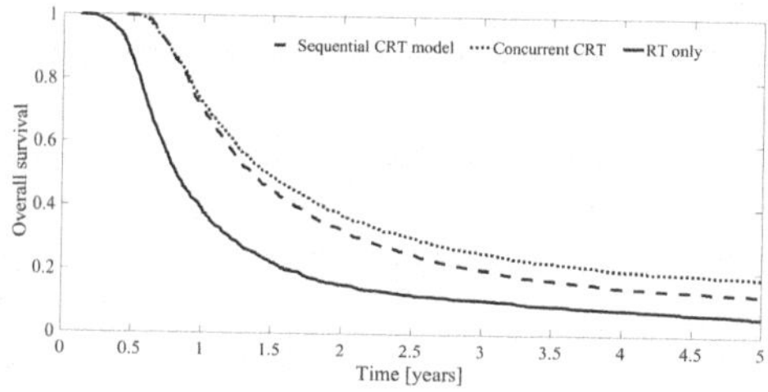

(b) Logistic model

Fig. 3. Kaplan Meier survival cures as a result of different therapy strategies: sequential chemo-radiotherapy, concurrent chemo-radiotherapy and only radiotherapy for (a) the Gompertz model and (b) the logistic model.

and logistic models lead to very similar Kaplan-Meier curves (Fig. 2(a)), there are significant differences between treatment outcomes, represented by survival curves, for different types of the cancer growth model.

When only radiotherapy was applied (Fig. 2(b)) the higher overall survival was predicted with the Gompertz growth model. However, when both radio- and chemotherapy were applied, it was the logistic model that led to a longer overall survival after first several months, both in the case of concurrent (Fig. 2(c)) and sequential (Fig. 2(d)) CRT therapy. The differences in Kaplan-Meier curves were negligible in short time after the treatment, but they increased in time, in particular between one and two and half years (Fig. 2c, d).

Secondly, we analyzed the effect of different treatment strategies on the survival for each model separately (Fig. 3). Concurrent CRT therapy gave the highest overall survival after 5 years for both models, where RT as the only treatment gave the lowest one. Sequential and concurrent CRT applied for Gompertz model gave the same treatment outcome, which suggests that this model is less sensitive to control. Results obtained with the logistic model suggest that application of sequential or concurrent CRT may have significant impact on overall survival of patients (log-rank test, p-value < 0.01).

There are some differences between overall survival of patients obtained by two different models. However, the shape of curves were similar and it may be related to tumor growth rate parameter value. With appropriate estimation of parameter values, any of presented models could reproduce clinical data with similar accuracy.

In all cases, the most effective treatment was concurrent chemo-radiotherapy. The logistic model was more sensitive to changes of therapy protocols than the Gompertz model.

4 Conclusions

Tumor growth and its treatment can be viewed and analyzed as an open-loop control system. Thus, optimal treatment protocols can be found using formalisms of optimization theory.

The particular form of optimal control in terms of the order in which chemo- and radiotherapy should be applied depends on including or neglecting pharmacokinetics of chemotoxic drugs. However, in both cases the necessary conditions suggest that both therapies should be applied concurrently, at least in some period of time. Incorporating or neglecting DNA repair mechanisms do not have the same impact on the form of necessary conditions of the theoretical optimal treatment protocol (in both cases the solution is qualitatively the same).

Furthermore, survival analysis, in which the actual clinical protocols for chemo- and radiotherapy were applied, confirms the above conclusion, showing that concurrent therapies yield significantly better survival prospects. This conclusion does not depend on the type of tumor growth that is used in the model.

The analysis shown above indicates that the following procedure should be used in an attempt to find best control strategies:

1. For a family of models, derive necessary conditions of optimal control in a single therapy cycle and use them to find optimal control, assuming realistic parameter values.
2. Compare control and state trajectories arising from each model. For those that are similar, choose arbitrarily one representative model.
3. For each model that yields significantly different outcomes and control trajectories found in the point 1., calculate survival curves, assuming a distribution of parameter values with mean equal to the parameter values used in the previous point.
4. Compare survival curves to those found in clinical data.

Acknowledgment. The Authors would like to thank Prof. Suwinski from the Maria Sklodowska-Curie Institute -Oncology Centre (MSCI), branch in Gliwice, for valuable comments on clinical applicability of the results obtained from modeling.

References

1. Swierniak, A., Kimmel, M., Smieja, J., Puszynski, K., Psiuk-Maksymowicz, K.: System engineering approach to planning anticancer therapies. Springer, Cham (2016). https://doi.org/10.1007/978-3-319-28095-0

2. Schaettler, H., Ledzewicz, U.: Optimal control for mathematical models of cancer therapies. An application of geometric methods. Springer, New York (2015). https://doi.org/10.1007/978-1-4939-2972-6
3. Geng, C., Paganetti, H., Grassberger, C.: Prediction of treatment response for combined chemo- and radiation therapy for non-small cell lung cancer patients using a bio-mathematical model. Sci. Rep. **7** (2017). https://doi.org/10.1038/s41598-017-13646-z
4. Curran Jr., W.J., Paulus, R., Langer, C.J., Komaki, R., Lee, J.S., Hauser, S., Movsas, B., Wasserman, T., Rosenthal, S.A., Gore, E., Machtay, M., Sause, W., Cox, J.D.: Sequential vs concurrentchemoradiation for stage III nonsmall cell lung cancer: Randomizedphase III trial RTOG 9410. J. Natl Cancer Inst. (JNCI) **103**(19), 1452–1460 (2011). https://doi.org/10.1093/jnci/djr325
5. Dolbniak, M., Kardynska, M., Smieja, J.: Sensitivity of combined chemo-and antiangiogenic therapy results in different models describing cancer growth. Discr. Conti. Dynam. Syst. B **23**, 145–160 (2018). https://doi.org/10.3934/dcdsb.2018009
6. Dudley, W.N., Wickham, R., Coombs, N.: An introduction to survival statistics: Kaplan-Meier analysis. J. Adv. Pract. Oncol. **7**(1), 91–100 (2016)
7. Bajgier, P., Fujarewicz, K., Swierniak, A.: Effects of pharmacokinetics and DNA repair on the structure of optimal controls in a simple model of radio-chemotherapy. In: Proceedings of the 23rd International Conference on Methods and Models in Automation and Robotics (MMAR), pp. 686–691 (2018). https://doi.org/10.1109/mmar.2018.8485901
8. Dolbniak M., Smieja, J., Swierniak, A.: Structural sensitivity of control models arising in combined chemo-radiotherapy. In: Proceedings of the 23rd International Conference on Methods and Models in Automation and Robotics (MMAR), pp. 339–344 (2018). https://doi.org/10.1109/mmar.2018.8486088
9. Skipper, H.E., Schabel, F., Wilcox, W.: Experimental evaluation of potential anticancer agents. XIII. On the criteria and kinetics associated with curability of experimental leukemia. Cancer Chemother. Rep. **35**, 1–111 (1964)
10. Fowler, J.F.: The linear-quadratic formula and progress in fractionated radiotherapy. Br. J. Radiol. **62**, 679–694 (1989)
11. Gerlee, P.: The model muddle: in search of tumor growth laws. Cancer Res. **73**(8), 2407–2411 (2013). https://doi.org/10.1158/0008-5472.CAN-12-4355
12. Lee, J.Y., Kim, M.-S., Kim, E.H., Chung, N., Jeong, Y.K.: Retrospective growth kinetics and radiosensitivity analysis of various human xenograft models. Lab. Anim. Res. **32**(4), 187–193 (2016). https://doi.org/10.5625/lar.2016.32.4.187

Challenges with Conventional Ventilation of Infants with Inhomogeneous Lungs

Barbara Stankiewicz(✉), Krzysztof Jakub Pałko, Marek Darowski, Maciej Kozarski, and Krystyna Górczyńska

Nalecz Institute of Biocybernetics and Biomedical Engineering, Polish Academy of Sciences, Warsaw, Poland
bstankiewicz@ibib.waw.pl

Abstract. Mechanical ventilation of infants with congenital diaphragmatic hernia (CDH) often is a challenge due to different elastic-resistive properties of both lungs. The first step of therapy requires choosing between conventional mechanical ventilation (CMV) and high frequency ventilation (HFV). This study's aim is to explore why CMV cannot always be applied in infants with ventilation inhomogeneity. The authors hypothesized that lung inhomogeneity is related to higher respiratory system impedance requiring higher respiratory pressure and work of breathing to obtain assumed minute ventilation. Real-time *in vitro* simulations using hybrid (numerical-physical) model of the respiratory system and a ventilator revealed the risk factors connected with inhomogeneous ventilation of lungs. The pressures and flows registered in the smaller, pinched lung exhibiting decreased compliance can reach dangerously high values and cause lung injury. Such effects were not observed with homogenous lungs as the same minute ventilation has been received at significantly lower pressure delivered to patient airway.

Keywords: Inhomogeneous ventilation of lungs · Congenital diaphragmatic hernia · Conventional mechanical ventilation · Hybrid respiratory simulator

1 Introduction

Small differences in elastic-resistive properties of lungs resulting from their natural maturation, are frequently seen in newborns and infants, and do not constitute a problem from a respiratory therapy point of view. In such cases, standard ventilation therapy procedures are implemented, similar to those of patients with homogenous lungs, based on general health condition and type of respiratory failure. However, if the difference between lungs is significant as it is in infants with congenital diaphragmatic hernia (CDH) [1], ventilation therapy requires more caution in choice of proper technique and ventilation parameters to avoid adverse effects of ventilation therapy such as risk of lung injury.

J. Korbicz et al. (Eds.): PCBEE 2019, AISC 1033, pp. 234–244, 2020.
https://doi.org/10.1007/978-3-030-29885-2_21

CDH is a defect in diaphragm muscle seen in 1 per 2000–5000 live births [2], with high mortality rate (20–50%) [3–5]. It is most frequently left side-localized (85%) and is more often found in male neonates (60%) [1]. The defect results in relocation of the digestive organs to the thoracic cage above the diaphragm and pressing against the lungs and heart. This result in the lungs' inability to develop normally, and the lung exhibiting the most pressure remaining smaller and more hypoplastic (Fig. 1) [1, 6]. This is root cause of lung inhomogeneity, and makes effective and safe ventilation therapy in CDH infants difficult [3, 7].

CDH severity is standardly assessed based on diaphragm defect position (left vs. right), liver position (abdomen vs. thorax), lung to head ratio (LHR) or lung to thorax transverse area ratio (LTR). For this purpose, roentgenography, computed tomography or magnetic resonance imaging are used [6, 8].

Lung inhomogeneity in CDH infants can be assessed using lung clearance index or moments calculated basing on the process of gas elution (N_2, SF_6, O_2) from the lungs. This can be measured using multiple-breath washout technique (MBW) [9–11], which unfortunately is not yet in standard clinical use.

As a first approach to performing ventilation therapy on an infant with CDH, conventional or high frequency ventilation is recommended [1]. Extracorporeal membrane oxygenation (ECMO) is preferred for the most severe CDH cases when the two above techniques are ineffective [1, 12].

We wondered why it is impossible to effectively ventilate some CDH infants using conventional method. The study's aim, therefore, was to examine the influence of lung inhomogeneity on the work of breathing as well as on pressures and flows inside the respiratory system. For that purpose, the infant hybrid (numerical-physical) respiratory simulator was used.

2 Methods

2.1 Numeric Model of Lungs

The numerical part of the simulator is a lung model consisting of a central and peripheral airway and a chest wall. The peripheral airway is divided into two compartments (two lungs). The central airway is described by resistance R_c and gas inertance L, each of lungs - by compliance C_1 or C_2 and resistance R_1 or R_2, and the chest wall - by compliance C_w. Physiological and electrical representations of the infant respiratory system are shown in Fig. 1.

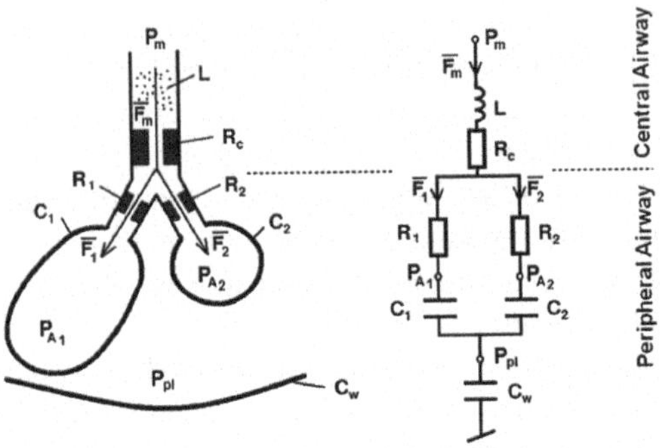

Fig. 1. Physiological and electrical schemes of the infant respiratory system numerical model. L, R_c – gas inertance and central airway resistance, P_m, F_m – pressure and flow at model inlet, respectively, C_1, C_2 – lung compliance in the 1st and 2nd compartments, respectively, R_1, R_2 – peripheral airway resistance in the 1st and 2nd compartments, respectively, P_{A1}, P_{A2} – alveolar pressure in the 1st and 2nd compartments, respectively, F_1, F_2 – flow in the 1st and 2nd compartments, respectively.

Operation of the respiratory model can be described by the following equation set 1:

$$\begin{cases} F_m(t) = \frac{P_m}{Z_m} \\ V(t) = \int F_m(t)\,dt \\ \frac{dV_1(t)}{dt}(R_1 + R_1) + V_1(t)\frac{C_1+C_2}{C_1 \cdot C_2} = F_m(t) \cdot R_2 + \frac{V(t)}{C_2} \\ V_2(t) = V(t) - V_1(t) \\ F_1(t) = \frac{dV_1}{dt} \\ F_2(t) = F_m(t) - F_1(t) \\ P_{A1}(t) = p_m(t) - L\frac{dF_m}{dt} - R_c F_m - R_1\frac{dV_1}{dt} \\ P_{A2}(t) = p_m(t) - L\frac{dF_m}{dt} - R_c F_m - R_2\frac{dV_2}{dt} \\ P_{pl}(t) = -\frac{V(t)}{C_w} \end{cases} \tag{1}$$

where: V – lung volume above fractional residual capacity (FRC), V_1, V_2 – gas volume, in the 1st and 2nd compartments, respectively.

Then, model impedance can be expressed by the equation set 2:

$$\begin{cases} |Z_m| = \sqrt{X^2 + R^2} \\ X = \omega \cdot L - \frac{1}{\omega \cdot C_w} - \frac{C_1 + C_2 + \omega^2 \cdot (R_1^2 \cdot C_1^2 \cdot C_1 + R_2^2 \cdot C_2^2 \cdot C_1)}{\omega^3 (R_1 \cdot C_1 \cdot C_2 + R_2 \cdot C_2 \cdot C_1)^2 + \omega \cdot (C_1 + C_2)^2} \\ R = R_C + \frac{R_1 \cdot C_1^2 + R_2 \cdot C_2^2 + \omega^2 \cdot R_1 \cdot C_1 \cdot R_2 \cdot C_2 \cdot (R_1 \cdot C_1 \cdot C_2 + R_2 \cdot C_2 \cdot C_1)}{\omega^2 (R_1 \cdot C_1 \cdot C_2 + R_2 \cdot C_2 \cdot C_1)^2 + (C_1 + C_2)^2} \\ \omega = 2\pi \cdot f \end{cases} \tag{2}$$

where: $|Z_m|$ is the model impedance of infant respiratory system, X and R are the imaginary and real part of impedance, respectively, ω – angular breath frequency, f – breath frequency.

2.2 The Hybrid Simulator of Respiratory System and Setup

The setup used in this study is presented in Fig. 2. It is composed of a Puritan Bennett 8400 Ventilator (Covidien Co., USA), NICO monitor (Respironics Corp., Murrysville, PA, USA) of respiratory parameters, tablet computer and a "Patient" represented by the simulator of the infant respiratory system. The simulator of a hybrid structure consists of numerical and physical parts. The work of both parts of the simulator and its interaction with a ventilator is managed and coordinated by LabVIEW Professional Development System 2013, with Real-Time Module (National Instruments Co., Austin, TX, USA). The numerical part of the simulator is lung model (Fig. 1). The physical part of the simulator is the impedance transformer (TR) of the piston-cylinder design driven by a servo-mechanical unit. The TR converts, back and forth, digital signals of pressure (P_m) and volume gas flow (F_m), obtained as the numerical solution of lung model equation set (Eq. 1) into physical signals of pressure and flow (P, F) to be the input signals to the simulator. The linear conversion, being a key property of TR, determines the numerical lung model to be transformed without distortions.

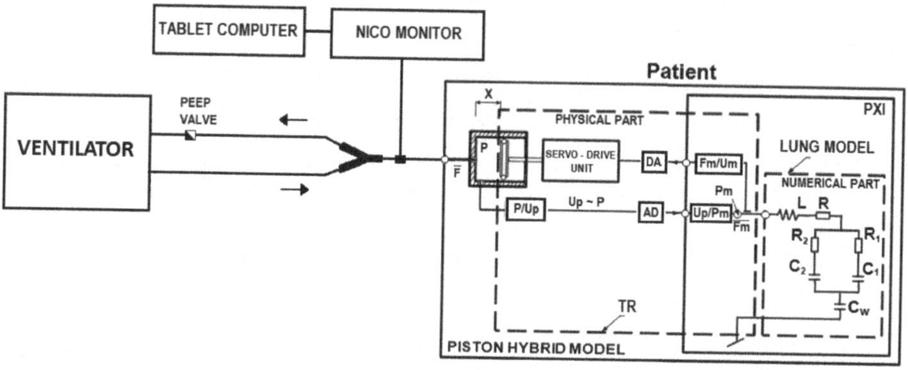

Fig. 2. Setup used in the study: ventilator, NICO monitor, tablet computer and "Patient" – piston hybrid (numerical-physical) simulator of respiratory system. Physical part of the simulator - impedance transformer (TR) with piston-cylinder construction driven by the servo-mechanical unit. PXI – PC-based platform for measurement and control systems. Numerical part with lung model: L, R_c – gas inertance and central airway resistance, P_m, F_m – pressure and flow at model inlet, respectively, C_1, C_2 – lung compliance in the 1st and 2nd compartments, respectively, R_1, R_2 – peripheral airway resistance in the 1st and 2nd compartments, respectively, C_w – chest wall compliance.

The operational principle of TR can be described by the following equation set 3:

$$\begin{cases} Z = k \cdot Z_m \parallel Z_x \\ Z = \frac{P}{F} \\ Z_m = \frac{P_m}{F_m} \end{cases} \quad (3)$$

where: k - a constant value, Z - the input impedance of the simulator, Z_m - the numeric model impedance, Z_x - the impedance of pneumatic capacitor representing gas compliance (C_x) in the piston chamber. C_x is a function of chamber volume ($V = A \cdot x$) (Fig. 2), atmospheric pressure and polytropic index ($n = 1.3 \pm 0.5$).

2.3 The Simulations

In the course of the study, ventilation of infants with homogeneous and inhomogeneous lungs was simulated. The parameters of infant respiratory system model used in the simulations are presented in Table 1.

During the tests, pressure-controlled ventilation (PCV) with inspiration to expiration ratio I:E of 1:1.5, respiratory rate RR of: 30, 60 and 90 bpm, and positive end expiratory pressure (PEEP) equal to 0.3 kPa were set to obtain minute ventilation MV of 1.2 L. The signals of pressures and flows in "infant" respiratory system, including: alveolar pressures (P_{A1} and P_{A2}) and flows (F_1 and F_2) in both lungs were registered. Besides, work of breathing (WOB) by ventilator and peak inspiratory pressure (PIP) were measured using NICO 7300 monitor fitted with a main stream sensor, neonatal flow adapter and a probe placed between the simulator and Y-piece of the ventilation circuit (Fig. 2.)

Table 1. Parameters of numeric model of infant respiratory system.

	R_c $kPa \cdot s \cdot l^{-1}$	R_1 $kPa \cdot s \cdot l^{-1}$	R_2 $kPa \cdot s \cdot l^{-1}$	C_1 $ml \cdot kPa^{-1}$	C_2 $ml \cdot kPa^{-1}$	C_w $ml \cdot kPa^{-1}$	L $kPa \cdot s^2 \cdot l^{-1}$
HL	6.8	3.4	3.4	20.5	20.5	124	0.002
IHL1	6.8	3.4	3.4	20.5	2.1	124	0.002
IHL2	6.8	34	3.4	20.5	2.1	124	0.002

HL - homogenous lungs: $R_1 = R_2$, $C_1 = C_2$; inhomogenous lungs: IHL1: $R_1 = R_2$, $C_1 = 10 \cdot C_2$, IHL2: $R_1 = 10 \cdot R_2$, $C_1 = 10 \cdot C_2$.

3 Results

The results of the simulations were presented in Figs. 3, 4 and 5. In Fig. 3, the signals of alveolar pressures in homogenous lungs (P_{A1} and P_{A2}) and gas flows (F_m, F_1, F_2) in "infant" respiratory system were presented. Similarly, in Fig. 4, the signals were presented for inhomogenous lungs.

As shown in Fig. 3, in the case of homogenous lungs, the amplitudes of alveolar pressures P_{A1} and P_{A2} were equal to 0.4 kPa. The intrinsic PEEP was 0.2 kPa. The pressure at inlet of patient airway (P_m), in reference to assumed minute ventilation (1.2 L) was 1.1 kPa. Whereas, in the case of inhomogenous lungs (Fig. 4), the pressure amplitude P_{A2} obtained from the lung of lowered compliance was several times higher than the pressure amplitude P_{A1} from the lung of normal compliance (3.4 kPa vs. 0.7 kPa, respectively). Auto-positive end expiratory pressures (auto-PEEP) were equal to 0.1 and 0.7 kPa, in the lungs of lowered and normal compliance, respectively. The pressure at patient airway (P_m) of 3.5 kPa was needed to obtain a minute ventilation 1.2 L in the infant with inhomogenous lungs. Whereas, in the patient with homogenous lungs, the pressure P_m of 1.1 kPa was sufficient.

Moreover, the gas flow values F_1 and F_2 received in the "patient" of homogenous lungs (Fig. 3) and in the patient of inhomogenous lungs (Fig. 4) differ significantly. In the last case, the gas flows F_1 and F_2 differ significantly in their peak inspiratory flow (PIF) and peak expiratory flow (PEF) values, but also in their flow patterns, which is more asymmetric in a shape (PEF >> PIF) with inhomogenous lungs (compare Figs. 3 and 4). Whereas, in the patient of homogenous lungs, the flows F_1 and F_2 have the same PIF and PEF values and flow pattern (Fig. 3).

Next we examined the impedance of infant respiratory system and work of breathing by ventilator obtained for "infant" of homogenous lungs and "infant" of inhomogenous lungs compared at different ventilator respiratory rate (Fig. 5). As we had previously noted, the same assumed minute ventilation (1.2 L) required a significantly higher pressure in the patient with inhomogenous lungs. The PIP values,

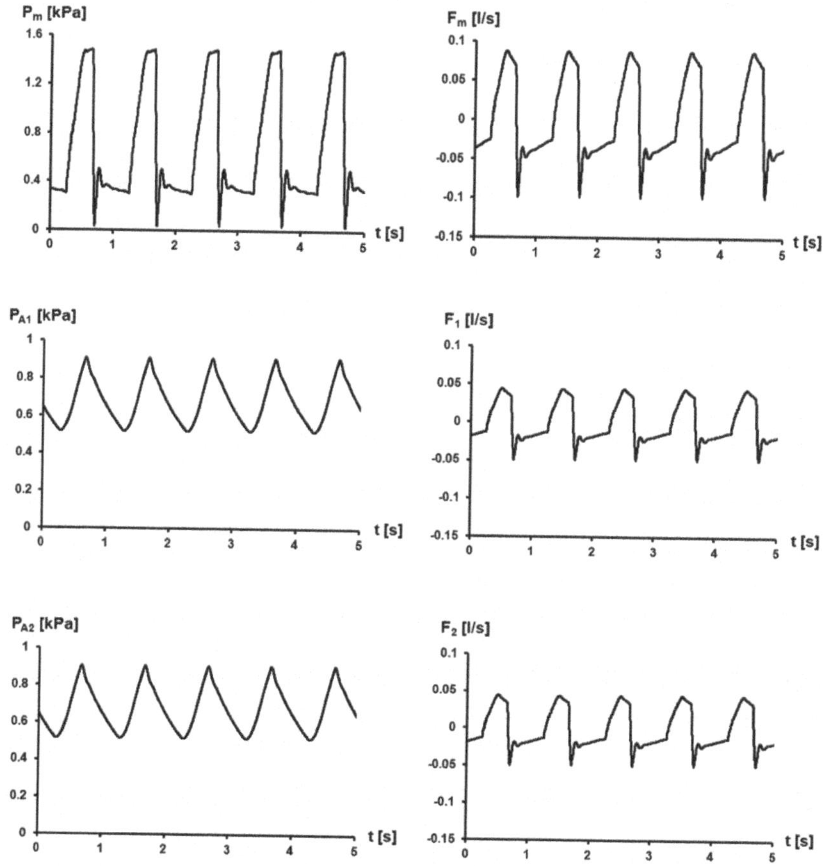

Fig. 3. Flow and pressure courses in infant respiratory system of homogenous lungs ($R_1 = R_2$, $C_1 = C_2$). P_m, F_m – pressure and flow at infant airway, respectively; P_{A1}, P_{A2} – alveolar pressure in the 1st and 2nd lungs, respectively; F_1, F_2 – flow in 1st and 2nd lungs, respectively.

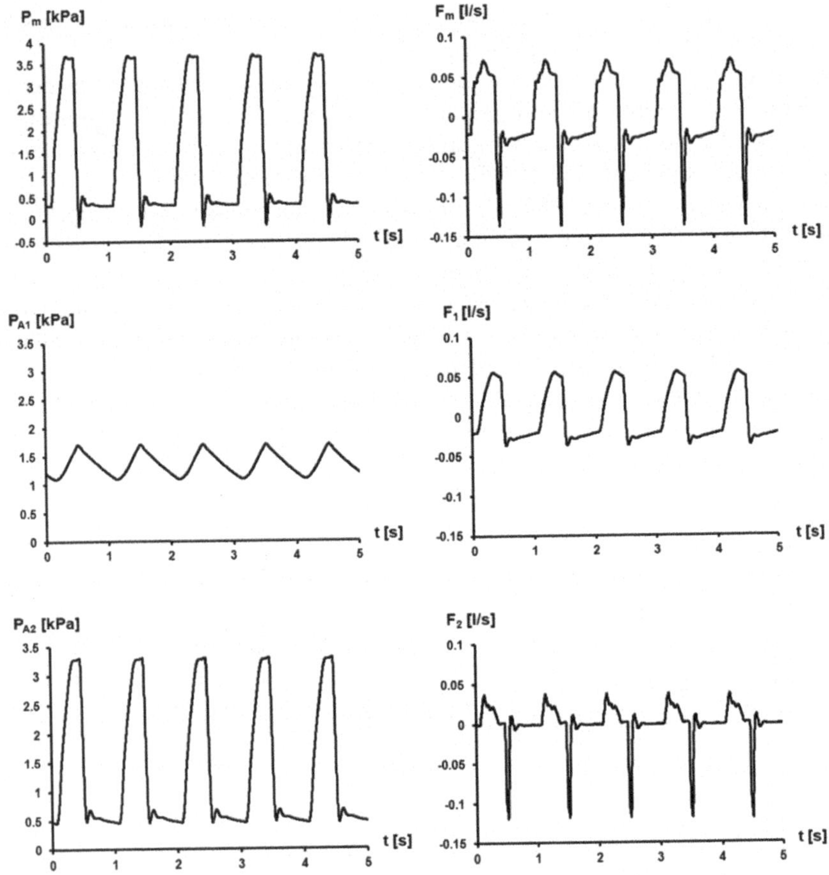

Fig. 4. Flow and pressure courses in CDH infant respiratory system with inhomogenous lungs ($R_1 = 10 \cdot R_2 =$, $C_1 = 10 \cdot C_2$). P_m, F_m – pressure and flow at infant airway, respectively; P_{A1}, P_{A2} – alveolar pressure in the 1st and 2nd lungs, respectively; F_1, F_2 – flow in 1st and 2nd lung, respectively.

at RR = 60 bpm, were: 3.8 kPa and 1.5 kPa, in the patients of inhomogenous and homogenous lungs, respectively (compare peak value of P_m from Figs. 3 and 4). In the case of the patient of inhomogenous lungs ($R_1 = 10 \cdot R_2$, $C_1 = 10 \cdot C_2$), the work of breathing (WOB) was higher ($2.83\,J \cdot l^{-1}$ at RR = 60 bpm) than in the patient with homogenous lungs ($1.1\,J \cdot l^{-1}$ for $R_1 = R_2$, $C_1 = C_2$) (Fig. 5). In both cases, the impedances of infant respiratory system were: $35.3\,kPa\,s\,l^{-1}$ and $9.9\,kPa\,s\,l^{-1}$, respectively (Fig. 5).

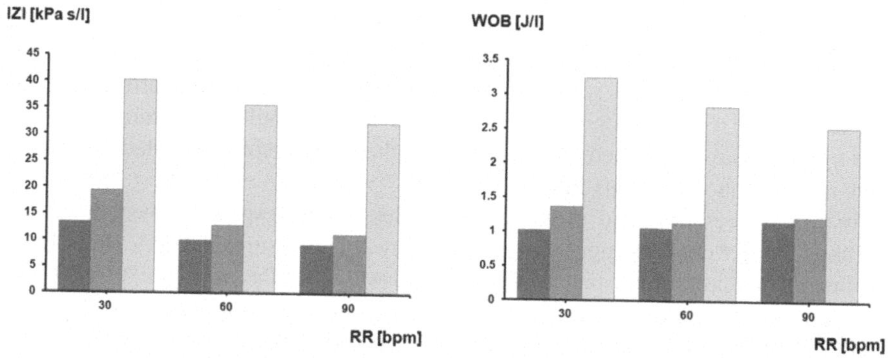

Fig. 5. Simulation results. Impedance of infant respiratory system and work of breathing during pressure-controlled ventilation. From left: homogeneous lungs ($R_1 = R_2$, $C_1 = C_2$), inhomogenous lungs with one lung severe restriction ($R_1 = R_2$, $C_1 = 10 \cdot C_2$) and inhomogenous lungs with severe obstruction ($R_1 = 10 \cdot R_2$) of one lung and severe restriction ($C_1 = 10 \cdot C_2$) of the other.

4 Discussion and Conclusions

The optimal initial management of CDH infants is still unknown. Practitioners often have to decide between two ventilation techniques, conventional mechanical ventilation (CMV) or high frequency ventilation (HFV), to best prepare the infants for surgical hernia repair. Clinical studies are also unequivocal, and there are very few randomized controlled trial (RCT) studies concerning the subject [1, 3, 7]. HFV is usually treated as a rescue therapy, in the case of hypoxia or increase of ventilation pressures, during CMV. It has also been shown that HFV can be used as initial therapy for preoperative stabilization of infants with CDH [3]. To date, a significant difference between the outcomes of the CMV and HFV techniques has not been noted. However, the international multicenter RCT study performed on 171 CDH infants [13] showed that in spite of insignificant differences between CMV and HFV techniques, in the combined outcomes of mortality and bronchopulmonary dysplasia (BPD) morbidity, in the CMV group, the shorter therapy times and less frequent use of ECMO were observed. According to the last statement of the CDH EURO Consortium [14] on postnatal management of CDH infants in Europe, CMV was indicated as the optimal initial ventilation strategy. However, it is not always possible to use CMV, and therefore other ventilation techniques like HFO or ECMO are implemented. We asked why CMV cannot be used to ventilate some patients with CDH. We hypothesized that CMV may be ineffective in some CDH cases because of the high impedance of respiratory system of inhomogenous lungs, requiring higher ventilation pressures to obtain assumed minute ventilation and proper gas exchange levels. We also supposed that CMV using high pressures delivered by ventilator to patient airway, when the lungs are very different in their elastic-resistive properties, can lead to adverse effects, making ventilation ineffective.

We perform real-time *in vitro* simulation of pressure-controlled ventilation of infant with inhomogenous and homogenous lungs to compare the work of breathing and mechanical parameters of ventilation. Seven-elements two-compartmental RLC numerical model of respiratory system enabled us to simulate pressures and flows inside of the respiratory system. The analysis of the study results indicated unfavorable effects in both inhomogenous lungs.

First, peak pressures and pressure amplitudes in the respiratory system of inhomogenous lungs were very high, in regard to allowable pressures in CMV (Fig. 4). The peak inspiratory pressure at patient airway (P_m) was 3.6 kPa, at set PEEP of 0.3 kPa whereas peak alveolar pressure P_{A2} in the lung of lowered compliance was 3.3 kPa at auto-PEEP of 0.2 kPa. Puligandla *et al.* [3] concluded based on review of several clinical studies, that the indication to switch from CMV to HFV was a PIP higher than 2.5 kPa. When using HFV, a difference between PIP and PEEP of 3–5 kPa was permissible [12]. Boloker *et al.* [15] found that none of the CDH infants ventilated with PIP exceeding 2.6 kPa survived. The last statement of the CDH EURO Consortium [14] is consistent with the above conclusions. According to them, the CMV should be changed to HFO if the PIP value is higher than 2.5 kPa. Therefore, in the considered case of lung inhomogeneity ($R_1 = 10 \cdot R_2,\ C_1 = 10 \cdot C_2$) this level was surpassed, but it was needed to reach the required minute ventilation.

Second, the high auto-PEEP value (0.8 kPa) was performed in the respiratory system of inhomogenous lungs, in the lung of normal compliance. Auto-PEEP is an adverse effect occurring in patients during mechanical ventilation. It is disadvantageous, because it can lead to WOB increase, barotrauma, hemodynamic instability and difficulty in ventilator triggering [16]. When auto-PEEP is created in lungs, the patient is unable to totally exhale the ventilator-delivered tidal volume. If the situation is repeated from breath to breath, air-trapping occurs, which can cause PIP to increase to a level that results in barotrauma, volutrauma, hypotension, patient-ventilator asynchrony, or even death [17]. Air-trapping was reported in 12 per 29 CDH infants by Wright *et al.* [8].

The simulation revealed that in the case of homogenous lungs, the situation was strikingly different – no pressure in respiratory system exceeded 1.5 kPa. The peak inspiratory pressure at patient airway (P_m) was 1.5 kPa, including set PEEP of 0.3 kPa. Both peak alveolar pressures (P_{A1} and P_{A2}) were equal to 0.9 kPa, including auto-PEEP of 0.2 kPa and external PEEP of 0.3 kPa.

It was also found, that CDH severity and lung inhomogeneity depth influence on respiratory system impedance; for homogenous infant lungs it was about $10\,\mathrm{kPa} \cdot \mathrm{s} \cdot \mathrm{l}^{-1}$, whereas for inhomogenous lungs - $15.2\,\mathrm{kPa} \cdot \mathrm{s} \cdot \mathrm{l}^{-1}$ ($R_1 = R_2,\ C_1 = 10 \cdot C_2$) and $35.5\,\mathrm{kPa} \cdot \mathrm{s} \cdot \mathrm{l}^{-1}$ ($R_1 = 10 \cdot R_2,\ C_1 = 10 \cdot C_2$), at f = 60 bpm.

Moreover, CDH severity and lung inhomogeneity depth influence the work of breathing - WOB at homogenous lungs was $1.1\,\mathrm{J} \cdot \mathrm{l}^{-1}$, and in inhomogenous lungs was increased to $1.5\,\mathrm{J} \cdot \mathrm{l}^{-1}$($R_1 = R_2,\ C_1 = 10 \cdot C_2$) and $2.8\,\mathrm{J} \cdot \mathrm{l}^{-1}$($R_1 = 10 \cdot R_2,\ C_1 = 10 \cdot C_2$), at f = 60 bpm. According to Mead *et al.* [18], the forces acting on lung tissue can be up to 4.5 times higher in inhomogeneous ventilation of lungs. Inhomogeneity therefore raises lung stresses and risk of ventilator-induced lung injury [10].

We conclude that inhomogeneity of lungs results in respiratory system impedance and increased work of breathing. Besides, a difference in elastic-resistive properties between lungs can lead to significantly increased PIP and alveolar pressure in the more hypoplastic lung, what can lead to lung injury, patient-ventilator asynchrony and other adverse effects. High value auto-PEEP in the less hypoplastic lung can compound this situation even more. In this context, a choice of HFO over CMV in these patients seems appropriate.

References

1. Wung, J.T.: Congenital diaphragmatic hernia multidisciplinary management. http://perinatal.com.br/8simposio/pdf/Aula7WungCDH.pdf. Accessed 21 Mar 2019
2. Jain, V., Agarvala, S., Bhatnagar, V.: Recent advances in the management of congenital diaphragmatic hernia. Indian J. Pediatr. **77**(6), 673–678 (2010)
3. Puligandla, P.S., Grabowski, J., Austin, M., et al.: Management of congenital diaphragmatic hernia: a systemic review from the APSA outcomes and evidence based practice committee. J. Pediatr. Surg. **50**, 1958–1970 (2015)
4. Guidry, Ch.A, Hranjec, T., Rodgers, B.M., et al.: Permissive hypercapnia in the management of congenital diaphragmatic hernia: our institutional experience. Am. College Surg. **214**(4), 640–647 (2012)
5. Guevorkian, D., Mur, S., Cavatorta, E., et al.: Lower distending pressure improves respiratory mechanics in congenital diaphragmatic hernia complicated by persistent pulmonary hypertension. J. Pediatr. **200**, 38–43 (2018)
6. Mielniczuk, M., Kusza, K., Brzeziński, P., et al.: Current management of congenital diaphragmatic hernia. Anesth. Intensive Care **44**(4), 259–264 (2012)
7. Van der Hout, L., Tibboel, D., Vijfhuize, D., et al.: The VICI trial: high frequency oscillation versus conventional mechanical ventilation in newborns with congenital diaphragmatic hernia: an international multicentre randomized controlled trial. BMC Pediatr. **11**, 98–105 (2011)
8. Wright, T., Filbrum, E., Bryner, B., et al.: Predictors of early lung function in patients with congenital diaphragmatic hernia. J. Pediatr. Surg. **49**, 882–885 (2014)
9. Fuchs, S.I., Gappa, M.: Lung clearance index: clinical and research applications in children. Paediatr. Respir. Rev. **12**(4), 264–270 (2011)
10. Bikker, I.G., Holland, W., Specht, P., et al.: Assessment of ventilation inhomogeneity during mechanical ventilation using a rapid-response oxygen sensor-based oxygen washout method. Intensive Care Med Exp. **2**(1), 14 (2014)
11. Latzin, P., Thamrin, C., Kraemer, R.: Ventilation inhomogeneities assessed by the multibreath washout (MBW) technique. Thorax **63**(2), 98–99 (2015)
12. Van der Hout, L.: A new light on lung disease in congenital diaphragmatic hernia. University Medical Center, Rotterdam (2011). https://repub.eur.nl/pub/23773/
13. Snoek, K.G., Capolupo, I., van Rosmalen, J., et al.: Conventional mechanical ventilation versus high-frequency oscillatory ventilation for congenital diaphragmatic hernia. Ann. Surg. **22**(10), 1–8 (2015)
14. Snoek, K.G., Reiss, I.K.M., Greenough, A., et al.: Standardized postnatal management of infants with congenital diaphragmatic hernia in Europe: the CDH EURO Consortium Consensus - 2015 Update. Neonatology **110**, 66–74 (2016)

15. Boloker, J., David, A., Bateman, D.A., et al.: Congenital diaphragmatic hernia in 120 infants treated consecutively with permissive hypercapnea/spontaneous respiration/elective repair. J. Pediatr. Surg. **37**(3), 357–366 (2002)
16. Reddy, V.G.: Auto-PEEP: how to detect and how to prevent - a review. Middle East J. Anaesthesiol. **18**(2), 293–312 (2005)
17. Medscape, What is intrinsic positive end-expiratory pressure (PEEP), or auto-PEEP, in mechanical ventilation? https://www.medscape.com/answers/304068-104803/what-is-intrinsic-positive-end-expiratory-pressure-peep-or-auto-peep-in-mechanical-ventilation. Accessed 21 Mar 2019
18. Mead, J., Takishima, T., Leith, D.: Stress distribution in lungs: a model of pulmonary elasticity. J. Appl. Physiol. **28**, 596–608 (1970)

Modelling the Influence of the Electromagnetic Field on a User of a Bone Conduction Hearing Medical Implant

Patryk Zradziński(✉), Jolanta Karpowicz, and Krzysztof Gryz

Central Institute for Labour Protection – National Research Institute (CIOP-PIB), Czerniakowska 16, 00-701 Warsaw, Poland
{pazra, jokar, krgry}@ciop.pl

Abstract. The electromagnetic field (EMF) may have a direct influence on the functions of human body, for example through the electrostimulation of the nervous system or tissue heating. The spatial distribution and the level of such influence, estimated by induced electric field strength (Ein) or specific energy absorption rate (SAR), are different in body tissues close to a medical implant, compared to the body without an implant during the same exposure. Numerical models of EMF sources (loop antennas) with dimensions typical for magnetotherapeutic applicators and high frequency radiofrequency identification (RFID HF) readers, and the user of a hearing implant (Bonebridge type – HI-BB) were worked out. The values of Ein and SAR were analyzed in the head of an HI-BB user exposed to EMF in the worst case scenario – at the shortest distance of 2 cm from EMF sources (at the side or at the edge in front of them). It was demonstrated that the use of HI-BB significantly increases the level of direct exposure effects caused by EMF – in low frequency EMF (Ein up to 3.4 times higher), and in radiofrequency EMF (SAR (1 g) up to 4.5 times higher). The level of such hazards depends on the dimensions of the EMF source and the location of the head against the source (i.e. the field polarization). It was shown that low and radiofrequency EMF is a direct environmental hazard for the hearing implant users, but the analysis of the individual sensitivity of particular users still required further studies involving more detailed models of implants and users' bodies, as well studies covering the indirect EMF influence on the electronic circuits of HI-BB, recognized as electromagnetic compatibility (EMC).

Keywords: Biomedical engineering · Environmental engineering · Induced electric field strength · Numerical simulations · Occupational exposure · Public health · Specific energy absorption rate

1 Introduction

Hearing loss can be compensated for by various types of hearing implants, including a partially implanted bone conduction prosthesis of the Bonebridge type (later referred as: hearing implant – Bonebridge (HI-BB)) located behind the user's ear. This is used in cases when the outer or middle ear is damaged, and sounds are unable to reach the inner ear effectively [1–3]. The internal (implanted) part of the HI-BB is a system that

J. Korbicz et al. (Eds.): PCBEE 2019, AISC 1033, pp. 245–255, 2020.
https://doi.org/10.1007/978-3-030-29885-2_22

uses bone conduction to transmit sound waves through the skull bone directly to the inner ear, where they are processed further as natural sound. The sound processor (the external part of HI-BB, fixed directly over the internal part by magnets) is responsible for processing the recorded signals and controlling the operation of the internal part.

The results of the influence of the electromagnetic field (EMF) include the induced electric field and current or energy absorption in exposed electrically conductive objects such as metals or living tissues. In the bodies of HI-BB users, a different distribution of the exposure effects compared to people without HI-BB should be considered. It was already identified that because of it, hearing implant users may experience: (i) heating malfunctions or even damage to tissue or electrical stimulation in the vicinity of the conductive elements of implants (caused by the direct EMF influence on tissue), (ii) problems with sound quality or (iii) damage to the implant's circuitry (caused by indirect EMF influence to the metal structure of the device and recognized as a problem with electromagnetic compatibility (EMC)), which reduces comfort or even safety, especially in the work environment [2, 3]. These effects depend on the frequencies of the environmental EMF, its level, polarization and distribution in time and space.

Legislation and guidelines regarding the safety of medical implants users include the requirements to individually evaluate hazards caused to them by EMF exposure [4–7]. Environmental EMF exposure and possible discomfort or hazards should also be considered when choosing what type of device will be implanted to a particular patient – by, among other things, analyzing his job and its environmental conditions.

Previous studies were focused on: (i) evaluating the direct effects of exposure to low frequency EMF in users of HI-BB and Bone Anchored Hearing Aid (BAHA) type implants in single exposure scenario with single numerical model of implant user, and (ii) inter-person variability of hazards to users of BAHA implants (with models of users of various dimensions and structure of head). They showed significantly higher electromagnetic hazards for hearing implant users compared to a person without an implant during the same exposure [8, 9].

The aim of this study was to evaluate the direct electromagnetic hazards for HI-BB users present near sources emitting low frequency (100 Hz) or radiofrequency (13.56 MHz) EMF. The study discussed in this paper extends the EMF frequency range and variability of EMF polarization in comparison to what was reported before [8, 9].

2 Material and Methods

2.1 Numerical Model of HI-BB

Simulations were carried out using developed numerical models imitating the construction of the HI-BB and its location in a simplified body model. The model of the HI-BB was developed based on technical data provided by the manufacturer of typical implants, MED-EL (Fig. 1) [1]. The HI-BB model takes into account the dimensions, shape and materials they are made from – of both the external and the internal parts, including the following structures (Table 1):

- external part: coil (transmitter), magnet fixing the external part over the internal part, sound processor, housing
- internal part (implanted): coil (receiver), magnet fixing the external part over the internal part, demodulator, the Bone Conduction – Floating Mass Transducer (BC-FMT) causing the vibration of skull bones, cover of all elements except BC-FMT.

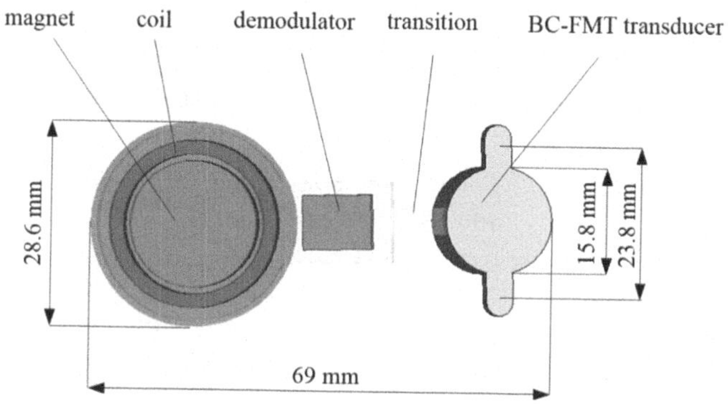

Fig. 1. Numerical model of HI-BB - with geometrical dimensions based on MED-EL data [1]

Table 1. Dielectric parameters of elements of the HI-BB model

Element of the implant model	Implant part including an element	Material	Relative permittivity	Conductivity, S/m
Transmission coils	External and internal	Copper	1.0	6.0×10^{7}
Magnet	External and internal	Steel	1.0	1.4×10^{6}
Housing	External	PET-polyester	3.5	1.0×10^{-15}
Demodulator	Internal	Copper	1.0	6.0×10^{7}
BC-FMT	Internal	Titanium alloy	1.0	5.6×10^{5}
Cover	Internal	Silicone	12	2.5×10^{-4}

PET – Polyethylene terephthalate, BC-FMT – Bone Conduction-Floating Mass Transducer

2.2 Numerical Model of an HI-BB User

The most important part of the HI-BB, which is responsible for transmitting sound waves directly to the inner ear, is fixed by screws to the temporal bone. The investigated user numerical model was limited to a phantom of the head, in which all the components of the implant were inserted. The HI-BB user head was modelled as a multi-layered ellipsoid with external dimensions corresponding to the head of a 50th percentile adult male of the Polish population: 24 cm along the long axis of the head

(HL); 16 cm along the head transverse (side to side) axis, and 19 cm along the sagittal axis of the head (HS) [10]. This model contains four layers of dimensions and dielectric parameters corresponding to the skin, subcutaneous layer of fat tissue, bone tissue (skull) and an internal part corresponding to the brain tissue (the average value of the white and grey matter) [9, 11]. The thicknesses of particular tissue layers: 4 mm for skin, 2 mm for fat and 9 mm for skull bone, referred to typical values in people being implanted with hearing prostheses reported in the literature. The values of the tissues' dielectric parameters (conductivity and relative permittivity) are varied by EMF frequency. Table 2 shows tissue parameters adopted in reported studies regarding exposure to EMF at 100 Hz and 13.56 MHz frequency [12].

Table 2. Dielectric parameters (conductivity and relative permittivity at 100 Hz and 13.56 MHz frequency) of tissues in head model of HI-BB user [12]

Tissue	Thickness	Conductivity, S/m		Relative permittivity	
		100 Hz	13.56 MHz	100 Hz	13.56 MHz
Brain	Internal part	0.07	0.25	2.8×10^6	208
Skin	4 mm	0.0004	0.38	4.5×10^4	177
Fat	2 mm	0.02	0.03	4.6×10^5	12
Bone	9 mm	0.02	0.13	5.9×10^3	59

2.3 Models of EMF Sources

The analysis of the EMF exposure effects in the HI-BB user concerned the low frequency (100 Hz) EMF emitted by models of magnetotherapy applicators. Such applicators are used in magnetic field-based therapeutic procedures (physiotherapy). The source of the magnetic field are usually coaxial coils (a volume set of coils) with various diameters, supplied by a regulated current ensuring the emission of the magnetic field at a magnetic flux density of 0–20 mT. The most typical cylindrical magnetotherapy applicators with diameters of 25 cm and 60 cm and a width of 20 cm were modelled [8, 9]. It was shown previously that results of modelling at frequency 100 Hz sufficiently represents the studied problem over a frequency range up to 40 kHz [13].

The analysis of the EMF exposure effects from radiofrequency EMF (13.56 MHz) in the HI-BB concerned the model of high frequency radio identification (RFID HF) reader. RFID HF readers are widely used (among other things) to protect tagged objects against its unauthorized movement (e.g. in shops or libraries), as access control to buildings or rooms, contactless payment cards, public transport cards, and so on. The source of EMF are planar loop (coil) antennas of various shape and dimensions. The main parameter of the RFID HF reader is reading range (up to 1.5 m when a device is compliant with ISO/IEC 15693-1:2018) – i.e. the distance from which it is possible to read the tag. This study concerned RFID HF desktop readers typically used in shops and libraries with a rectangular shape and dimensions 25 × 35 cm.

Both EMF sources have similar dimensions. Numerical models of EMF sources were validated by measurements of magnetic field distribution near real magnetotherapy applicators and RFID HF readers carried out using Narda (Narda Safety Test Solutions, Germany) meters (ELT 400 and NBM-550) with probes dedicated for the investigated frequencies.

2.4 Characteristic of Exposure to EMF

A physiotherapist operating a magnetotherapy device is responsible for laying the patient inside an applicator fixed to the bed or the table, and setting the parameters of the EMF exposure (shape of variability in time, frequency, level and exposure duration – depending on the appropriate treatment), and starts its generation using the control panel of the generator. Sometimes it is necessary to approach the patient and to monitor the treatment procedure. Doing so, the physiotherapist may lay down by the active applicator. Similarly, the patient may move the upper part of the body into the applicator, e.g. looking for treatment staff. Both movements may expose their head to a strong, heterogeneous EMF near the applicator cover.

The popular radiofrequency identification of an object with the attached RFID tag is made by using planar loop (coil) antennas of various dimensions, typically supplied by a current of 13.56 MHz frequency (RFID HF readers). Readers may be fixed to furniture, walls or a special frame (gate), or may be manually movable. The head of employees (e.g. at a shop or library) or their customers may be exposed to strong heterogeneous EMF near RFID HF reader at various locations.

All the investigated exposure scenarios refer to the worst cases of exposure. In all exposure scenarios, the surface of the HI-BB user's head model and source were spaced at a distance of 2 cm, to be exposed to the highest and the most heterogeneous EMF. In the AR-HL (Fig. 2a), and AR-HS exposure scenarios, the head model was located to the side of the source, along its radius (AR line), while in the AE-HL and AE-HS exposure scenarios it was located in front of the source, at its edge, along the axis (AE line). Furthermore, in the AR-HL and AE-HL exposure scenarios (Fig. 2a)

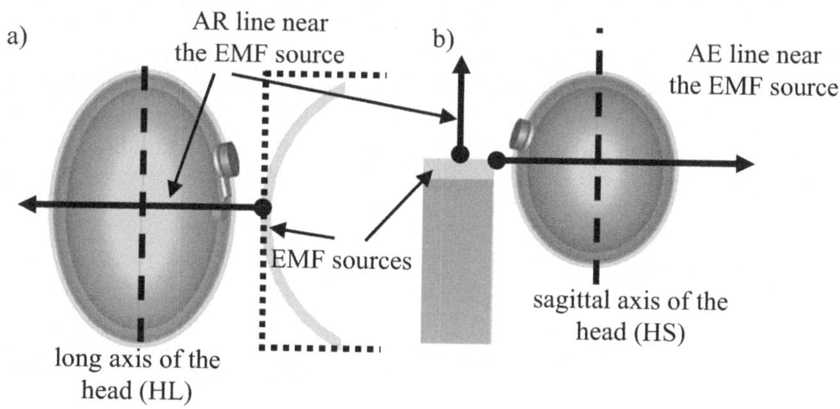

Fig. 2. The scheme of investigated exposure scenarios (a) AR-HL and (b) AE-HS

the long axis (from top to bottom) of the head model (HL) was perpendicular to the AR or AE lines respectively, while in AR-HS and AE-HS exposure scenarios (Fig. 2b), the sagittal axis (from face to back) of the head model (HS) was perpendicular to them.

2.5 Numerical Simulations

The complexity of the virtual models was 10-18 million voxels, with a resolution of approx. (0.25–0.50) mm in the vicinity of the HI-BB model. Specialized software to analyze the EMF problems – CST STUDIO SUITE 2016 – was used.

The standard uncertainty of the results of the numerical simulations was estimated at approximately ±30%, covering: discretization of the model, field source model, the calculation method used, and the values of dielectric parameters of particular elements of the numerical model (compliant with EN 50413:2008).

2.6 Criteria for Assessing Electromagnetic Hazards

In the investigated exposure scenarios, the level of EMF exposure was evaluated against reference levels (RL) limits provided for the general public (GP) or occupational exposure (OE) [5, 6]. The direct exposure results (characterized by the induced electric field strength, Ein, in volts per metre (V/m) or specific energy absorption rate, SAR, in watts per kilogram (W/kg)) in the head of the HI-BB user were analyzed in the context of basic restrictions (BR). The analysis omitted BR limits related to Ein in the central nervous system in the head, because HI-BB elements are not in contact with the brain tissues.

3 Results

Figure 3 compares the spatial distribution of the magnetic field evaluated by numerical simulations along AR and AE lines (Fig. 2), as the normalized values in relation to point values evaluated in the centre of each loop antenna (applicators or RFID HF reader).

The magnetic field level (evaluated by point calculated values) at a distance of 2–20 cm from the cover of the applicator (where the head locations were defined in the analyzed exposure scenario) decreases more slowly with distance from an applicator of 60 cm-diameter (5-times) than compared to the field distribution near an applicator of 25 cm-diameter (7-times) and near an RFID HF reader of 25 × 35 cm dimensions (36-times) (Fig. 3).

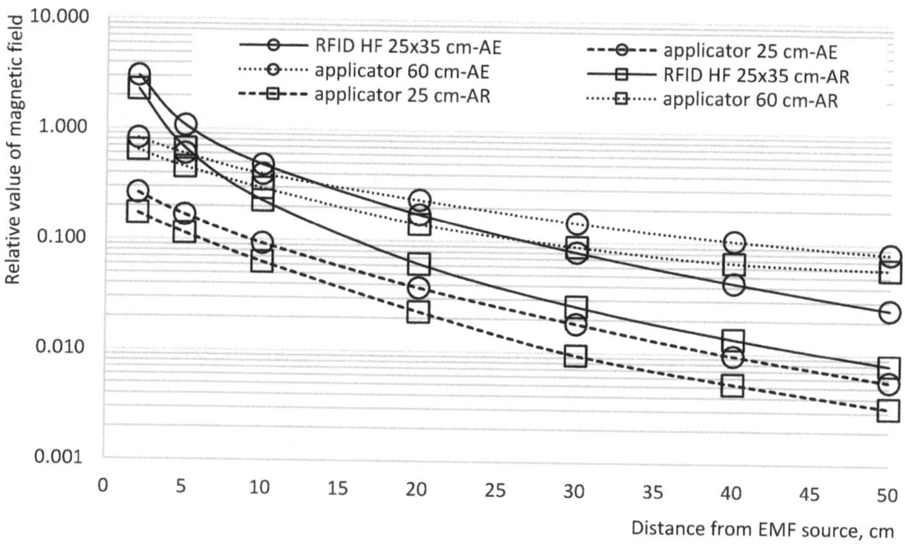

Fig. 3. Magnetic field spatial distribution (results of simulations), along the axial direction at the edge (AE) or along radius (AR) of an RFID HF reader with dimensions of 25x35 cm and magnetotherapy applicators with a diameter of 25 cm or 60 cm (reference value – magnetic field value in the centre of particular sources)

According to Appendix A of EN 50527-1:2010, active implantable medical devices manufactured for use in the European Union should be resistant to electromagnetic interferences up to the exposure level equal to the general public limits provided by the Council of the European Union Recommendation, 1999/519/EC [6, 14]. This limit for a magnetic field of 100 Hz frequency is equal to 40 A/m, i.e. 4-times lower than ICNIRP's general public RL, while for a magnetic field of 13.56 MHz both limits are equal to 0.073 A/m. The range of distances from the investigated EMF sources where exposure exceeds 1999/519/EC and ICNIRP's general public RL limit depends on the intensity of the magnetic field set at the magnetotherapy device control panel, or the reading range of RFID HF readers. When the maximum settings in a typical magnetotherapy device are applied (5 mT in a 60 cm-applicator, or 20 mT in a 25 cm-applicator) the limit of 40 A/m is exceeded at distances of approximately 80 cm and 60 cm, respectively. At maximum settings, the ICNIRP's general public RLs are exceeded at distances of approximately 40 cm and 30 cm respectively, while occupational RLs are exceeded only near the smaller applicator at a distance of approximately 10 cm.

In the case of RFID HF readers, the limit of 0.073 A/m is exceeded at distances of approximately 40 cm, 65 cm and 135 cm for 30, 50, 100 cm reading ranges respectively. These distances are approximately 35% higher than the reading ranges. Occupational RLs are exceeded at distances equal to the reading ranges of readers. Additionally, according to EN 60601-1-2:2015 on EMC, in EMF at a frequency from 80 MHz to 2.7 GHz, medical equipment (including medical implants) should be manufactured to be resistant to electromagnetic interference from an electric field up to

3 or 10 V/m. It can be assumed that for lower frequencies the relevant limit may be similar. The EMC requirements for the implant manufacturer are more restrictive than the requirements from EN 50527-1:2010 [2].

Tables 3 and 4 show the calculation results of the induced electric field strength (Ein) and localized specific energy absorption rate (SAR(1 g)) averaged over any 1 g of tissues inside the HI-BB user's head model (1 g averaging of tissue was chosen because of the small dimensions of analyzed parts of the implant), normalized to the values obtained in the same head model but without the implant at the same exposure (reference model - RM).

Table 3. The strength of electric field (Ein) induced inside the head model of HI-BB user by EMF near magnetotherapy applicators (100 Hz)

Tissue	The ratio: Ein (HI-BB)/Ein (RM), value ± uncertainty of simulations			
	Exposure scenarios near the applicator of 25 cm (60 cm) diameter			
	AR-HL	AR-HS	AE-HL	AE-HS
Skin	3.4 ± 1.4 (3.2 ± 1.3)	2.6 ± 1.1 (2.7 ± 1.1)	2.6 ± 1.1 (2.0 ± 0.8)	2.0 ± 0.8 (1.8 ± 0.8)
Fat	1.9 ± 0.8 (2.4 ± 1.0)	2.0 ± 0.8 (2.5 ± 1.0)	1.3 ± 0.5 (1.2 ± 0.5)	1.3 ± 0.5 (1.2 ± 0.5)
Bone	1.6 ± 0.7 (2.3 ± 1.0)	2.7 ± 1.1 (2.8 ± 1.2)	2.0 ± 0.8 (1.1 ± 0.5)	1.9 ± 0.8 (1.3 ± 0.5)

Ein (HI-BB) – strength of electric field induced in the head model of HI-BB user
Ein (RM) – strength of electric field induced in the head model without HI-BB (reference model)
AR-HL – long axis of the head model (HL) perpendicular to AR line
AR-HS – sagittal axis of the head model (HS) perpendicular to AR line
AE-HL – long axis of the head model (HL) perpendicular to AE line
AR-HS – sagittal axis of the head model (HS) perpendicular to AE line

Table 4. Specific energy absorption rate (SAR) inside the head model of HI-BB user by EMF near RFID HF readers (13.56 MHz)

Tissue	The ratio: SAR (1 g) (HI-BB)/SAR (1 g) (RM), value±uncertainty of simulations			
	Exposure scenarios near RFID HF reader of 25 × 35 cm dimensions			
	AR-HL	AR-HS	AE-HL	AE-HS
Skin	2.8 ± 1.2	2.0 ± 0.8	1.5 ± 0.6	1.4 ± 0.6
Fat	2.7 ± 1.1	1.9 ± 0.8	1.7 ± 0.7	1.5 ± 0.6
Bone	3.0 ± 1.3	2.3 ± 1.0	4.5 ± 1.9	1.8 ± 0.8

SAR (1 g) (HI-BB) – localized specific energy absorption rate averaged in any 1 g of contiguous tissue in the head model of HI-BB user
SAR (1 g) (RM) – localized specific energy absorption rate averaged in any 1 g of contiguous tissue in the head model without HI-BB (reference model)
Other abbreviations as in Table 3

HI-BB elements are electrically insulated from head tissues with a silicon coat – except the BC-FMT in which the conductive metal structure is in contact with the tissues – skin, fat and bone. The values of Ein and SAR (1 g) in the brain tissue are not modified by the presence of the HI-BB because its elements are not in contact with it. In an evaluation of the hazards associated with the effects of interaction between EMF and an HI-BB user, the results of numerical simulations of Ein in skin and subcutaneous fat tissue may be evaluated against BR limits, because they contain receptors and nerves of the peripheral nervous system (PNS).

The EMF influence on an HI-BB user may be a factor increasing a health hazard in the workplace in comparison to the sensitivity to EMF exposure of other workers [3]. The conclusions from conducted studies are useful for implementing European directive 2013/35/EU concerning electromagnetic hazards in the workplace, which indicates the need to individually consider these hazards with respect to EMF exposure of "workers at particular risk" – in an evaluation of hazards, planning protection measures, workers training and health surveillance [7].

The Ein and SAR values depend on the geometric dimensions of the human body. The presented results concern a head with the dimensions of a 50th percentile male. The results of other studies indicate 10–60% differences in Ein values in the head for models with different geometrical dimensions [9, 15]. This requires further study.

The unequivocal assessment of whether the use of an implant HI-BB may be a contraindication to the employment of a particular employee requires individual studies using more detailed models of implants and users (anatomical models), as well as studies involving volunteers. The conclusions of these studies may be helpful in the analysis of the effects of exposure to EMF from other sources that cause exposure of a similar profile to that analyzed in this study.

4 Conclusions

It was found that the EMF should not cause a malfunction in an HI-BB at a distance longer than 80 cm from the magnetotherapy applicators, and approximately 35% longer than RFID HF readers reading range.

It was demonstrated that the use of HI-BB significantly increase the level of direct exposure effects caused by low frequency and radiofrequency EMF (Ein up to 3.4 times and SAR (1 g) up to 4.5 times, respectively). The level of such hazards is a function of the dimensions of field source and location of the head against the source (i.e. the field polarization).

It was shown that low and radiofrequency EMF is an environmental direct hazard for the hearing implant users, but the analysis of the individual sensitivity of particular users requires further studies involving more detailed models of implants and users' bodies, as well as studies covering the indirect influence on the function of implants, recognized as electromagnetic compatibility (EMC), involving volunteers or phantoms in laboratory tests.

Acknowledgments. This paper has been based on the results of a research carried out within the scope of the National Programme "Improvement of safety and working conditions" (tasks II. P.14, II.N.19) partly supported in Poland – within the scope of research and development – by the Ministry of Science and Higher Education/National Centre for Research and Development. Main coordinator: CIOP-PIB.

References

1. Med-EL. http://www.medel.com
2. Zradziński, P., Karpowicz, J., Gryz, K., Leszko, W.: Evaluation of the safety of users of active implantable medical devices (AIMD) in the working environment in terms of exposure to electromagnetic fields – practical approach to the requirements of European Directive 2013/35/EU. Int. J. Occup. Med. Environ. Health **31**(6), 795–808 (2018)
3. Hansson Mild, K., Hocking, B.: Guidance note: risk management of workers with medical electronic devices and metallic implants in electromagnetic fields. Int. J. Occup. Saf. Ergon. (JOSE) **14**(2), 217–222 (2008)
4. International Commission on Non-Ionizing Radiation Protection: Guidelines for Limiting Exposure to Time-Varying Electric and Magnetic Fields (1 Hz–100 kHz). Health Phys. **99** (6), 818–836 (2010)
5. International Commission on Non-Ionizing Radiation Protection: Guidelines for Limiting Exposure to Time-Varying Electric, Magnetic and Electromagnetic Fields (up to 300 GHz). Health Phys. **74**(4), 494–522 (1998)
6. Council of the European Union Recommendation of 12 July 1999 on the limitation of exposure of the general public to electromagnetic fields (0 Hz to 300 GHz), 1999/519/EC, Official Journal of the European Communities, L 199/59
7. Directive 2013/35/EC of the European Parliament and of the Council of 26 June 2013 on the minimum health and safety requirements regarding the exposure of workers to the risks arising from physical agents (electromagnetic fields) (20th individual Directive within the meaning of Article 16(1) of Directive 89/391/EEC) and repealing Directive 2004/40/EC. Off J Eur Union. 2013;L 179/1
8. Zradziński, P., Karpowicz, J., Gryz, K., Leszko, W.: Ocena zagrożeń wynikających z oddziaływania pola magnetycznego emitowanego przez aplikator magnetoterapeutyczny dla użytkowników protez słuchu wykorzystujących przewodnictwo kostne. Med. Pr. **68**(4), 469–477 (2017)
9. Zradziński, P.: Evaluation of the inter-person variability of hazards to the users of BAHA hearing implants caused by exposure to a low frequency magnetic field. Int. J. Radiat. Biol. **94**(10), 918–925 (2018)
10. Gedliczka, A.: Atlas of human body measures – Data sheets for ergonomic and evaluation. Central Institute for Labour Protection, Warszawa (2001)
11. Ruan, J., Prasad, P.: The effects of skull thickness variations on human head dynamic impact responses. Stapp. Car Crash J. **45**(12), 395–414 (2001)
12. Gabriel, S., Lau, R.W., Gabriel, C.: The dielectric properties of biological tissues: II. measurements in the frequency range 10 Hz to 20 GHz. Phys. Med. Biol. **41**(11), 2251–2269 (1996)

13. Zradziński, P., Karpowicz, J., Gryz, K.: In Silico modelling of influence from low or intermediate frequency magnetic fields on users of wearable insulin pumps. Int. J. Radiat. Biol. **94**(10), 926–933 (2018)
14. European Committee for Electrotechnical Standardization (CENELEC). Procedure for the assessment of the exposure to electromagnetic fields of workers bearing active implantable medical devices—Part 1: General. (Standard No. EN 50527-1:2016), CENELEC, Brussels (2016)
15. Dimbylow, P.J.: Development of the female voxel phantom, NAOMI, and its application to calculations of induced current densities and electric fields from applied low frequency magnetic and electric fields. Phys. Med. Biol. **50**(6), 1047–1070 (2005)

Measurement of Dynamic Properties of Animal Liver to Identify the Material Model

Monika Palmerska[1] and Tomasz Klekiel[2(✉)]

[1] Biomedical Engineering, University of Zielona Gora, Licealna 9 Street, 65-547 Zielona Gora, Poland

[2] Faculty of Mechanical Engineering, University of Zielona Gora, Szafrana 4 Street, 65-516 Zielona Gora, Poland

t.klekiel@ibem.uz.zgora.pl

http://www.zib.uz.zgora.pl

Abstract. The paper is focused on the problem of liver characterization. The dynamic properties include very important information about behavior of tissue under impact loading. The work presents the methodology for testing the dynamic properties of the liver and parameter identification of Mooney-Rivlin hyperelastic material model. A measuring tip was hammered into the perfused liver with different deformation rates. The displacements of the liver surface were recorded using the ARAMIS system and the results were analyzed in GOM Correlate software. The experimental test was modeled and simulated using a finite element method (FEM). The obtained experimental data was used to validation a numerical model. Both experiments shows similar results for the same load and boundary conditions, what suggested that the numerical model was correct assumed.

Keywords: Biomechanics · Liver tissues · Finite element method · Mooney-Rivlin · Material model · Model identification

1 Introduction

Blunt abdominal injury in result of impact loading is one of an important factors causing the dead [1]. Studies show that due to car accidents, the liver, spleen and kidneys suffer the most serious injuries right after the brain. The liver is partially surrounded by the chest skeleton, but despite its relatively safety and protected position, it often suffers injuries as a result of a blunt injury. Under impact loading the liver is subjected to blast wave, which can cause the its damage [2]. The high deformation rate contribute to the disintegration of the liver parenchyma, vascular damage and intraperitoneal hemorrhage [3–5].

The surgical techniques allow to reduce the mortality associated with a liver damage, but injury prevention remains the challenge. The automotive industry require the knowledge about passenger safety, and in consequences, the information about influence of impact generated by the vehicles acting on the human body is needed to understand the critical loads and to predict the range of

J. Korbicz et al. (Eds.): PCBEE 2019, AISC 1033, pp. 256–264, 2020.
https://doi.org/10.1007/978-3-030-29885-2_23

injury. Usually the special dummies are used to assess the risk of injury, but they are not adopted to represent injuries for individual abdominal organs [3,4]. Understanding behavior of the liver under deformities for different strain rates is useful to develop a vehicle safety measures. Possibilities of a virtual simulators, that will predict abdominal injuries during accidents are developed, however it is depend on the predictive models used to representing a mechanical properties of tissues. Finite element models (FEM) are therefore an integral tool to modeling the relations between the load, deformation and tissue damages [4,6–9]. The liver is highly vascularized and one of its main functions is a blood reservoir. It usually contains about 10% of the total body's blood volume, an average of about 450 ml, but it can expand to stay up to 1 l. The pressure generated by the blood is of great importance for liver mechanical response tests. Perfused organs exhibit similar properties to in vivo conditions [10]. Due to the limited availability of human organs, often the tests are carried out on an animal organs.

In this investigation the pork liver has been acting by the measuring tip with diameter equal 4 mm. The liver was perfused by the fluid with temperature and pressure similar to in vivo conditions. The port liver is recognized as similar to human, but some differences should be highlighted. The pig's liver is divided into 5 lobes, it is thinner than human and has more pronounced divisions. Therefore, it is expected that the pig's liver will be stiffer than the human one [10,11].

2 Materials and Methods

The liver was loaded by the measuring tip in one axis. The measurement station (Fig. 1) was built based on the testing machine Zwick and consists the beaker with force sensor (1) on which the liver sample was laid, measuring tip (2). The reservoirs (3) and (4) are used to set the suitable pressures in two fluid circuits, respectively: h1 = 97 5 mmHg to the hepatic artery and h2 = 8 2 mmHg to the portal vein.

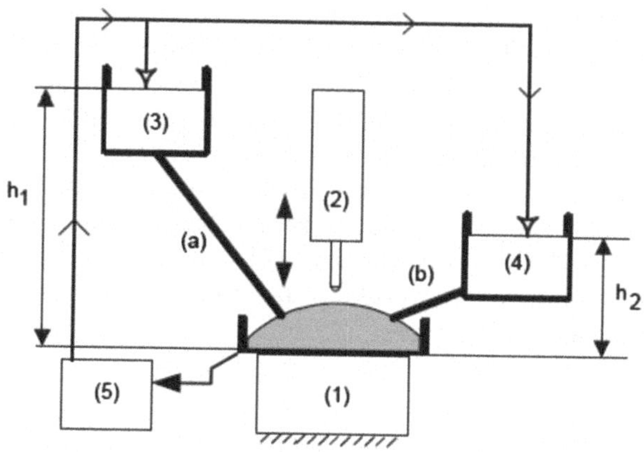

Fig. 1. Measurement station: (1) - force sensor, (2) - piston with measuring tip, (3) reservoir for the hepatic artery circuit, (4) - reservoir for the portal vein circuit, (5) - pump with the heater.

The pump with heater (5) stabilized the fluid temperature and pumped the fluid to the containers (3) and (4). From container (3) the fluid gravitationally was flowed to the hepatic artery (a) and from (4) the fluid gravitationally was flowed to the portal vein.

The fluid was prepared as simulate body fluid (SBF). The force sensor measure the reaction force generated after acting the measurement tip on the liver. The sample liver was placed on the beaker plate attached to the force sensor of the testing machine. The measuring tip was mounted in upper jaw of testing machine. During tests the liver was perfused with constant fluid pressure at the temperature equal 300 K. The fluid getting out from the liver returned to the pump container and was heated again. The connection with the liver was made by inserting wires into appropriate vessels. The pressure inside the hepatic artery was equivalent to the hydrostatic pressure for height about 1.2 m and for the portal vein the hydrostatic pressure was equivalent to about 12 cm.

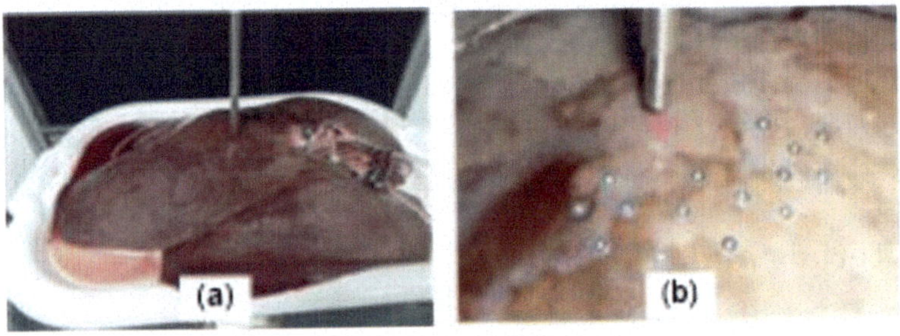

Fig. 2. The liver on the measurement station (a), the measurement trip and marker localizations (b).

The deformations were measured based on the markers placed on the liver surface, recognized by the Aramis system (http://www.gom.com). The Aramis system was calibrated and placed in front of the testing machine with cameras directed to the surface with markers (Fig. 2b). For correct measurement process, the marker positions are important. They must be visible in the field of view of both cameras, before and after deformation. In result, it is ensured that the object is located in the measuring area of the system.

Before test, the perfusion process was made to reach a physiological temperature. It lasted about 20 min. In result the liver shape was changed and the tissue height was increased. Throughout the test, the surface of the liver was flowed to maintain suitable humidity. The results of reaction force measurements indicate that the speed increases.

3 Results

Tests was carried out by the testing machine, in which the measuring tip recess was repeated three times for speeds of 5 mm/min, 20 mm/min, 50 mm/min, 100 mm/min and 500 mm/min, respectively. The test was repeated in two places distant about 3 cm from each other. The results are similar to previous studies which indicate that for higher deformation rate the organ is stiffer [10, 11].

The Fig. 3 presents the relationship between force reaction and deformation rate. For slow deformation the reaction force is in range from 1 to 1.2 N. For faster deformations the reaction force is increased asymptotically to 3.68 N.

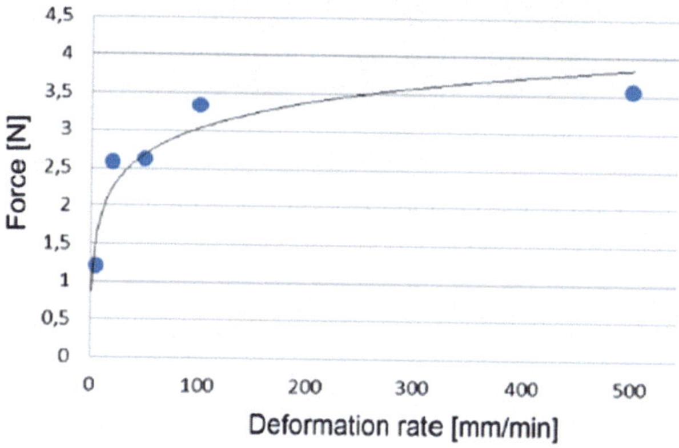

Fig. 3. The force reaction in relation to deformation rate.

The GOM Correlate software was used to the deformation measurements. Three selected markers were used in each measurement series. A special coordinate system was established, which passed through the central point, being the reference point of the displacements of the remaining markers along the Z axis placed perpendicular to the liver surface defined by the markers shown on Fig. 5.

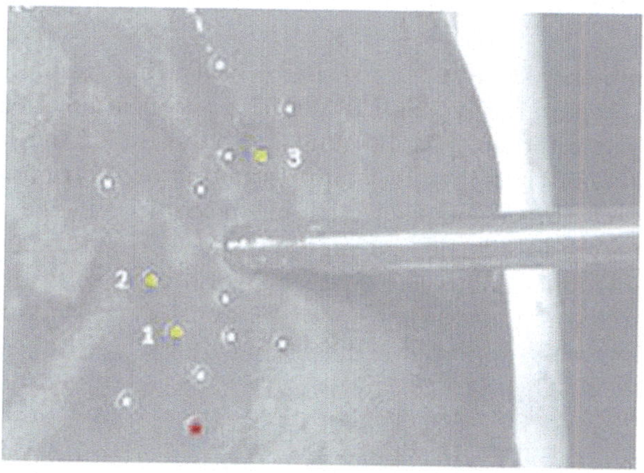

Fig. 4. Deflections for different deformation rates.

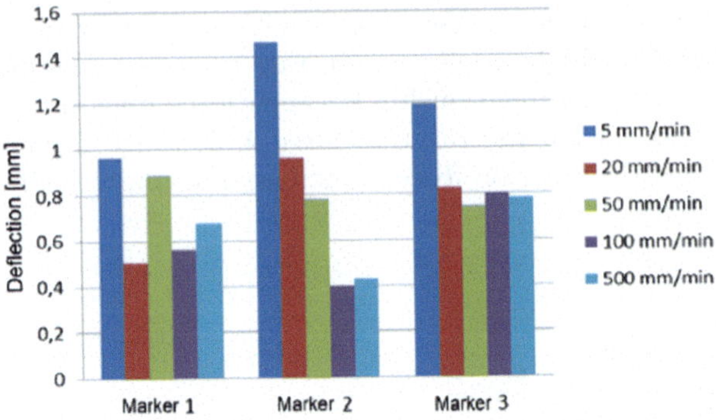

Fig. 5. Selected matrices on the liver surface

The deformation measured by the markers are not entirely reliable because the fluid flowed on the liver surface are distorted in cameras. The markers are shifted slightly but the Aramis system not recognized any errors. The Fig. 4 presents the deflection of the surface markers for different deformation rates. For the rate of 5 mm/min the deformations are higher than for each other rates. The displacements are differ significantly from the rest which are similar to each other. These results indicate on the creep process inside the liver. For higher deformation rates, the displacements around the acting point decreases what suggesting a higher deformation resistance of the tissue.

The experimental results was used to material model identification. The numerical model of the liver was prepared as the solid body based on the sphere shape with the radius calculated based on the marker positions getting before deformation experiment.

4 Model Identifiaction

To perform a numerical analyzes, a real-size measuring tip and the liver solid geometry were prepared using SOLIDWORKS application. For the liver the surface radius was appointed as 150 mm and a height of the liver was equal 75 mm. The radius of the sphere was approximated as the radius of the circles designated for each three points selected and recognized in the Aramis system markers. The analyze was carried out in the GOM Correlate software. The solid geometry was introduced into the ANSYS Workbench software. The contact between liver surface and measuring tip was established as frictionless. The lower section of the liver was supported as fixed (Fig. 6).

Fig. 6. FEM model of the liver and measuring tip.

Table 1. Material properties assumed for liver.

No.	Material type	Parameters	Ref.
1	Linear isotropic model	Density = 1170 kg/m^3	[4]
		Young modulus = 160 MPa	
		Poisson ratio 0.49	
2	Mooney-Rivlin material model	Density = 1170 kg/m^3	[4]
		C10 = 22,2 kPa	
		C01 = 21,4 kPa	
		C20 = 21,245 kPa	
		C11 = 26,01 kPa	
		C02 = 26,72 kPa	
		D1 = 1.0 Pa-1	
		Shear modulus = 600 Pa	
3	Mooney-Rivlin	C10 = 11,1 kPa	-
		C01 = 10,5 kPa	
		C20 = 10,6 kPa	
		C11 = 13 kPa	
		C02 = 13,3 kPa	
4	Money-Rivlin	C10 = 44,4 kPa	-
		C01 = 42 kPa	
		C20 = 42,5 kPa	
		C11 = 52 kPa	
		C02 = 52 kPa	

The load was defined as the tip displacements along its axis by 8 mm. Analyzes were carried out for two basic material models: linear and hiperelastic Mooney-Rivlin model. The material properties are shown in Tables 1 and 2.

Table 2. Markers position.

Marker number	Displacement [mm]	Distance from the contact point [mm]
1	0.5–0.95	17.5
2	0.4–1.42	16.5
3	0.75–1.2	18.5

Table 3. Results for experiment and numerical model.

Number of material Property	Displacement range in the numerical model [mm]	Displacement range in the experiment [mm]
2	0.5–1.7	0.1–1.4
3	1.25–1.6	0.1–1.4
4	2.3–3.3	0.1–1.4

The loads were simulated for different materials defined in Table 1. The changes of the Mooney-Rivlin model parameters were made based on the assumption that the stiffness of the liver is depends on the state of the tissues. Frozen or after long time from preparation, the liver tissues are changed and the mechanical properties can be different. The changes of the model parameters in positions 4 and 5 of Table 1 is used to determining a material sensitivity of the. The linear model using to an analysis suggests that the displacement distribution is concentrated to 15 mm around the load point (Fig. 7). Marker distances from the center point of the measuring tip is placed from 15 to 25 mm. Marker movements range from approx. 0.1 mm to approx.1.4 mm. Displacements in suitable distances in the numerical model for all series are mixed in the range of about 0.64–1.59 mm. The simulation results obtained for linear isotropic material model is similar to the experimental ones.

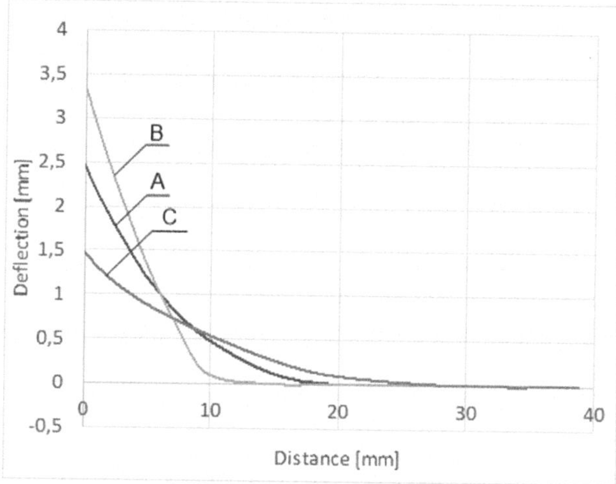

Fig. 7. The deformations for three cases of Mooney-Rivlin models: – basic model, B - flexible liver material, C - stiffer liver material.

The results for different types of Mooney-Rivlin model was simulated for the same boundary conditions. The distribution of the displacements for the basic model for which the parameters were chosen according to Umale et al. [4]. The deformations along path from the contact center for selected material parameters of Mooney-Rivlin model are shown in Fig. 7. The Table 3 presents the range of displacements for the distances similar to experiments.

The results obtained for the basic model (Table 1, position 2) reach the upper limit of the experimental values. The twice stiffer liver material (Table 1, position 4) is beyond of range getting from the experiment. The more flexible liver material model (Table 1, position 3) allowed to obtain the values closest to experimental conditions. The displacements obtained from simulations for both models are close to the maximum experimental values. The displacement distribution obtained for hyperelastic material parameters better defined real conditions. The stiffness reduction allowed to obtain the values closest to experimental results. The numerical model with Mooney-Rivlin material model shown that the displacements include wider area and passes more smoothly than the linear isotropic material model.

5 Conclusion

The paper presents a method for numerical model identification based on the measuring of liver dynamic properties. The test results shown that the mechanical response of the organ is depend on the deformation rate. If the deformation rate increased, the values of reaction forces increased too. At the same time the liver tissues on the surface are less deformed. For the rate 5 mm/min, the marker displacements were significantly larger than for higher speeds. It indicates the creep phenomena inside material. The experimental results were used to matching the numerical model in which the liver material is defined as Mooney-Rivlin model. Comparison with experimental results indicate that the selected material generate similar reactions. The presented method allows to the preparation of a numerical model, which giving correct mechanical response of the tissues.

References

1. Ramasamy, A., Hill, A.M., Clasper, J.C.: Improvised explosive devices: pathophysiology, injury profiles and current medical management. J. R. Army Med. Corps **155**, 265–272 (2009)
2. Klekiel, T.: Biomechanical analysis of lower limb of soldiers in vehicle under high dynamic load from blast event. Ser. Biomech. **29**(2–3), 14–30 (2015)
3. Pervin, F., Chen, W.W., Weerasooriya, T.: Dynamic compressive response of bovine liver tissues. J. Mech. Behav. Biomed. Mater. **4**(1), 76–84 (2011)
4. Umale, S., et al.: Modeling and validation of the human liver and kidney models. The International Research Council on Biomechanics of Injury Conferance (2013)
5. Uravić, M., et al.: Penetrating liver war injury: a report on 172 cases. Mil. Med. **168**(5), 419–421 (2003)
6. Augenstein, J., Bowen, J., Perdeck, E., Singer, M., Stratton, J., Horton, T., Rao, A., Digges, K., Malliaris, A., Steps, J.: Injury Patterns in Near–Side Collisions. SAE Technical Paper Series 2000–01–0634, pp. 1–9 (2000)
7. Elhagediab, A.M., Rouhana, S.W.: Patterns of abdominal injury in frontal automotive crashes. In: 16th ESV Conference, Windsor, Canada (1998)
8. Kemper, A.: Multi-scale biomechanical characterization of human liver and spleen. In: Proceedings of the 22nd Enhanced Safety of Vehicles Conference, vol. 11 (2011)
9. Sparks, J.L., Dupaix, R.B.: Constitutive modeling of rate dependent stress-strain behavior of human liver in blunt impact loading. Ann. Biomed. Eng. **36**(11), 1883 (2008)
10. Kerdok, A.E.: Characterizing the nonlinear mechanical response of liver to surgical manipulation. Havard University, Diss (2006)
11. Snedeker, J.G., Niederer, P., Schmidlin, F.R.: Strain-rate dependent material properties of the porcine and human kidney capsule. J. Biomech. **38**, 1101–1021 (2005)

A Novel Real-Time Computer Model of the Coronary System Including a Physiological Feedback Between the Left Ventricular Muscles and Coronary Perfusion

Krzysztof Zieliński(✉), Krystyna Górczyńska, Maciej Kozarski, Piotr Okrzeja, Marek Darowski, and Gianfranco Ferrari

Nalecz Institute of Biocybernetics and Biomedical Engineering, Polish Academy of Sciences, Ks. Trojdena 4, 02109 Warsaw, Poland
kzielinski@ibib.waw.pl

Abstract. Coronary dysfunction is one of the most common coronary diseases in people living in highly developed regions of the world, where heart infarction and sudden cardiac death often result from stenosis. Besides symptomatic treatment of ischaemic heart diseases, invasive methods are also used. An effective form of aid in the case of ischaemic heart diseases resulting from coronary stenosis and/or heart infarction is mechanical heart assistance provided by an intra-aortic balloon pump (*IABP*). Therefore, the main goal of this paper was to elaborate a sophisticated computer model of the coronary circulation (*CCM*) to enable coronary system pathological states modeling such as coronary stenosis. This coronary model must be executed in the real-time and be suitable to be used in the hybrid (hydraulic-numerical) cardiovascular simulator. *CCM* consists of 4 main coronary arteries modeled by series of resistance-inertance-compliance compartments: left main artery (*LCA*), right main artery including intraventricular hind branch, left main circumflex artery, left anterior descending artery. The peripheral coronary circulation was modeled by an intramyocardial pump model. An innovative feature introduced in *CCM* is a physiological model between the coronary circulation and the end systolic pressure volume relationship function of the ventricles. *CCM* was integrated with the whole, closed-loop model of the cardiovascular system. Exemplary simulation results at rest and myocardial hyperemia conditions as well as for *LCA* stenosis are consistent with the physiology and literature. As a part of the hybrid cardiovascular simulator *CCM* can be exploited in simulation scenarios of *IABP* assistance by real, physical balloon.

Keywords: Coronary circulation model · IABP assistance · Modeling and simulation

J. Korbicz et al. (Eds.): PCBEE 2019, AISC 1033, pp. 265–276, 2020.
https://doi.org/10.1007/978-3-030-29885-2_24

1 Introduction

The role of the coronary system is the distribution of oxidized blood and nutritive products to the heart muscle, and removing metabolic waste from the heart. A man at rest typically has a coronary flow of 220–250 ml/min, which makes up about 5% of cardiac output (*CO*).

Coronary vessels undergo different kinds of pathological states, one of them is coronary stenosis i.e. established ischaemic heart disease, caused by a substantial reduction of the coronary arteries cross sectional area. Coronary stenosis causes a temporary or permanent unbalance between the heart muscle's oxygen demand and supply. Coronary dysfunction is one of the most common coronary diseases in people living in highly developed regions of the world, where heart infarction and sudden cardiac death often result from stenosis.

Besides symptomatic treatment of ischaemic heart diseases, invasive methods are also used. One such method, percutaneous coronary intervention, is used to unblock occluded coronary blood vessels. Percutaneous coronary intervention is completed with stents as well as a heart bypass. An effective form of aid in the case of ischaemic heart diseases resulting from coronary stenosis and/or heart infarction is mechanical heart assistance provided by an intra-aortic balloon pump (*IABP*) which functions on the principle of counterpulsation. This type of therapy aims to improve heart muscle perfusion through diastolic augmentation, resulting in substantial coronary flow increase. Independently of the above mentioned ways of treatment, the ischaemic heart disease and other cardiovascular heart diseases connected with oxygen balance caused the new studies on the coronary system modeling to be undertaken.

The main goal of this paper was to elaborate a novel, sophisticated computer model of the coronary circulation to enable coronary system pathological states modeling such as coronary stenosis. The main problem we faced was the choice of an appropriate mathematical model to enable pressure-flow data analysis - it is made conditional on specific terms and at least three requirements:

- Description of a coronary elastic behavior with accuracy that makes the estimated value of a pericardial compliance to be near to a real value.
- The results from different experiments estimated by the model should be consistent and resistant to external noise as well as to random and systemic noise.
- A model should be relatively simple and the results should be obtained as soon as possible – it is very important when applied in hybrid simulator of a total circulatory system working in the real-time.

The last requirement is important due to the fact that the developed coronary system model will be implemented in a hybrid (hydraulic-numerical) simulator of the cardiovascular system [1] developed jointly by the Nalecz Institute of Biocybernetics and Biomedical Engineering of the Polish Academy of Sciences and the Institute of Clinical Physiology of the Italian National Council of Research.

2 Materials and Methods

2.1 The Model of the Coronary System

The developed model of the coronary circulation consists of four main parts: an "anatomic" model of four main coronary arteries, a peripheral coronary systemic circulation model, a model of a coronary vessel with advanced stenosis and a physiological model between the left ventricular muscles and coronary perfusion.

"Anatomic" Model of the Coronary Vessels
It considers four sections of the main coronary vessels (Fig. 1):

- Left main artery (*LCA*).
- Right main artery (*RCA*) including intraventricular hind branch.
- Left main circumflex (*LCX*).
- Left anterior descending (*LAD*).

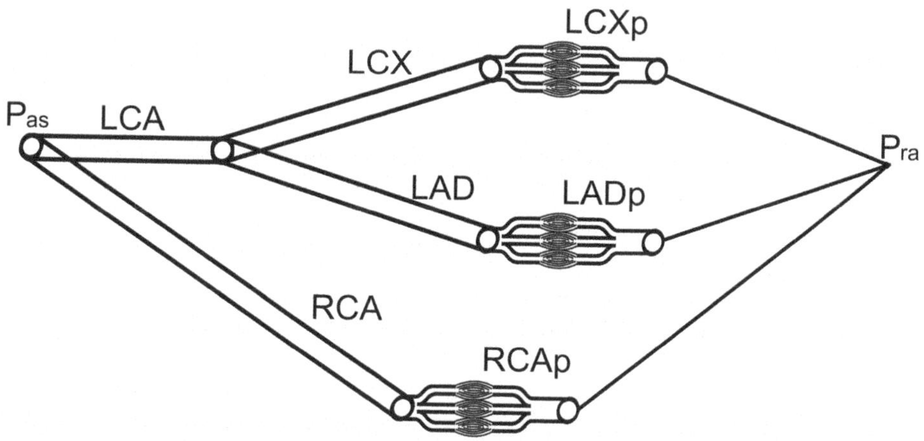

Fig. 1. "Anatomic" model of the coronary system. P_{as} – aortic pressure, P_{ra} – right atrial pressure

Two main sections (*LCA, RCA*) are connected to the aorta – their input value is aortic pressure (P_{as}). Three sections (*LCX, LAD, RCA*) are ended with peripheral modules *LCXp, LADp, RCAp*, respectively, representing arterial distal branches, small and large veins, directly connected to the model of the right atrium.

This model schematic was chosen for the following reasons:

- proper simulation of the anatomy of the human coronary arterial system;
- the four sections of the main coronary vessels chosen to represent coronary circulation are easily visible during coronarography and therefore carefully

investigated. This model can be easily verified and personalized due to the large amounts of clinical data available on the subject;

- this model allows for coronary stenosis modeling as well as the observation and verification of stenotic changes in ventricular hemodynamics and energetics in the real-time;
- the simplicity of this model makes it working in the real-time regime and allows it to include hybrid (hydro-numerical) modelling of the circulatory system.

Model of a Coronary Vessel Segment

A coronary artery is modeled as a concatenation of *n RLC* compartments representing the resistive (*R*), inertial (*L*) and capacitive (*C*) properties of its segment (Fig. 2). The R_n and L_n values are calculated from the Navier-Stokes equations. They can be also directly calculated from the Hagen-Poiseuille law (*R*) and from the Second Newton's law (*L*) [2]. The C_n values are calculated from the Law of Laplace [2]. In order to use these formulas, some assumptions such as blood behaving like a Newtonian fluid and blood flow being laminar were made:

$$R = \frac{128 \cdot \mu}{\pi \cdot D^4} l \tag{1}$$

$$L = \frac{4 \cdot \rho}{\pi \cdot D^2} l \tag{2}$$

$$C = \frac{3 \cdot \pi \cdot D^3}{16 \cdot E \cdot h} l \tag{3}$$

where μ, ρ are blood dynamic viscosity ($4 \cdot 10^{-3} \mathrm{kg} \cdot \mathrm{m}^{-1} \cdot \mathrm{s}^{-1}$) and density ($10^3 \mathrm{kg} \cdot \mathrm{m}^{-3}$), *D* is a vessel's diameter, *E* is a vessel's Young's modulus (13 MPa), *h* is a vessel's wall thickness (0,1 mm) and *l* is a length of the vessel modeled by one compartment.

A coronary artery stenosis can be simulated by increasing the resistance for the particular compartments. If the simulated vessel consists of more compartments, then the stenosis can be simulated more realistic, including different resistance values for different compartments – i.e. ununiform narrowing of the blood vessel.

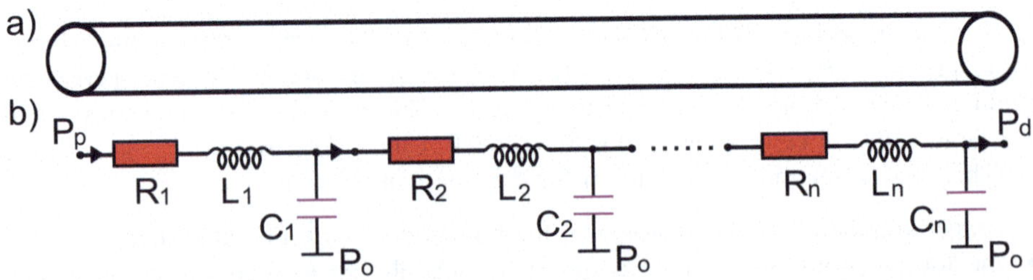

Fig. 2. Model of a segment of a coronary arterial vessel. P_p – proximal pressure, P_d – distal pressure, P_o – intrapericardial pressure (assumed to be equal to an intrathoracic pressure)

Model of the Peripheral Coronary Circulation

Model of a whole peripheral coronary circulation (Fig. 3) is based on the model of the intramyocardial pump [3–5]. Due to the degree of its complexity and difficulties associated with clinical verification, a single layer version was used.

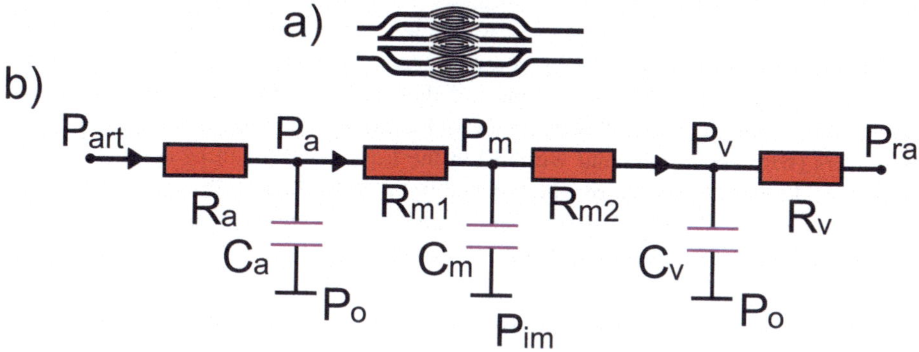

Fig. 3. Model of the peripheral coronary circulation; (a) symbol, (b) general outline: R_a, R_v and R_{m1}/R_{m2} – coronary arterial, venous and capillary resistance; coronary arterial (C_a), venous (C_v) and capillary (C_m) compliance, coronary arterial (P_a), venous (P_v) and capillary (P_m) pressure, P_{art} – distal pressure for *LAD*, *LCX* or *RCA*, P_{ra} – right atrial pressure

In the model R_a and C_a represent small arteries resistance and compliance respectively, R_{m1}/R_{m2} and C_m represent the resistive and compliant properties of the myocardium and R_v and C_v simulate the resistance and compliance of all coronary veins. An intramyocardial pressure (P_{im}) is proportional to the pressure in the left/right ventricle.

Model of the Physiological Feedback Between the Left Ventricular Muscles and the Coronary Perfusion

When developing a physiological feedback model between the ventricular muscle represented by end systolic pressure-volume relationship (*ESPVR*), described by end-systolic elastance E_{max} and unstressed volume V_0 parameters, and coronary circulation (Q_{cor}) represented by peripheral flow throw R_{m1} resistance (see Fig. 3), it was assumed that this relation can be described by the following formulas:

$$V_0' = (V_0)^{BASE} + \left((V_0)^{MAX} - (V_0)^{BASE}\right) \cdot \left(1 - \frac{(Q_{cor})'}{(Q_{cor})^{BASE}}\right) \tag{4}$$

$$E_{max}' = (E_{max})^{BASE} - \left((E_{max})^{BASE} - (E_{max})^{MIN}\right) \cdot \left(1 - \frac{(Q_{cor})'}{(Q_{cor})^{BASE}}\right) \tag{5}$$

where $(V_0)^{BASE}$ and $(E_{max})^{BASE}$ are default *ESPVR* parameters for physiological condition of the ventricle, while $(Q_{cor})^{BASE}$ is default peripheral coronary flow for the same condition. $(V_0)^{MAX}$ and $(E_{max})^{MIN}$ are connected with diastolic pressure-volume relationship (*DPVR*) of the ventricle. V_0' and E_{max}' are current *ESPVR* parameters for current $(Q_{cor})'$ caused by coronary stenosis. In the model it was assumed that *LCA* supplies the left ventricular myocardium and *RCA* supplies the right ventricular one.

If the above equations are adapted to the left ventricle, then Q_{cor} is the sum of peripheral flows of *LADp* and *LCXp* compartments. *LCA* stenosis in the model results in decreased *LAD* and *LCX* circulation and then ischemic left ventricular disease is simulated by changing its *ESPVR* parameters (V_{0l} and E_{maxl}). Note, that the V_0 and E_{max} parameters (with apostrophe and without) in the Eqs. 4 and 5 can be applied for both left and right ventricle, therefore when related to the left ventricle, they are indicated by *l*-index (V_{0l}, E_{maxl}, V_{0l}', E_{maxl}'). If the Eqs. 4 and 5 are adapted to the right ventricle, then Q_{cor} corresponds to the peripheral flow of *RCAp*.

2.2 The Coronary Model Parametrization

To evaluate the lengths and diameters of the main coronary arteries (*LCA, RCA, LCX, LAD*) and their parameters *R*, *L* and *C* (Eqs. 1–3), data (Table 1) was taken from literature [6, 7]. According to this data, the total resistance of the four main coronary vessels was approximately 2.04 mmHg · s/ml.

To assess the values of the peripheral resistance of coronary vessels (*LCXp, LADp, RCAp* moduls) an information on this subject was estimated that flow, used for coronary perfusion makes up about 5% of the total cardiac output in the blood circulatory system of a man at rest. Assuming that in a healthy patient the mean pressure in the aorta is P_{as} = 100 mmHg and cardiac output is *CO* = 5, 5 l/min, the total mean peripheral coronary resistance (R_c) is about 22 mmHs/ml/s, if we neglect venous coronary pressure.

Basing on the model from paper [8], the percentage R_c division for *LCXp, LADp, RCAp* modules was determined to be 40%; 40% and 20%, respectively. Next, according to the paper [5], the following resistance distribution for coronary arteries, veins and capillars was assumed: 25%, 7%, 68% respectively. Using the determined R_c value and all above information, the all resistance parameters values were calculated.

Making use of paper [5], we analogically determined the value of total coronary compliance C_c = 0.27 ml/mmHg (assuming that the left ventricular tissue's mass is 200 g). For this C_c value, the contributions of arteries, veins and capillars were determined to be 2%, 33% and 65% [5], respectively. This gives the following values for coronary arterial (C_a), venous (C_v) and capillary compliance (C_m): 0.0054; 0.0891 and 0.1755 ml/mmHg. According to the model in [8] the compliance distribution for the *LCXp, LADp* and *RCAp* modules was calculated to be 40%, 40%, 20% respectively. All of this data was next applied to a coronary peripheral model.

Table 1. The default parameters of the main coronary arteries

	Diameter [mm]	Length [mm]	R [mmHg/ml/s]	L [mmHg/ml/s^2]	C [ml/mmHg] $\times 10^{-3}$
LCA	8.7	17.7	0.038	0.0022	0.7
RCX	3.6	130	0.94	0.095	0.3
LAD	3.6	140	1.02	0.1	0.4
RCA	3	92	1.39	0.097	0.15

Values of parameters of the peripheral coronary circulation are presented in Table 2.

Table 2. The default parameters of the coronary peripheral vessels, resistance (*R*) values are expressed in *mmHg/ml/s*, compliance (*C*) values are expressed in *ml/mmHg*

	R_a	R_{m1}	R_{m2}	R_v	$C_a \times 10^{-3}$	$C_m \times 10^{-3}$	$C_v \times 10^{-3}$
LCXp	13.35	18.7	18.7	3.85	1.5	70.2	35.6
LADp	13.35	18.7	18.7	3.85	1.5	70.2	35.6
RCAp	26.7	37.4	37.4	7.7	0.77	31.5	17.8

To simulate the myocardial hyperemia, the all resistance values for *LCXp, LADp, RCAp* were changed. According to the literature data [5] the resistance distribution for coronary arteries, veins and capillaries is 42%, 31%, 27% respectively. Comparing this distribution with the one for the rest condition, it was approximated that the arterial, venous and capillary resistance must by decreased by about 65%, 0%, 95% respectively. The calculated values are presented in Table 3. The resistances of main arteries were not changed for hyperemia.

Table 3. The parameters of the coronary peripheral vessels for hyperemia, resistance (R) values are expressed in *mmHg/ml/s*, compliance (C) values are expressed in *ml/mmHg*

	R_a	R_{m1}	R_{m2}	R_v	$C_a \times 10^{-3}$	$C_m \times 10^{-3}$	$C_v \times 10^{-3}$
LCXp	4.05	1.87	1.87	3.85	1.5	70.2	35.6
LADp	4.05	1.87	1.87	3.85	1.5	70.2	35.6
RCAp	8.11	3.74	3.74	7.7	0.77	31.5	17.8

In Table 4 the parameters for the physiological feedback model (necessary for Eqs. 4 and 5) were presented.

Table 4. The parameters of the physiological feedback model, *l* – index connected with the left ventricle

Parameter	Value	Unit
$(V_{0l})^{BASE}$	5	ml
$(V_{0l})^{MAX}$	180	ml
$(E_{maxl})^{BASE}$	3.5	mmHg/ml
$(E_{maxl})^{MIN}$	2.625	mmHg/ml

2.3 The Coronary Model Implantation in the Whole Numerical Cardiovascular Model

The developed numerical model of the coronary system was integrated with the closed-loop model of the cardiovascular system. This model consists of 6 functional modules: left heart, systemic arterial circulation, venous systemic circulation, right heart, pulmonary arterial circulation, pulmonary venous circulation. Each circulation module is simulated by one compartment, 4-element Windkessel model. A Windkessel model reproduces resistive, inertial and elastic properties of the simulated vessels. The atriums are simulated by passive, linear compliances. The heart ventricles systole is described by means of the time-varying elastance model by Suga and Sagawa. Ventricular feeling phase is simulated by the sum of exponentials. The coronary circulation model was added as a next, 7th module. *LCA* was modeled by one *RLC* compartment (see Fig. 2), while *LCX, LAD* and *RCA* were modeled by *10 RLC* compartments. The whole numerical cardiovascular model was executed and successfully tested in the hard real-time regime and is ready to be part of the hybrid cardiovascular simulator. More details about the used numerical circulatory model as well as about the hybrid simulator were described in [1, 9].

3 Results and Discussion

In the first step, a simulation for default coronary model parameters (presented in Tables 1 and 2) as well as for default parameters of the whole cardiovascular model was performed. It simulated a physiological condition of the circulatory system at rest. Simulation results were presented in Fig. 4.

Simulation results present coronary arterial flow decreasing during systole and increasing during diastole, what is consistent with physiology. Moreover, a small negative left coronary artery flow (*Qlca*) is visible during systolic phase and a coronary venous flow (*Qcv*) is in counterphase in relation to the coronary arterial flow. Both effects are well observed in clinical practice [4]. A peak *Qlca* about 6 ml/s and mean total coronary flow about 250 ml/min (4.4% of the simulated *CO* = *i*) as well as *Qlca* and *Qrca* flow patterns are also consistent with the physiology. Thus, the coronary flow shapes reproduction by the presented coronary model is correct from physiological point of view.

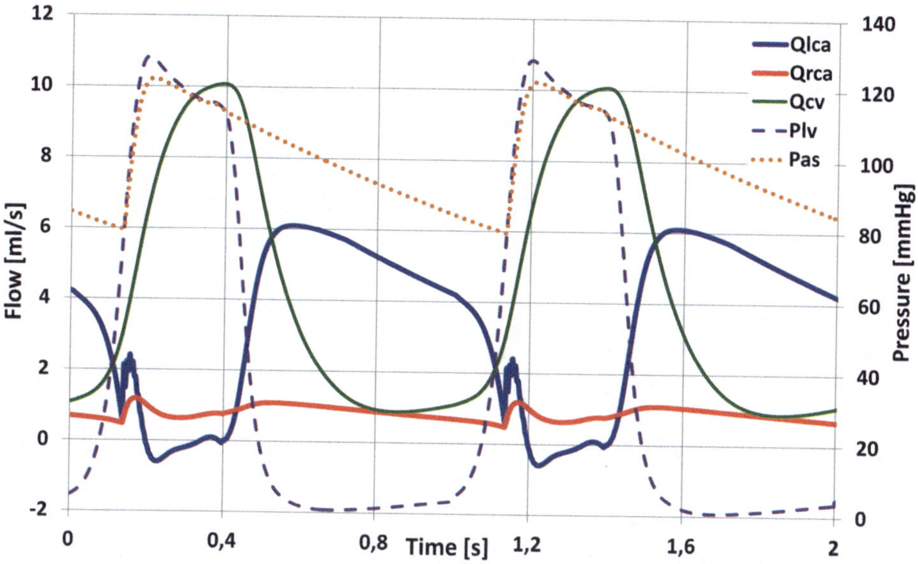

Fig. 4. Simulation results for physiological condition at rest, *Plv* – left ventricular pressure, *Pas* – arterial pressure, *Qlca* – left coronary artery flow, *Qrca* – right coronary artery flow, *Qcv* – venous coronary flow

In the second step the hyperemia of the myocardium was simulated, like in case of a physical activity condition. To simulate this condition, the peripheral coronary resistances were reduced (Table 3) to increase the coronary flow. The simulation results were presented in Fig. 5.

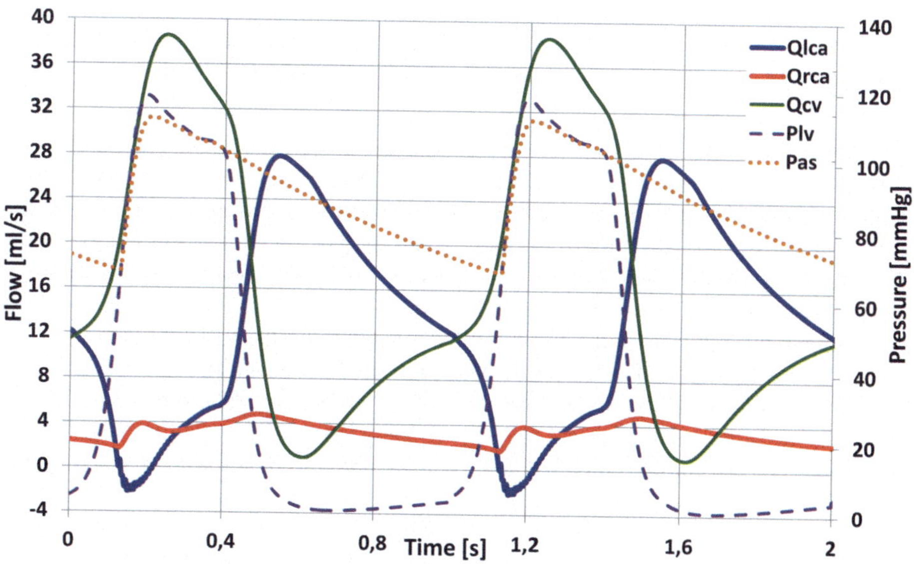

Fig. 5. Simulation results for the hyperemia of the myocardium, *Plv* – left ventricular pressure, *Pas* – arterial pressure, *Qlca* – left coronary artery flow, *Qrca* – right coronary artery flow, *Qcv* – venous coronary flow

A mean total coronary flow was about 0.99 l/min, what is about 17% of the simulated CO = 5.82 l/min, thus in comparison to the situation in Fig. 4, the coronary flow was increased four times. It is known from the physiology that during the physical activity, the coronary flow can be increased 3–4 times. The simulations are agreed with it then.

In the last step the physiological feedback between the *ESPVR* and coronary perfusion was introduced. During the simulation *LCA* resistance was increasing and the left ventricular *ESPVR* displacement was recorded. The results are presented in Fig. 6.

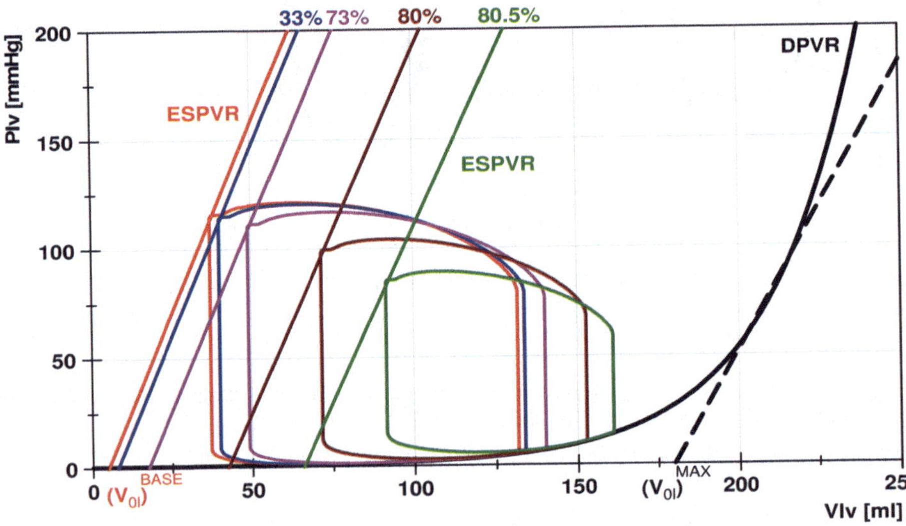

Fig. 6. Pressure-volume loops for the left ventricle for the selected *LCA* stenosis levels (expressed in %), *ESPVR* – end systolic pressure-volume relationship, *DPVR* – diastolic pressure-volume relationship

Increasing *LCA* resistance limits *LCX* and *LAD* coronary peripheral flows, what effects on the left ventricular *ESPVR* by its automatically shifting towards right and decreasing the slope (according to the Eqs. 1 and 2). The pressure-volume loop of the left ventricle is smaller and smaller as a result of reduced flow in the myocardium supplied by *LCA*, smaller oxygen supply and muscle ischaemia. The simulation results are in agreement with the study presented by Sunagawa [10] and Little [11].

This study has several limitations. The presented coronary model is still linear, especially the submodel of the peripheral coronary circulation. There is also a lack of collateral circulation in the model. Moreover, only one layer of the myocardium is considered. The last limitation is that *ESPVR* is assumed to be a linear function also in pathological heart's condition.

4 Conclusions

The described coronary model is much more advanced than popular coronary models [3, 4, 12] like Waterfall, pure intramyocardial pump or models basing on time-varying elastance concept. It is designed to simulate the coronary stenosis diseases and cardiac muscle ischaemia/infraction with affecting on the contraction of the ventricles what is the novel feature. The model can be personalized by e.g. coronarography data. It is suitable to be used in the hybrid simulator of the cardiovascular system (working in the real-time) to simulate the assistance scenarios by means of real *IABP*, what is the main goal of this coronary model exploitation.

Compliance with Ethical Standards

Funding
The study received financial support from the Nalecz Institute of Biocybernetics and Biomedical Engineering, Polish Academy of Sciences

Ethical approval
This article does not contain any studies with human participants or animals performed by any of the authors.

References

1. Darowski, M., Kozarski, M., Ferrari, G., Zieliński, K., Górczyńska K., Szczepanowski, A., Pałko, K.J., Fresiello, L., Di Molfetta, A.: A new hybrid (hydro-numerical) model of the circulatory system. Bull. Pol. Ac.: Tech. **61**(4), 993–1003 (2013)
2. Westerhof, N., Stergiopulos, N., Noble, M.I.M.: Snapshots of Hemodynamics: An Aid for Clinical Research and Graduate Education. Springer, Boston (2005)
3. Spaan, J.A., Breuls, N.P., Laird, J.D.: Diastolic–systolic coronary flow differences are caused by intramyocardial pump action in the anesthetized dog. Circ. Res. **49**(3), 584–593 (1981)
4. Westerhof, N., Boer, C., Lamberts, R.R., Sipkema, P.: Cross-talk between cardiac muscle and coronary vasculature. Physiol. Rev. **86**(4), 1263–1308 (2006)
5. Bovendeerd, P.H.M., Borsje, P., Arts, T., Van de Vosse, F.N.: Dependence of intramyocardial pressure and coronary flow on ventricular loading and contractility: a model study. Ann. Biomed. Eng. **34**(12), 1833–1845 (2006)
6. Miklaszewska, D., Gawlikowska-Sroka, A., Czerwiński, F.: Morphometric study of the left coronary artery. Ann. Acad. Med. Stetin. **56**(2), 12–17 (2010)
7. Douglas, P.S., Fiolkoski, J., Berko, B., Reichek, N.: Echocardiographic visualization of coronary artery anatomy in the adult. J. Am. Coll. Cardiol. **11**(3), 565–571 (1988)
8. Kim, H.J., Vignon-Clementel, I.E., Coogan, J.S., Figueroa, C.A., Jansen, K.E., et al.: Patient-specific modeling of blood flow and pressure in human coronary arteries. Ann. Biomed. Eng. **38**(10), 3195–3209 (2010)

9. Ferrari, G., Kozarski, M., Zieliński, K., Fresiello, L., Di Molfetta, A., Górczyńska, K., Pałko, K.J., Darowski, M.: A modular computational circulatory model applicable to VAD testing and training. J. Artif. Organs **15**(1), 32–43 (2012)
10. Sunagawa, K., Maughan, W.L., Sagawa, K.: Effect of regional ischemia on the left ventricular end-systolic pressure-volume relationship of isolated canine hearts. Circ. Res. **52** (2), 170–178 (1983)
11. Little, W.C., O'Rourke, R.A.: Effect of regional ischemia on the left ventricular end-systolic pressure-volume relation in chronically instrumented dogs. J. Am. Coll. Cardiol. **5**(2 Pt 1), 297–302 (1985)
12. Downey, J.M., Kirk, E.S.: Inhibition of coronary blood flow by a vascular waterfall mechanism. Circ. Res. **36**(6), 753–760 (1975)

Other Bioengineering Studies

Mechanical Properties of Porcine Aorta – Influence of Specimen Taken Orientation

Sylwia Łagan(✉) and Aneta Liber-Kneć

Cracow University of Technology, Warszawska 24, 31-155 Cracow, Poland
slagan@mech.pk.edu.pl
http://pk.edu.pl

Abstract. The aim of the work was a characteristic of anisotropic mechanical properties of a pig's aorta. The isolated ascending, descending and thoracic aortas were taken. The longitudinal and transversal orientations of specimens taken were considered. Tissue strips were cut in the longitudinal orientation and in the transversal orientation as rings. Part of the rings was cut and indicated as the transversal strip samples, the second part in the form of rings was indicated as the transversal ring samples. The prepared samples were tested in an uniaxial tensile test. The values of the maximal force, the elongation, the ultimate tensile strength and strain at the UTS, Young's moduli and energy of destruction were estimated. The influence of orientation of samples taking was estimated. The highest values of the maximal force and the UTS were determined for the transversal ring samples and the lowest for the longitudinal strip samples.

Keywords: Porcine aorta · Anisotropy · Tensile test

1 Introduction

The clinical and literature data show that the mechanical properties of the arteries depend on physical and chemical parameters of the physiological environment, such as temperature, osmotic pressure, pH, partial pressure of oxygen and carbon dioxide, and ion concentration [1]. In order to perform the basic mechanical characteristics, the aorta is treated as an inflated, thin-walled tubular structure. Arterial blood pressure is transferred as a biaxial mechanical stress. Structural components of the aortic wall are specifically oriented which indicates the local anisotropy of the wall. This causes peripheral and axial stresses in the aorta while maintaining other stiffness values for these directions. Although the arteries contain an intracellular and extracellular substance that has the ability to flow both within and outside the vessel (as a result of pressure), they cannot be considered incompressible, because they experience isochoric movements, i.e. they do not change their volume, even under different load conditions. In addition, the arteries are not homogeneous, their properties change with location, which is also confirmed by their layered structure. Very often, for the purpose of research, homogeneity of mechanical properties is assumed within one wall

J. Korbicz et al. (Eds.): PCBEE 2019, AISC 1033, pp. 279–287, 2020.
https://doi.org/10.1007/978-3-030-29885-2_25

layer, but this results in its non-linear, global behavior. Therefore, it is necessary to assess the influence of both local and global aortic wall heterogeneity with respect to mechanical parameters [2,3]. The aorta, like most soft tissues, exhibits viscoelastic behavior. During cyclic loading, hysteresis, as well as stress relaxation in conditions of constant elongation and creep under the influence of constant load can be observed. Repetitive reaction to subsequent loading cycles after initial conditioning has also been shown. Detailed tests should also take into account the presence of residual stresses and voltage even in the absence of external loads. However, their measurements are only possible for ex vivo cut tissues and intended for analysis in vivo [2,4]. In this study, a porcine aorta was tested under uniaxial tensile load with respect to the orientation of specimens taken. The main goal was an estimation of differences between specimens cut as rings in the transversal orientation and then divided into two groups: the transversal strip samples and the transversal ring samples. Additionally, the influence of sample measurement length for the longitudinal orientation on mechanical parameters was estimated.

2 Materials and Methods

For this study, samples of a porcine ascending, descending and thoracic aortas were isolated as a food waste from a local abattoir. Polish large white pigs, 5 males and 4 females, weighting ca. 116 ± 7 kg were used. Before the test tissue specimens were frozen for 12 days in $-18\,^{\circ}$C degrees in polypropylene boxes without immersion liquid. Next, the specimens were heated to the room temperature directly before tests. The tensile tests were conducted with the rate of 5 mm/min until broken of specimens, in the room temperature, without medium. The tensile machines: Instron 4465 with load cell 5 kN and MTS Insight 50 with 1kN load cell were used. The biological material was divided into four groups of specimens. Tissue strips were cut in the longitudinal orientation (short and long samples) and in the transversal orientation as rings. Part of the rings was cut and indicated as the transversal strip samples, and the second part in the form of rings was indicated as the transversal ring samples. Two different lengths of specimens were considered with 10:1 and 3:1 coefficient of length in relation to the width of specimens (Table 1). The examples of samples used in tests were shown in Fig. 1. Registered force (F) - elongation (Δl) curves were recalculated into stress (σ) - strain (ε) curves with the use of equations (1–2) and the average experimental curves were determined for a pig aorta. The ultimate tensile strength (UTS), strain at the UTS and the energy of deformation (W) as the area under elongation curve were calculated. The Young's modulus values (E) were calculated according to the formula (3) for three different ranges: physiological, hypertensive and maximal [8].

$$\sigma = \frac{F_{max}}{A} \tag{1}$$

$$\varepsilon = \frac{\Delta l}{l_0} \tag{2}$$

$$E = \frac{\partial \sigma}{\partial \varepsilon} \tag{3}$$

where: $\partial\sigma$- increase of stress, $\partial\varepsilon$- increase of strain.

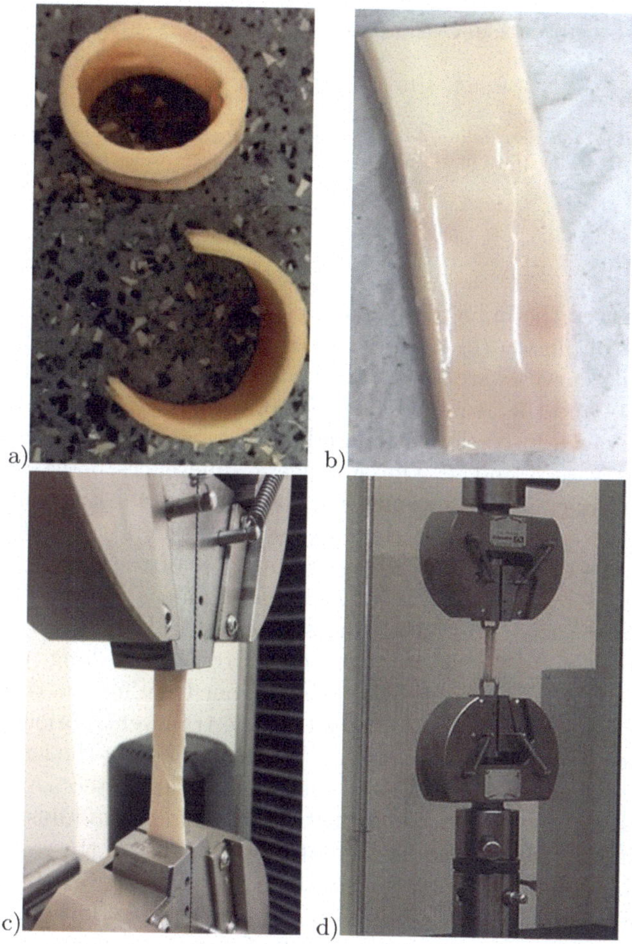

Fig. 1. The example of aorta samples: (a) the transversal ring and strips samples, (b) the longitudinal strip samples, (c) the strip samples in strength machine grips, (d) the transversal ring samples in strength machine grips

Table 1. Groups of samples used in tests (X ± SD mm)

Orientation	Group	Location	Thickness [mm]	Width [mm]	A [mm^2]	l_0 [mm]
Transversal (strips)	GI	A & DA	2,37 ± 0,28	12,65 ± 0,82	30,01 ± 4,61	30,00 ± 0,10
Longitudinal (short)	GII	A & DA	1,88 ± 0,21	13,07 ± 1,30	24,83 ± 5,41	30,00 ± 0,10
Transversal (rings)	GIII	T & abA	1,58 ± 0.02	10,01 ± 0,02	31,50 ± 4,38	45,03 ± 4,61
Longitudinal (long)	GIV	T & abA	1,15 ± 0,06	9,95 ± 0,19	11,48 ± 0,61	86,68 ± 9,44

A&DA- Ascending and descending aorta T& abA-Thoracic and abdomen aorta, GI & GII(6 m–th, n = 9), GIII & GIV(9–12 m–th, n = 6)

3 Results

The results showed the influence of samples taking orientations on the mechanical parameters of pig's aorta. In order to compare individual groups of samples, averaged tensile curves were determined. Examples of tensile curves registered during the tests and the averaged curve for one group of samples were shown in Fig. 2. The averaged force-elongation and stress-strain curves for all four groups were shown in Figs. 3 and 4. The significant differences between stress-strain curve of the transversal ring samples and the longitudinal strip samples of aorta can be seen. Also specimens cut in the transversal orientation, but distinguished as rings or strips of tissue had different character of the tensile curves. These differences can be compared quantitatively based on the parameters listed in the Table 2. In case of the transversal ring samples, the highest values of Young's modulus and ultimate tensile strength compared to the longitudinal orientation were obtained. These values were also higher than for the transversal strip samples. The material's ability to deform in the transversal orientation was the highest for the transversal strip samples. When considering the length of the longitudinal strip samples, differences in mechanical parameters can be seen. Significantly higher values of tensile strength and Young's modulus were obtained for long specimens. This indicates a significant influence of the research methodology on the obtained results. It is also connected with limitations related to the dimensions of isolated animal samples taken for the examination. To characterize the physiologic and hypertensive moduli of tested specimens, pressure ranges were adopted according to [8,9]. In Fig. 5, the comparison of average stress-strain curves of aorta samples with marked ranges of pressure included in the moduli values calculation were shown. The values of elastic modulus calculated for physiologic (PYM) and hypertensive (HYM) range, as well as the maximum elastic modulus (MYM) were compared in Fig. 6 and Table 2.

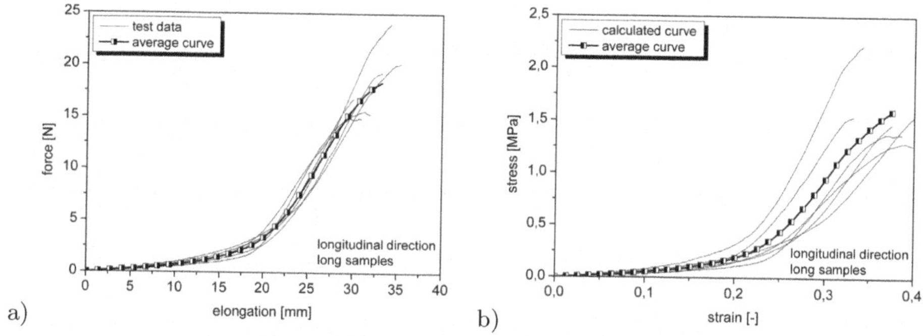

Fig. 2. The example of registered tensile curves and the averaged curve for GIV: (a) load-elongation curves, (b) stress-strain curves

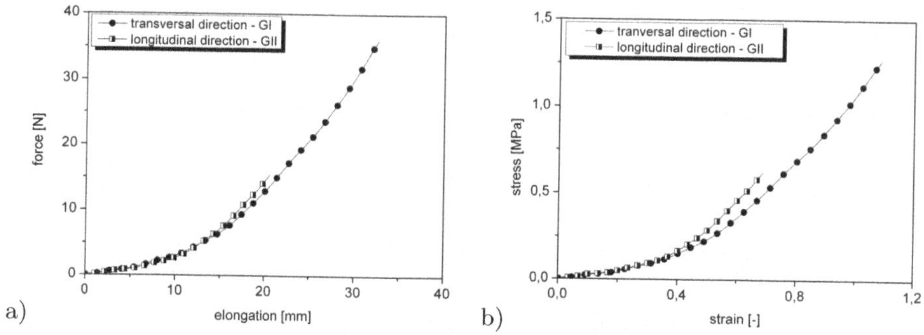

Fig. 3. The comparison of characteristic (a) force-elongation curves, (b) stress-strain curves for the transversal strip samples and the longitudinal (short) strip samples

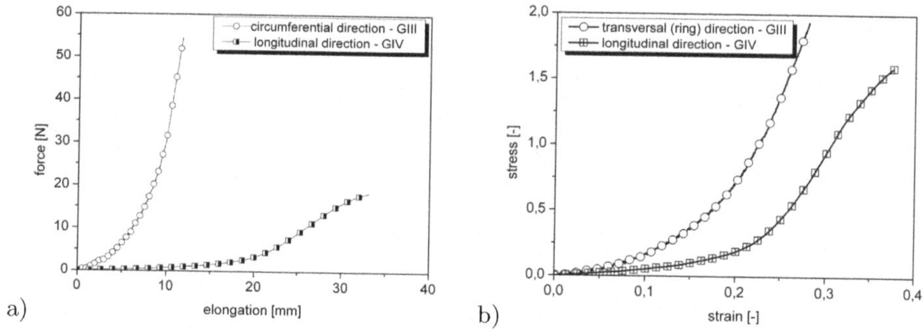

Fig. 4. The comparison of characteristic: (a) force-elongation curves, (b) stress-strain curves for the transversal ring samples and the longitudinal (long) strip samples

Table 2. The tensile test results (X ± SD mm)

Orientation	Fmax [N]	lmax [mm]	UTS [MPa]	Strain at UTS [–]	W [J]	PYM [MPa]	HYM [MPa]	MYM [MPa]
Transversal (strips)	34,79 ± 5,96	32,62 ± 6,58	1,16 ± 0,17	1,08 ± 0,22	0,33 ± 0,13	0,66 ± 0,10	0,95 ± 0,18	2,4 ± 1,02
Longitudinal (short)	15,43 ± 4,90	21,04 ± 1,84	0,62 ± 0,18	0,70 ± 0,06	0,10 ± 0,03	0,93 ± 0,25	1,35 ± 0,35	1,67 ± 0,63
Transversal (rings)	57,71 ± 20,11	12,93 ± 2,31	1,78 ± 0,45	0,29 ± 0,04	0,13 ± 0,03	3,31 ± 0,44	3,75 ± 0,66	13,27 ± 1,54
Longitudinal (long)	18,88 ± 3,51	31,37 ± 2,23	1,66 ± 0,35	0,36 ± 0,02	0,16 ± 0,04	2,09 ± 0,28	4,53 ± 1,25	10,05 ± 1,04

PYM/HYM/MYM physiological/hypertensive/maximal Young modulus

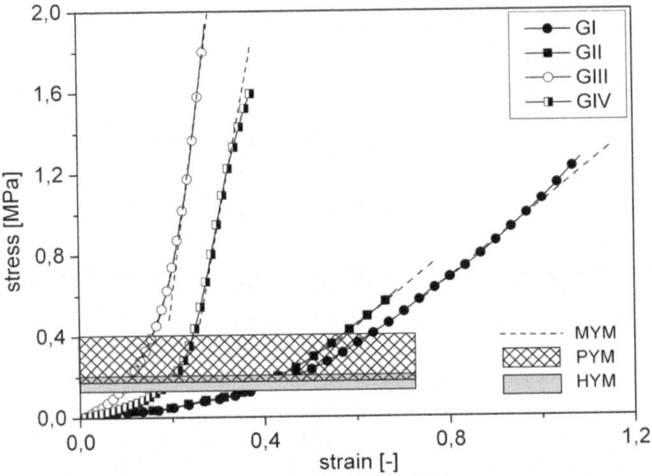

Fig. 5. The comparison of stress-strain curves for GI–GIV with marked ranges of physiological (PYM), hypertensive (HYM) and maximal (MYM) pressure

The elastic moduli in the physiological range was considerably higher in the transversal ring samples than in the transversal and longitudinal strip samples, whereas smaller differences were found for the elastic moduli in the hypertensive range between the transversal and longitudinal strip samples.

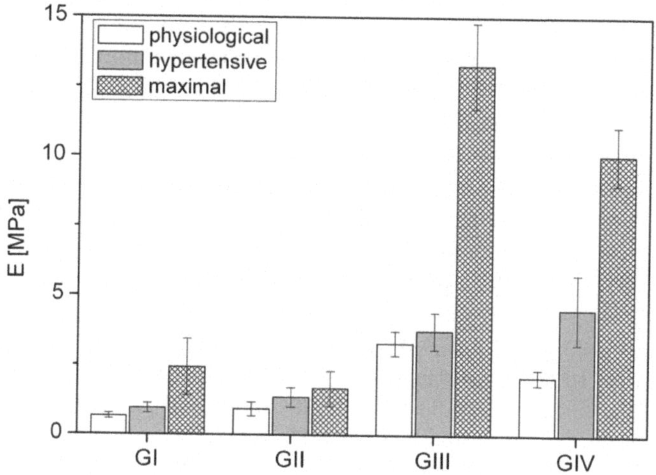

Fig. 6. The comparison of Young modulus in range of physiological (PYM), hypertensive (HYM) and maximal (MYM) range of pressure

4 Discussion

It is common practice to study arteries mechanical behavior paying particular attention to species, age, state of health and location along the vascular tree. Furthermore, a significant factor that affects the mechanical properties considered is orientation of samples taken. Also the elastic moduli is one of a key factors in assessing changes in elasticity of the aortic wall when designing of aortic endografts. Deficiency in human test materials causes, that most investigations on the mechanical properties of coronary arteries have been done on animals [3,4]. Higher values of mechanical parameters of aorta obtained in this study for the transversal orientation are consistent with the biomechanical function of the vessel, which in physiological conditions is subjected to a much greater forces and circumferential strains [1,2]. The values of mechanical parameters for porcine aorta are in the range given by Bartkowiak-Jowsa et al. [5] for samples of porcine coronary arteries for the longitudinal and transverse direction. Gzik-Zorska et al. [6] reported the value of maximum force equal to 15,6 N (±3,55 N) for porcine left anterior descending artery, tested as a whole fragment in the longitudinal direction. It is in good compliance with our results for the longitudinal samples. Hence animal arteries differ from human ones, e.g. there is no process of growth of intimal layer of arterials throughout life of an animal, the axial prestretch is much less for aged human arteries than for animal arteries, from the mechanical point of view some similarities can be found. For the strip samples from adventitia, media and intima of the human artery, the ultimate tensile strength was equal to 1,43/0,44/0,39 MPa in circumferential direction and 1,30/0,41/0,39 MPa in the longitudinal direction [3]. These results are comparable with those obtained from this study. The investigation of Jankowska et al. [7] reported the value of the UTS for longitudinal human artery strip samples equal to 2,14 MPa, higher than in our results. Experiment conducted by Karimi et al. [4] for human arteries (tested non separated arteries) showed the value of E = 1,55 ± 0,26 MPa, UTS = 1,44 ± 0,87 MPa and strain equal to 0,54 ± 0,25. If we compare these results with the longitudinal porcine strips samples, we can observe compliance only for modulus of elasticity for the longitudinal short samples. The analysis of the elastic moduli for the physiological and hypertensive range of pressure showed good agreement with results obtained for the human artery strips by Khanafer et al. [8], but only for two groups of samples: the transversal rings and the longitudinal strips (the longer one). For the transversal orientation, the value of elastic moduli in the physiological pressure range was reported between 2,61 and 3,07 MPa, and for the longitudinal orientation, it was 2,35 to 3,72 MPa. In the hypertensive range of pressure, the value of moduli was 3,37 to 4,21 MPa and 2,83 to 4,33 MPa, respectively for the transversal and longitudinal orientation [8]. The values obtained in our measurement are close to these values. The value of maximum moduli reported in [8] for human artery was considerably different form our results for pig's artery. When comparing only strip samples, we observed much lower values of elastic moduli. An important point to be noted is that presented investigations on the mechanical properties of arteries had different methodological approaches (strips of tissue or fragment of the artery), different spices, age and type of the artery.

The presented study showed problems occurring during this type of research and the directions of future methodological modifications. One of the issues associated with in vitro characterization of biomechanical behavior of soft tissue is the absence of the internal pressure of the vessel. The second of limitation in vitro tissue-testing is the simplification of the forces being applied to the tissue in uniaxial direction, what is not appropriate to reproduce the physiological condition of stress, and to catch the anisotropic behavior of biological tissue. However, due to the simplicity of execution, uniaxial tensile tests still represent a testing method widely used by researchers. Next issue, is the effect of the strain rate on the tensile properties of tissues. In this study, strain rate was 5 mm/min and the majority of experimental investigations on vessels mechanics used non-physiological strain rates (i.e. 10 mm/min) in a tensile tests. For better assessment of the mechanical properties, tensile test machines should be able to operate with strain rate corresponding to the physiology of aortic wall deformation. Tensile testing techniques should be revised in order to minimize slippage of the tissue and supplemented by visual detection such as digital image correlation (DIC) to measure full-field displacement. Another problem in the present study, was the storage of tissue samples before the test and the need of frozen them to collect sufficient number of specimens. Comparison of results with the consideration of sex or age of individual animal was impossible to carry out. In order to achieve statistically significant results, it was necessary to put together data from different animals within the same group.

5 Conclusions

Particular attention was paid to the examination of the overall response of an aorta walls to loading in the longitudinal and transversal directions. The stress-strain characteristics were identified on the base of experimental data obtained from uniaxial tensile tests. Diversification of research results depending on the anisotropy and methodological approaches with consideration of strips or ring samples in the transversal orientation, as well as different measurement lengths of the longitudinal samples showed that it is still necessary to improve methodology of experimental research of the arteries. In the next step, a constitutive modeling approach will be applied to predict the response of aorta under load. Such constitutive relations provide a basis for the formulation of boundary value problems and possible further theoretical analysis or computer simulations of the tissue behavior under loading.

References

1. Humphrey, J., Holzapfel, G.: Mechanics, mechanobiology, and modeling of human abdominal aorta and aneurysms. J. Biomech. **45**, 805–814 (2012)
2. Holzapfel, G., Gasser, T., Stadler, M.: A structural model for the viscoelastic behavior of arterial walls: continuum formulation and finite element analysis. Eur. J. Mech. A/Solids **21**, 441–463 (2002)
3. Holzapfel, G., Sommer, G., Gasser, C., Regitnig, P.: Determination of layer-specific mechanical properties of human coronary arteries with nonatherosclerotic intimal thickening and related constitutive modeling. Am. J. Physiol.-Heart Circulatory Physiol. **289**, 2048–2058 (2005)
4. Karimi, A., Navidbakhsh, M., Shojaei, A., Faghihi, S.: Measurement of the uniaxial mechanical properties of healthy and atherosclerotic human coronary arteries. Mater. Sci. Eng. C **33**, 2550–2554 (2013)
5. Bartkowiak-Jowsa, M., Bedziński, R.: Mechanical and structural properties of porcine coronary arteries. In: Proceedings of 9th YSESM: Youth Symposium on Experimental Solid Mechanics, Italy (2010)
6. Gzik-Zroska, B., Joszko, K., Wolański, W., Gzik, M.: Development of new testing method of mechanical properties of porcine coronary arteries. In: Information Technologies in Medicine, Advances in Intelligent Systems and Computing, vol. 472, pp. 289–297. Springer, Cham (2016)
7. Jankowska, M., Bartkowiak Jowsa, M., Bedziński, R.: Experimental and constitutive modeling approaches for a study of biomechanical properties of human coronary arteries. J. Mech. Behav. Biomed. Mater. **50**, 1–12 (2015)
8. Khanafer, K., Duprey, A., Zainal, M., Schlicht, M., Williams, D., Berguer, R.: Determination of the elastic modulus of ascending thoracic aortic aneurysm at different ranges of pressure using uniaxial tensile testing. J. Thorac. Cardiovasc. Surg. **142**(3), 682–686 (2011)
9. Duprey, A., Khanafer, K., Schlicht, M., Avril, S., Williams, D., Berguer, R.: In vitro characterization of physiological and maximum elastic modulus of ascending thoracic aortic aneurysms using uniaxial tensile testing. Eur. J. Vasc. Endovasc. Surg. **39**(6), 700–707 (2010)

Experimental and Constitutive Approaches for a Study of Mechanical Properties of Animal Tendons

Aneta Liber-Kneć(✉) and Sylwia Łagan

Cracow University of Technology, Warszawska 24, 31-155 Cracow, Poland
aliber@pk.edu.pl
http://pk.edu.pl

Abstract. The aim of the work was an analysis of the seven hyperelastic material models ability to capture tendon (from a sheep and domestic pig) response during quasi-static tensile loading. In the first step, animal tendons were tested under tensile loading; then the neo-Hookean, Mooney-Rivlin, Ogden, Humprey, Martins, Veronda-Wenstmann and Yeoh material models were fitted to tendons tensile data. Three different approaches to a modeling procedure were used: (1) models were fitted to tensile data for all tested specimens and coefficients of models were averaged, (2) on the base on registered tensile curves, one average stress-stretch curve was determined and then models were fitted, and (3) for two above described variants, the tensile data were limited to the first and second phase of elongation during fitting procedure and extrapolated to the third phase. Models sensitivity to limitation of experimental data and possibility of predicting tensile behavior in full range of elongation were analyzed. The range of experimental data used to fit the model was crucial factor for the predictive ability of each model.

Keywords: Fitting hyperelastic material model · Animal tendon · Tensile test

1 Introduction

Tendons form an integral part of a musculoskeletal unit. Their primary function is to transmit force between muscles and the skeleton but their elastic deformation allows them to store potential energy [1,2]. Tendons are composed of collagen and elastin fibers, organized at different length scales. Such structure influences on mechanical behavior of tendons. A number of recent studies have characterized a non-linear, anisotropic, inhomogeneous and viscoelastic properties of tendons for different species and for different anatomical locations [1,3,4]. An analysis of the mechanical behavior of tendons under tensile loads delivers the primary information for the understanding the injury mechanism and the healing potential of tendons, and can provide necessary data for developing appropriate model formulations. The constitutive modeling of tendon behavior under different loading conditions can function as complementary modules to

J. Korbicz et al. (Eds.): PCBEE 2019, AISC 1033, pp. 288–297, 2020.
https://doi.org/10.1007/978-3-030-29885-2_26

evaluate its mechanical properties under normal function, predicts the mechanism of injury and healing, and delivers data for finite element analyses of tissues and joints [5–7]. On the tissue scale, mathematical models of tendon mechanics have been widely used [8–12]. These phenomenological approaches use mathematical formulation that can closely match the mechanical response of tendon but the parameters of the models often have no clear physical formulation (e.g. the neo-Hookean model). Soft tissues, due to their complex hierarchical structure, exhibit the properties of orthotropic heterogeneous materials, but for the purposes of mathematical descriptions, simplification as assumption of their anisotropy or isotropy are made. The analysis of the modeling results presented in the literature indicates that the scope of application of described material models should be carefully determined, due to the possibility of obtaining divergent results depending on the data range used. Especially, when it is difficult to clearly define the boundaries of small and large deformation ranges for the examined tissues and the elastic limit of the biological material. In modeling approaches, the boundary conditions and the limits of the applicability of hyperelastic constitutive material models should be strictly defined [13,14]. The objective of the work was an analysis of the hyperelastic material models ability to capture tendon (from a sheep and domestic pig) response during quasi-static tensile loading. We hypothesize that experimental data limited to the first and second phase of uniaxial loading can be used to accurately predict the full stress-stretch response of tendons. For comparison, calculations for the full range of experimental data were also performed. Seven hyperelastic material models: neo-Hookean (n-H), Mooney-Rivlin (M-R), Ogden (O), Humprey (H), Martins (M), Veronda-Wenstmann (V-W) and Yeoh (Y) were used. Additionally, the modeling procedure was conducted twice: for the average experimental stress-stretch curve and also for all experimental curves; then the results of fitting were averaged.

2 Materials and Methods

Tendons taken from legs of a domestic pig and a sheep obtained from a local abattoir were used in this investigation. Both species were male, pigs weighted ca. 110–130 kg and were 8–10 months old. Sheep weighted ca. 25–30 kg and were 12 months old. Eight samples from a sheep and six samples from a pig were taken. Until the time of examination (no longer than 24 h), samples were stored in 0.9% saline solution at 4 °C. The average cross-section area was 20 mm^2 for pig tendons and 29 mm^2 for sheep tendons. The uniaxial tensile tests were conducted on the MTS Insight 50 tensile machine at a constant rate of 10 mm/min. The length of the measurement base (l_0) was 30 mm for each species. Registered force (F)-elongation (Δl) curves were recalculated into stress (σ)-stretch (λ=1+Δl/l_0) curves. In the next step, the average experimental curves were determined based on experimental tensile curves and the values of Young's modulus were calculated. Obtained stress-stretch characteristics were used for constitutive modeling with the use of several hyperelastic material models. The process of fitting the tensile experimental data was made with the use of the Origin Pro 7.5 software. The fitting procedure was conducted in three variants:

1. models were fitted to experimental data registered for all tested specimens and the values of models coefficients were averaged;
2. the experimental data were averaged and one stress-stretch curve was determined for each group of tendons, then models were fitted to the average curve;
3. for two above described cases, the experimental data were limited to the first and second phase (I and II) of elongation and during the fitting procedure prediction of tensile behavior in the third (III) phase was made.

A hyperelastic material model is characterized by the definition of the strain energy density function which formulation depends on the material. If the material is isotropic, the strain energy density function (W) depends upon three invariants of the Cauchy-Green tensor I_1, I_2, I_3 (2):

$$W_{isotropic} = W(I_1, I_2, I_3) \tag{1}$$

$$I_1 = \lambda_1^2 + \lambda_2^2 + \lambda_3^2,\ I_2 = \lambda_1^2\lambda_2^2 + \lambda_2^2\lambda_3^2 + \lambda_3^2\lambda_1^2,\ I_3 = \lambda_1^2\lambda_2^2\lambda_3^2 \tag{2}$$

where λ_1, λ_2, λ_3 are the principal stretches.

If the material is also incompressible (I_3=0), the expression of the invariants of the stretch tensor of deformation becomes simplified and can be written as the principal stretch (3):

$$\lambda_1 = \lambda,\quad \lambda_2 = \lambda_3 = \frac{1}{\sqrt{\lambda}} \tag{3}$$

Seven hyperelastic constitutive models described in the literature were used to determine the stress-stretch relationship for tested tendons. Chosen models differ in type and number of quantities (principal stretches or Cauchy-Green tensor) they depend on. The distribution of density function for these models was presented in Table 1.

Table 1. The distribution of density function [7,8]

Model	Density function
neo-Hookean	$W = C_1(I_1 - 3)$
Mooney-Rivlin	$W = C_1(I_1 - 3) + C_2(I_2 - 3)$
Humphrey	$W = C_1(e^{C_2(I_1-3)} - 1)$
Martins	$W = C_1(e^{C_2(I_1-3)} - 1) + C_3(e^{C_4(\lambda-3)} - 1)$
Veronda-Westmann	$W = C_1(e^{C_2(I_1-3)} - 1) - \frac{C_1C_2}{2}(I_2 - 3)$
Yeoh	$W = C_1(I_1 - 3) + C_2(I_1 - 3)^2 + C_3(I_1 - 3)^3$
Ogden	$W = \frac{2C_1}{C_2{}^2}(\lambda^{C_2} + \lambda^{-\frac{C_2}{2}} - 3)$

To analyze result of fitting, the coefficient of determination (R^2) and the reduced chi-square χ^2_{red} were calculated. The coefficient of determination R^2 is defined by the formula (4):

$$R^2 = \frac{\sum_{n=1}^{n}(\hat{y_i} - \bar{y})^2}{\sum_{n=1}^{n}(y_i - \bar{y})^2} \quad (4)$$

where: y_i is the actual variable value, $\hat{y_i}$ is the theoretical variable value obtained from modeling, $\bar{y}$ is the arithmetic mean value of the experimental variable value. The reduced chi-square χ^2_{red} was calculated according to (5):

$$\chi^2_{red} = \frac{\chi^2}{d} \quad (5)$$

where: χ^2_{red} is chi-square and d is the number of degrees of freedom.

3 Results

On the basis of the tensile test, force-elongation curves for both pig and sheep tendons were obtained (Fig. 1). For each group, the average curve was determined and an analysis of result was performed. In Fig. 2, the average stress-stretch curves for pig and sheep tendons were shown. The typical for soft tissues shape of these curves with three characteristic phases can been seen [14,15]. In the first part (I), the stress increases exponentially with increasing strain. In the second part (II), the relationship is fairly linear. In the third part (III), the relationship is nonlinear and ends with break. The "toe" region (I) is usually the physiological range in which the tissue normally functions. The Young's modulus value for pig tendon was 77 ± 22 MPa and 156 ± 57 MPa for sheep tendon. The Young's modulus value is twice as high for sheep's tendons in comparison to the swine ones. The average ultimate stress for pig tendon was 8.4 ± 3.4 MPa at stretch of

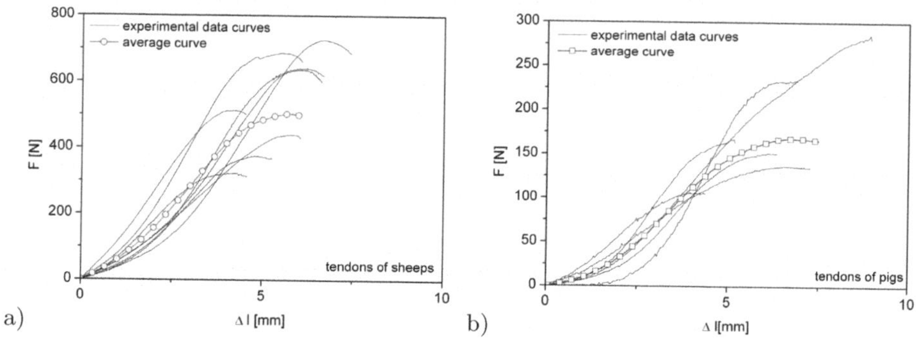

Fig. 1. Comparison between all experimental curves and the average curve obtained from uniaxial tensile tests for sheep tendon (a) and pig tendon (b)

1.2 and 17.5 ± 7.4MPa at stretch of 1.2 for sheep tendon. Each obtained stress-stretch curves and the average curve were used for constitutive modeling with the use of models presented above. Additionally, both for pig and sheep tendons, the range of experimental data used for obtaining fitting curves was limited to phases I and II of stress-stretch curves (Fig. 2), but fitting curves were extrapolated to the third phase of elongation (III). Stress-stretch curves obtained after procedure of fitting to experimental data were presented in Fig. 3a for sheep tendons and in Fig. 3b for pig tendons. In order to estimate model curves for whole range of stretch, the exploration of models to stretch value at 1.2 was made (for sheep Fig. 4a and b, for pig: Fig. 5a and b). In this way, the extrapolation of models solutions in the third phase (III) was performed. Two from seven models, the Yeoh and the Martins fitted the data well for the full range of experimental data with average $R^2 = 0.99$ for sheep tendon (Fig. 3a) and $R^2 = 0.99$ for pig tendon (Fig. 3b). For sheep tendon, in case of experimental data limited to stretch level of 1.12, six from seven models closely fitted the experimental data ($R^2 = 0{,}99$) (Fig. 4a). Above this stretch level, a highly accurate fit between two models (Yeoh, Martins) and experimental data can be seen (Fig. 4b). For sheep tendon, the average experimental stress value was 17.5 MPa at stretch of 1.2, whereas this value calculated with the use of Martins model was 13.6 MPa at stretch of 1.14. For Yeoh model it was 15 MPa at stretch of 1.16. Similar tendency was observed for pig tendon for the stretch level below 1.12 (Fig. 5a). Broadening the scope of analysis to stretch level of 1.2, caused good ability of stress prediction only for Yeoh model (Fig. 5b). The maximum theoretical value of the stress was 5 MPa at stretch of 1.17. In experimental studies this value was 8 MPa at stretch of 1.20 (Tables 2 and 3).

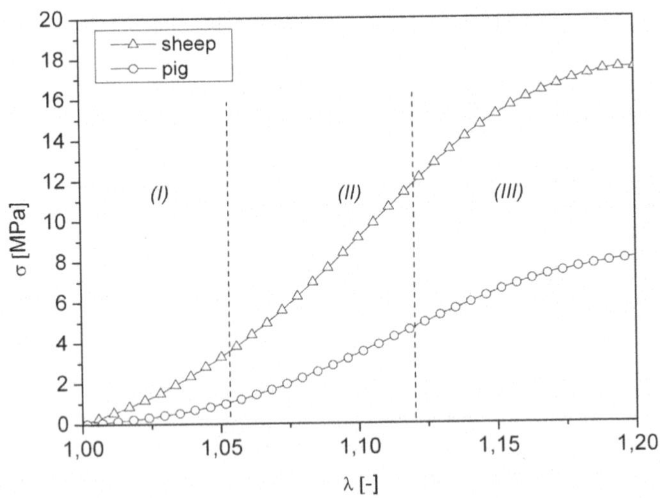

Fig. 2. The average experimental stress-stretch curves for tested tendons with distinguished three phases

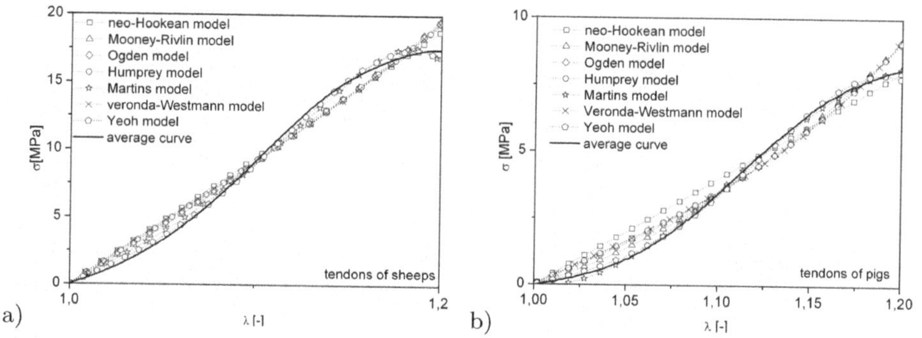

Fig. 3. Fitting curves in the full range of stretch (experimental data from phases (I–III)): (a) for sheep tendon, (b) for pig tendon

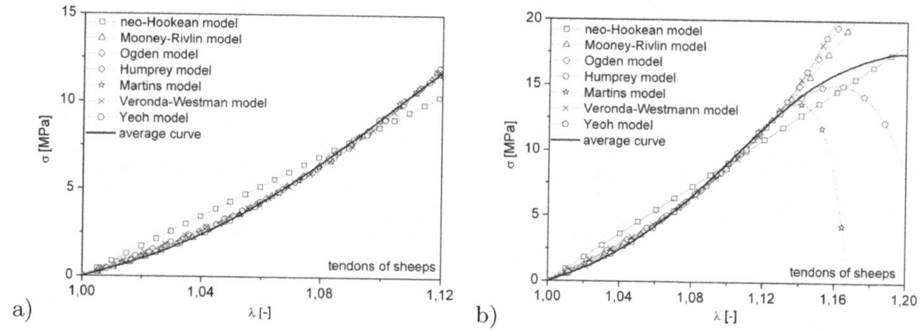

Fig. 4. Fitting curves for sheep tendons: (a) the range of stretch in phases (I–II), (b) the extrapolation of fitting curves to third phase (III)

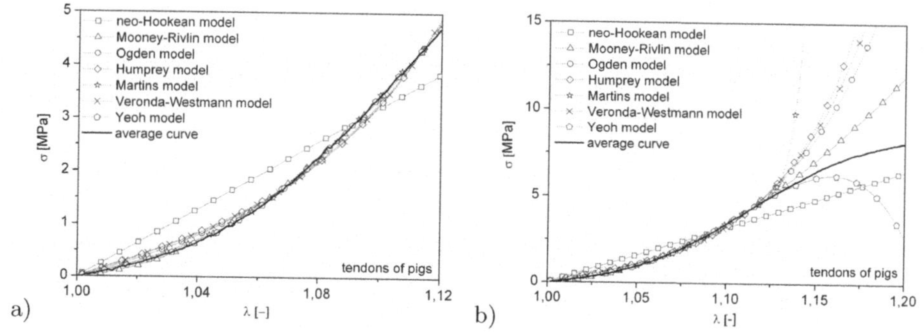

Fig. 5. Fitting curves for pig tendons: (a) the range of stretch in phases (I–II), (b) the extrapolation of fitting curves to third phase (III)

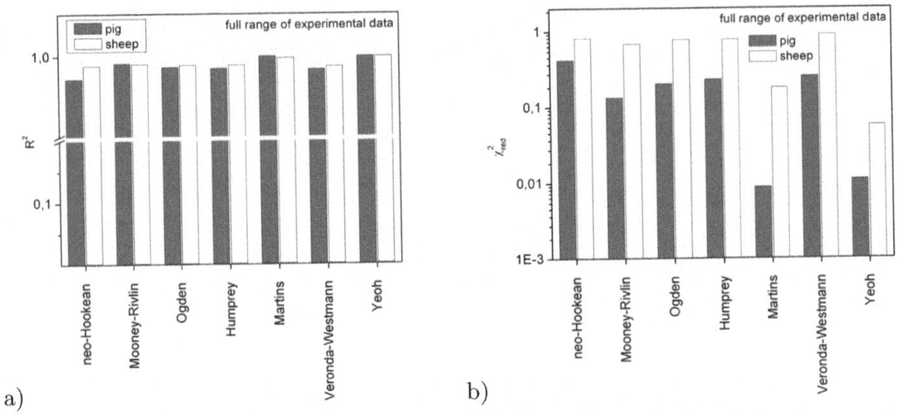

Fig. 6. The coefficient of correlation (R^2) and the reduced chi-squared (χ^2_{red})

Table 2. The values of model coefficients for sheep tendons calculated for all experimental curves and given as mean value and for the average curve (Mean-mean value, Average-Value for average curve)

	C_i	sheep (I)(II)		sheep (I)(II)(III)	
		Mean ± SD	Average ± error of fitting	Mean ± SD	Average ± error of fitting
n-H	C_1	16,23 ± 3,22	14,36 ± 0,18	18,94 ± 3,04	15,59 ± 0,10
M-R	C_1	136,59 ±33,35	97,79 ± 0,51	60,86 ± 40,04	33,09 ± 2,49
	C_2	−95,53 ± 95,68	−91,09 ± 0,56	−48,25 ± 44,22	−20,08 ± 2,86
H	C_1	0,93 ± 0,41	0,93 ± 0,04	9,83 ± 11,39	13,00 ± 2,78
	C_2	14,79 ± 4,56	11,40 ± 0,34	3,18 ± 1,54	1,11 ± 0,22
M	C_1	66,43 ± 15,91	54,03 ± 17,36	33,83 ± 10,05	20,95 ± 8,28
	C_2	−0,63 ± 0,37	−0,03 ± 0,07	−3,37 ± 2,479	−0,15 ± 0,38
	C_3	160,02 ± 48,65	65,92 ± 16,74	83,69 ± 27,10	22,45 ± 10,05
	C_4	0,95 ± 0,37	0,50 ± 0,01	4,57 ± 3,44	2,00 ± 0,09
V-W	C_1	2,32 ± 0,86	3,04 ± 0,12	32,47 ± 28,53	29,89 ± 3,57
	C_2	10,52 ± 2,24	7,19 ± 0,21	1,97 ± 1,07	1,05 ± 0,12
Y	C_1	9,03 ± 2,21	8,58 ± 0,03	9,95 ± 3,29	10,21 ± 0,11
	C_2	238,45 ± 106,44	166,95 ± 1,51	177,46 ± 51,14	104,18 ± 2,05
	C_3	−2420,22 ± 2641,68	−1191,35 ± 20,79	−1153,13 ± 518,53	−547,89 ± 11,26
O	C_1	19,86 ± 5,14	2,60 ± 0,06	31,71 ± 6,97	29,26 ± 0,61
	C_2	19,00 ± 2,37	14,35 ± 0,20	7,61 ± 3,34	5,99 ± 0,34

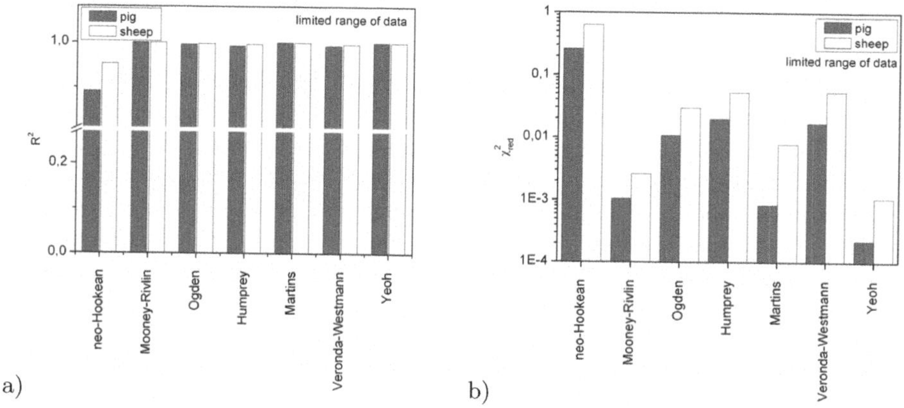

Fig. 7. The influence of upper limit of the elastic strain on models matching results: (a) the coefficient of correlation (R^2), (b) the reduced chi-squared (χ^2_{red})

Table 3. The values of model coefficients for pig tendons calculated for all experimental curves and given as mean value and for the average curve (Mean-mean value, Average-Value for average curve)

	C_i	pig (I)(II)		pig (I)(II)(III)	
		Mean ± SD	Average ± error of fitting	Mean ± SD	Average ± error of fitting
n-H	C_1	$5{,}40 \pm 0{,}82$	$5{,}33 \pm 0{,}11$	$17{,}24 \pm 22{,}85$	$6{,}60 \pm 0{,}06$
M-R	C_1	$66{,}19 \pm 21{,}82$	$56{,}17 \pm 0{,}31$	$22{,}41 \pm 27{,}10$	$28{,}44 \pm 1{,}12$
	C_2	$-66{,}28 \pm 24{,}26$	$-55{,}56 \pm 0{,}34$	$-21{,}78 \pm 25{,}64$	$-25{,}09 \pm 1{,}29$
H	C_1	$0{,}19 \pm 0{,}18$	$0{,}17 \pm 0{,}01$	$3{,}04 \pm 3{,}29$	$1{,}60 \pm 0{,}17$
	C_2	$24{,}56 \pm 10{,}07$	$18{,}64 \pm 0{,}60$	$3{,}78 \pm 2{,}01$	$3{,}29 \pm 0{,}28$
M	C_1	$97{,}84 \pm 26{,}49$	$81{,}04 \pm 1{,}23$	$5{,}98 \pm 3{,}41$	$3{,}91 \pm 2{,}34$
	C_2	$-0{,}31 \pm 0{,}32$	$-0{,}23 \pm 0{,}01$	$-28{,}61 \pm 14{,}25$	$-30{,}47 \pm 14{,}56$
	C_3	$257{,}96 \pm 72{,}51$	$211{,}99 \pm 3{,}41$	$12{,}29 \pm 8{,}82$	$7{,}38 \pm 5{,}95$
	C_4	$0{,}37 \pm 0{,}38$	$0{,}29 \pm 0{,}01$	$46{,}13 \pm 22{,}73$	$47{,}80 \pm 32{,}81$
V-W	C_1	$0{,}59 \pm 0{,}60$	$0{,}53 \pm 0{,}03$	$7{,}25 \pm 6{,}73$	$4{,}77 \pm 0{,}44$
	C_2	$17{,}36 \pm 8{,}04$	$12{,}12 \pm 0{,}38$	$2{,}69 \pm 1{,}29$	$2{,}33 \pm 0{,}16$
Y	C_1	$1{,}80 \pm 1{,}72$	$1{,}93 \pm 0{,}01$	$2{,}67 \pm 2{,}72$	$2{,}68 \pm 0{,}05$
	C_2	$101{,}14 \pm 39{,}88$	$90{,}79 \pm 0{,}65$	$69{,}94 \pm 42{,}58$	$62{,}68 \pm 0{,}92$
	C_3	$-776{,}35 \pm 859{,}85$	$-559{,}55 \pm 8{,}75$	$-381{,}16 \pm 280{,}72$	$-283{,}04 \pm 4{,}99$
O	C_1	$5{,}04 \pm 3{,}06$	$5{,}13 \pm 0{,}08$	$11{,}54 \pm 2{,}82$	$9{,}76 \pm 0{,}26$
	C_2	$24{,}85 \pm 7{,}25$	$22{,}16 \pm 0{,}29$	$7{,}50 \pm 4{,}65$	$9{,}71 \pm 0{,}34$

4 Discussion

The hyperelastic material models are commonly used to describe behavior of soft tissues in an uniaxial load conditions [7,8,10]. A wide variety of models has been proposed to describe tendon tissue behavior from both viscoelastic [10,14,16] and hyperelastic point of view [10,16]. In this paper, we evaluated the ability of hyperelastic models to predict stress-strain behavior of animal tendons. The predictive ability of each model was evaluated for two ranges of experimental data (full and limited range of stretch). These analysis showed that prediction of mechanical behavior of tendons in the first and second phase of elongation is possible with the use of several hyperelastic material models but the best results were obtained for the Yeoh and the Martins models. In the case of the third phase prediction, only two models were appropriate (the Yoeh and Martins). Previous studies evaluating the usefulness of hyperelastic models for predicting tendon behavior showed good fitting for different models. Money-Rivlin model showed good correlation between the experimental data and the fitted curve for human ACL at strain of 0.1 [16]. De Frate et al. [5] used Money-Rivlin model to fit experimental data of ACL in different strain levels. It showed quite good prediction ability, especially as the strain level used to fit the model increased. In literature, it is clearly shown that tendons cannot be regarded as a homogenous structure [14]. It has to be taken into account when creating a numerical model of the tendon. To better known behavior of tendons a three-dimensional constitutive framework is necessary. For accurate stress predictions, the transverse isotropy must be considered to describe tendon material symmetry. The physical interpretation of transversely isotropic symmetry is that of a matrix reinforced with a single fiber group. The material behavior results from matrix properties as well as fiber-matrix and fiber-fiber interactions. In aim to precise represent tendons mechanics, an uniaxial tensile tests should be supplemented by the multi-axial quasi-static and viscoelastic examinations [4,11,12].

5 Conclusions

The constitutive modeling conducted in this study was based on polynomial forms of the strain energy functions because of their simplicity, and therefore calculation efficiency. Hyperelastic approach in modeling tendon behavior under uniaxial tensile loading showed that several models provide good correlation between the experimental and predicted stress-strain curves for animal tendons. Among the number of used models, Yeoh and Martins models were found to be the most suitable choice. In addition, this analysis revealed a number of aspects related to the sensitivity of models to experimental data. The range of experimental data used to fit the model was crucial factor for the predictive ability of each model. The most of used hyperelastic models closely fitted the experimental data at low strain levels. The extension of analyzed strain level revealed poor prediction ability for the most of them, reducing the number of good fitted models to two (Yeoh and Martins). It should be considered that the material parameters are computed by fitting the experimental data to the model and, thus, can vary greatly, depending not only on the specimens and the experimental conditions but also on the so-called fitting parameter. The comparison of modeling on the base on the average stress-stretch curve and all stress stretch curves showed similar results, thus using average experimental data is sufficient.

References

1. Fung, Y.C.: Biomechanics. Mechanical Properties of Living Tissues. Springer, USA (1993)
2. Maganaris, C.N., Narici, M.V.: Mechanical properties of tendons. In: Maffulli, N., Renström, P., Leadbetter, W.B. (eds.) Tendon Injuries, pp. 14–21. Springer, London (2005)
3. Abraham, A., Moyer, J., Villegas, D., Odegard, G., Donahue, T.: Hyperelastic properties of human meniscal attachments. J. Biomech. **44**(3), 413–418 (2011)
4. Pena, E., Calvo, B., Martinez, M.A., Doblare, M.: An anisotropic visco-hyperelastic model for ligaments at finite strains. Formulation and computational aspects. Int. J. Solids Struct. **44**, 760–778 (2007)
5. De Frate, L.E., Li, G.: The prediction of stress relaxation of ligaments and tendons using the quasi-linear viscoelastic model. Biomech. Model. Mechanobiol. **6**, 245–251 (2007)
6. Pena, E., Pena, J.A., Doblare, M.: On modelling nonlinear viscoelastic effects in ligaments. J. Biomech. **41**, 2659–2666 (2008)
7. Safshekan, F., Tafazzoli-Shadpour, M., Abdouss, M., Shadmehr, M.B.: Mechanical characterization and constitutive modeling of human trachea: age and gender dependency. Materials (Basel) **9**(6), 456 (2016)
8. Martins, P., Jorge, R.N., Ferreira, A.: A comparative study of several material models for prediction of hyperelastic properties: application to silicone-rubber and soft tissues. Strain **42**, 135–147 (2006)
9. Chagnon, G., Rebouah, M., Favier, D.: Hyperelastic energy densities for soft biological tissues: a review. J. Elast. **120**, 129–160 (2015)
10. Fang, F., Lake, S.P.: Modelling approaches for evaluating multiscale tendon mechanics. Interface Focus (2016). https://doi.org/10.1098/rsfs.2015.0044
11. Guo, Z.Y., Peng, X.Q., Moran, B.: A composites-based hyperelastic constitutive model for soft tissue with application to the human annulus fibrous. J. Mech. Phys. Solids **54**, 1952–1971 (2006)
12. Weiss, J., Gardiner, J., Ellis, B., Lujan, T., Phatak, N.: Three-dimensional finite element modeling of ligaments: technical aspects. Med. Eng. Phys. **27**, 845–861 (2005)
13. Polak, K., Czyż, M., Ścigała, K., Jarmundowicz, W., Będziński, R.: Biomechanical characteristics of the porcine denticulate ligament in different vertebra levels of the cervical spine - preliminary results of an experimental study. J. Mech. Behav. Biomed. Mater. **34**, 165–170 (2014)
14. Johnson, G., Livesay, G., Woo, S., Rajagopal, K.: A single integral finite strain viscoelastic model of ligaments and tendons. J. Biomech. Eng. **118**, 221–226 (1996)
15. Janjic, N., Ninkovic, S., Harhaji, V., Stankovic, M., Savic, D., Milankov, M.: Biomechanical properties of porcine tendon. ICET-2013;T.1–1.9:1–4 (2013)
16. Chakraborty, S., Mondal, D., Motalab, M.: Constitutive modeling of the human anterior cruciate ligament (ACL) under uniaxial loading using viscoelastic prony series and hyperelastic five parameter Mooney-Rivlin model. In: AIP Conference Proceedings (2016). https://doi.org/10.1063/1.4958358

A Workspace Analysis for a Planar Model of a Tibiofemoral Joint - A Preliminary Study

Jakub Gałuszka and Adam Ciszkiewicz(✉)

Institute of Applied Mechanics,
Cracow University of Technology, Cracow, Poland
jakub.galuszka@interia.pl, adam.ciszkiewicz@pk.edu.pl

Abstract. The aim of this study was to compute and analyze the workspace of a tibiofemoral joint. To facilitate this research, an established, planar multibody model the joint was used. The model was composed of two rigid bodies corresponding to the tibia/fibula complex and the femur. These bodies were connected by a system of four nonlinear cables representing the ligaments and two Hertzian contact pairs, which modeled the cartilage of the knee. The workspace was computed by iteratively modifying the location of the tibia/fibula segment, specified by two linear coordinates and one angular coordinate. At each location, custom software prepared in Python iterated over the six elements of the joint and computed the loads that they generated. These loads were then compared to the maximal safe loads taken from published experimental studies. The obtained workspace of the tibiofemoral joint was moon-shaped with varying thickness. The largest workspace area was observed for a partially bent knee at 40.00°. Furthermore, significant reductions in the workspace were noted for hyperextension and deep flexion.

Keywords: Multibody System Method · Ligament · Bone contact · Nonlinear system · Knee

1 Introduction

The knee joint is one of the most important structures in the human body. It consists of two subjoints: tibiofemoral (TF) and patellofemoral (PF). The TF joint is responsible for defining the relative motion of the bones, while the PF mostly transfers the forces from the extensor muscles to the bones and is sometimes considered as the drive part of the knee. The TF joint contains the femur and the tibia/fibula, which are connected by a layer cartilage and a set of four ligaments: Anterior Cruciate Ligament (AC), Posterior Cruciate Ligament (PC), Medial Collateral Ligament (MC), Lateral Collateral Ligament (LC). The cartilage in the knee transfers mostly compressive loads, while the ligaments are active only when elongated.

Most elements of body joints exhibit nonlinear material behavior, which makes the problem of modeling them is a very complex one. Currently the main methods used for this purpose are:

- Finite Element Method (FEM) with examples including: [1–7],
- Multibody System Method (MBS) with examples including [8–14].

J. Korbicz et al. (Eds.): PCBEE 2019, AISC 1033, pp. 298–307, 2020.
https://doi.org/10.1007/978-3-030-29885-2_27

The FEM models have been mainly used for accurate analysis of the stress-strain state within the elements of the joint. On the other hand, the MBS models, due to their low numerical complexity, can be used effectively to study the joints over their full range of motion as in gait analysis [15, 16]. In the MBS approach each element of the joint (i.e. ligament, bone, cartilage) is replaced with its mechanical counterpart, such as a rigid body, a cable or a damper. This makes the structure of the model simple. Furthermore, the results are also easier to interpret and draw conclusions from.

Many MBS models of the knee have been proposed in the literature. These models differ in structure and can be divided into three major groups. The first group contains simplified representations of the joint - the knee replaced with a revolute or spherical kinematic pair [15, 17, 18]. In the second group the models are based on more complex parallel platform mechanisms with rigid links [9, 12, 19]. In the last group the models are again similar to parallel robots, nevertheless some of their links are modeled as flexible [8, 20–26]. The referenced models have been applied in various types of simulations including muscle force estimation, analysis of the knee's passive motion, knee's dynamic responses and more. Nevertheless, to the best of our knowledge, a problem of knee workspace has never been considered before. The results of such analysis could be employed in systems for minimally-invasive knee surgery [1, 27].

The aim of this research was to compute and analyze the workspace of the knee in the sagittal plane using a planar MBS model of the TF joint. Our main contributions to the existing research were in proposing a method for iteratively computing the workspace for the TF joint and analyzing the workspace for the assumed joint model. The problem was solved using a custom iterative algorithm prepared in Python. The methodology and the results were included in the following paragraphs.

2 Method

2.1 The Model of the TF Joint

In order to obtain the workspace of the TF joint a planar model of the TF joint presented in [8] was used. Based on [8], the following assumptions were employed:

- the model contained two rigid bodies: the first rigid body represented the tibia and the fibula, while the second one substituted the femur,
- the tibia and the fibula experienced no relative displacements - the ligaments and the cartilage, which connected these bones were not modeled,
- two reference frames were established - the tibia/fibula reference frame and the femur reference frame,
- the femur reference frame was assumed to be stationary, while the tibia/fibula reference frame could be displaced with regards to the tibia reference frame,
- four ligaments connected the bones - AC, PC, MC, LC,
- the ligaments were substituted with nonlinear, quadratic cables,
- the condyles of the tibia were idealized as spheres,
- the condyles of the femur were idealized as planes,
- the contact of the condyles was modeled with Hertz contact mechanics [28].
- the meniscus was not included in the model.

The geometrical data for the model was obtained mainly from a 3D scan of the knee [1]. The material parameters for the bone contact and the ligaments were assumed after [8]. Furthermore, the free lengths of the ligaments were obtained in the neutral location of the joint, which was set in accordance to [8]. The model in its neutral location was presented in Fig. 1.

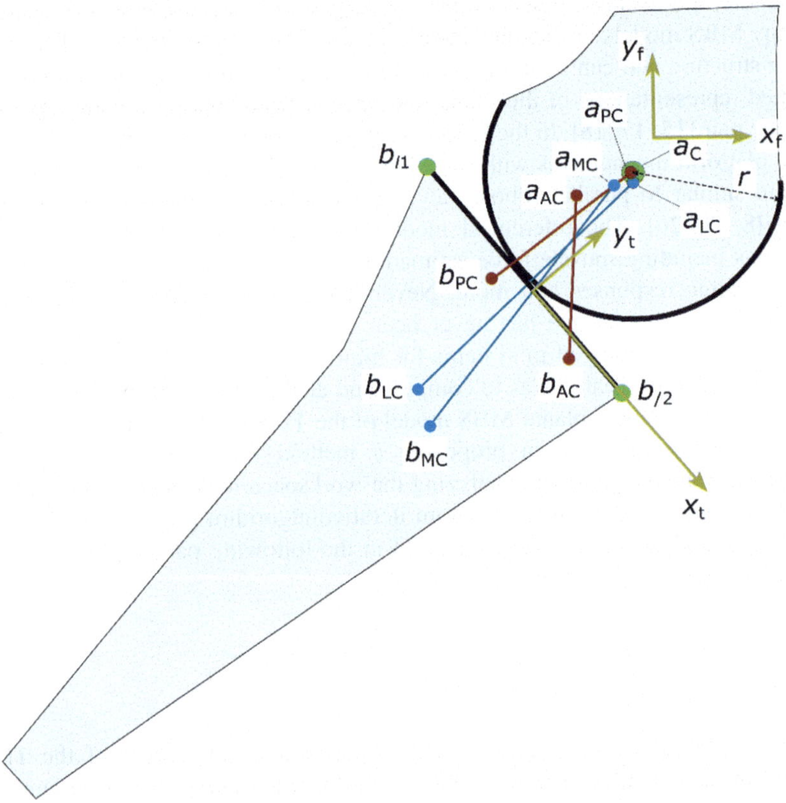

Fig. 1. The assumed model of the TF joint.

In this state, the model allowed for a fairly accurate prediction of the ligament loads and simplified estimation of the stress state in the bone-contact area of the TF joint.

To specify the location of the tibia/fibula in the tibia reference frame three variables were used:

- two linear coordinates p_x and p_y, which constituted the translation vector $\boldsymbol{p}$ between the reference frames of the femur and the tibia/fibula,
- one angular coordinate α, which defined the rotation matrix $\boldsymbol{R}$ between the frames; for the joint in full extension $\alpha = 0.00°$; in hyperextension $\alpha > 0.00°$; normal flexion occurred between $-120.00° < \alpha < 0.00°$, while deep flexion corresponded to $\alpha < -120.00°$ [29].

2.2 The Workspace

The workspace of the joint was defined as a set of locations of the tibia/fibula with regard to the femur specified by $\boldsymbol{p}$ and α, in which all the elements of the TF joint (the ligaments and the cartilage) were loaded below their maximal safe loads. In other words, the workspace contained the locations, which the tibia/fibula could safely achieve. The safe loads for the ligaments were assumed after [8], while the load for the cartilage was estimated based on a body weight of an adult male.

In order to compute the workspace, custom software was prepared in Python with Numpy [30] and Matplotlib. The software iterated over multiple locations of the tibia/fibula. At each location, the load was computed for every element in the joint. The obtained load was then compared to the maximal allowed load for the element. If the obtained load was smaller than the allowed one in all of the elements, the location was added to the list of safe locations. Finally, the list was visualized with Matplotlib.

To compute the force generated by a nonlinear cable at the location of the joint defined by $\boldsymbol{p}$ and α, the following procedure was employed. In the first step, the tibial/fibial attachment of the cable b_i was transformed to the femur reference frame:

$$\boldsymbol{b}_i = \boldsymbol{R}(\alpha)\boldsymbol{b}_i^t + \boldsymbol{p}, \tag{1}$$

where: $\boldsymbol{R}$ - the rotation matrix, $\boldsymbol{b}_i$ ($\boldsymbol{b}_i^t$) - the position vector of the tibial/fibial attachment of the cable i with regards to the tibia/fibula (femur) reference frame.

Then, the deformation Δl_i of the nonlinear cable was computed:

$$\Delta l_i = |\boldsymbol{b}_i - \boldsymbol{a}_i| - l_{free,i}, \tag{2}$$

where: $\boldsymbol{R}$ - the rotation matrix, $\boldsymbol{a}_i$ - the position vector of the femoral attachment of the cable i with regards to the femur reference frame, $l_{free,i}$ - the free length of the cable i, computed based on the joint's neutral position.

If the deformation Δl_i was larger than 0, the value of the force generated by the cable F_i was obtained, as per [8]:

$$F_i = k(\Delta l_i)^2, \tag{3}$$

where: k - the stiffness coefficient of the cable i.

To compute the force generated by the Hertzian contact pair the line $b_{l1}b_{l2}$ representing the cartilage was transformed to the femur reference frame. Then the distance between the center of the circle a_c and the line $b_{l1}b_{l2}$ was obtained. Based on this distance the relative penetration δ of the contact elements was computed:

$$\delta = l_{l1l2-c} - r, \tag{4}$$

where: l_{l1l2-c} - the distance between the center of the circle a_c and the line $b_{l1}b_{l2}$, r - the radius of the circle a_c representing the femoral condyle.

The obtained penetration δ was used to compute the contact force value:

$$F_{c,i} = \frac{2E\sqrt{r}}{3(1 - \nu^2)}\delta^{3/2}, \tag{5}$$

where: E, ν - the Young's modulus/Poisson's ratio of the elements in contact

Based on the contact force, the contact stress was computed, as per [31]:

$$\sigma_{\max,i} = \frac{1}{\pi}\sqrt[3]{\frac{3}{2}\frac{F_{c,i}E^2}{r^2\left(1-\nu^2\right)^2}}. \tag{6}$$

3 Results and Discussion

3.1 Testing the Procedure: A One-Ligament and a Contact-Only Model

In order to verify the proposed software, the initial tests were carried out on two simplified models: a single-ligament model and then a contact-only model. In both cases α was fixed to 30.00°, while p_x and p_y were modified - see Fig. 2.

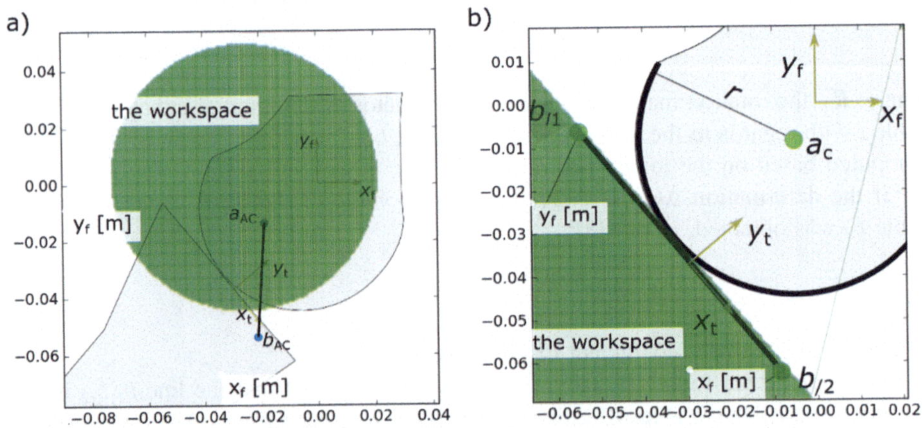

Fig. 2. The workspace for: (a) a single-ligament model, (b) a contact-only model, units: [m].

The workspace for these simplified models can be obtained analytically. For the first model, the analytical workspace has a circular shape - the obtained numerical result was correct, see Fig. 2a. For the contact-only model with plane-sphere Hertzian pairs, the analytical workspace is a half plane. Again, the proposed algorithm for iterative workspace computation returned correct results - Fig. 2b.

3.2 Simulations for the Complete TF Model

After the verification of the model's components, the workspace simulations were performed using the complete 6-element TF model and over the full range of the joint angle $\alpha = -150.00° : 20.00°$ with a step of 10.00°. The step the for the linear coordinates was set to 0.36 mm. In total 720000 locations of the tibia/fibula were checked to compute the workspace.

Selected snapshots from the simulations at different joint angles were presented in Figs. 3 and 4. The workspace in these cases resembled a disc with a flattened side towards the femur. The femur side was determined by the Hertzian contact pairs as seen in the previous test for the bone-contact only model - Fig. 2b. The shape of the rounded side - towards the tibia/fibula - reflected the relative geometry of the nonlinear cables representing the ligaments.

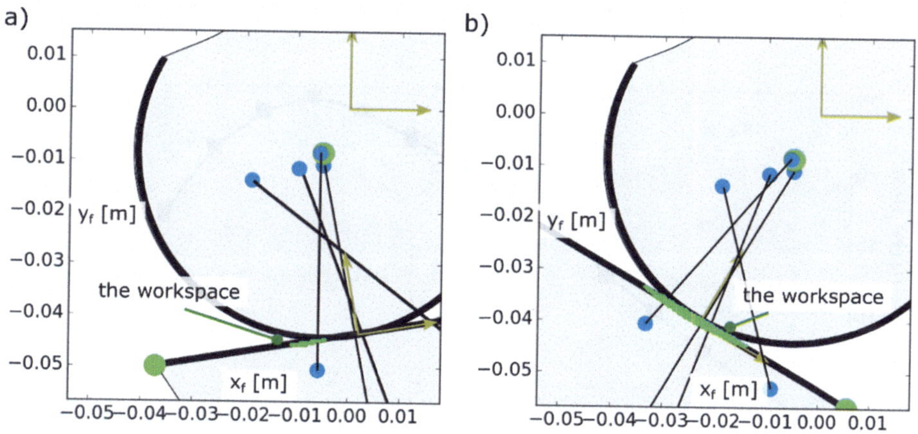

Fig. 3. The workspace for the TF model at α equal to: (a) 10.00°, (b) −30.00°, units: [m].

It is worth noting that while the model allowed for hyperextension of 20.00°, the workspace in the hyperextension phase, for $\alpha > 0.00°$, was significantly reduced when compared to the normal flexion range for α in (−120.00°; 0.00°), see Figs. 3 and 4. The reduced workspace in the hyperextension was more pronounced when considering the workspace area at different angles - see Fig. 5. After the hyperextension, a nonlinear increase of the workspace was observed for α in range of (−40.00°; 0.00°). The peak workspace area of 24.22 mm^2 was noted for $\alpha = -40.00°$. Further flexion of the joint resulted in a nearly linear decrease of the workspace area.

As seen in Fig. 5, the performed simulations also covered the deep flexion phase - $\alpha < -120.00°$. In this range a linear decrease of the workspace area was noted.

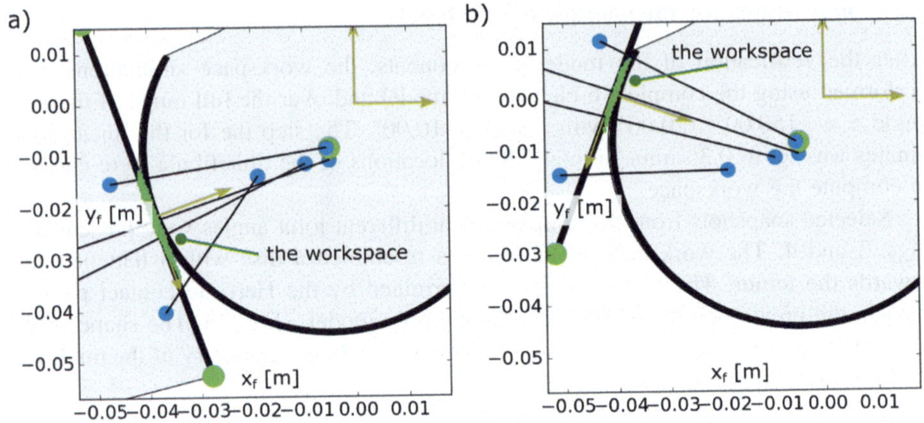

Fig. 4. The workspace for the TF model at α equal to: (a) −70.00°, (b) −110.00°, units: [m].

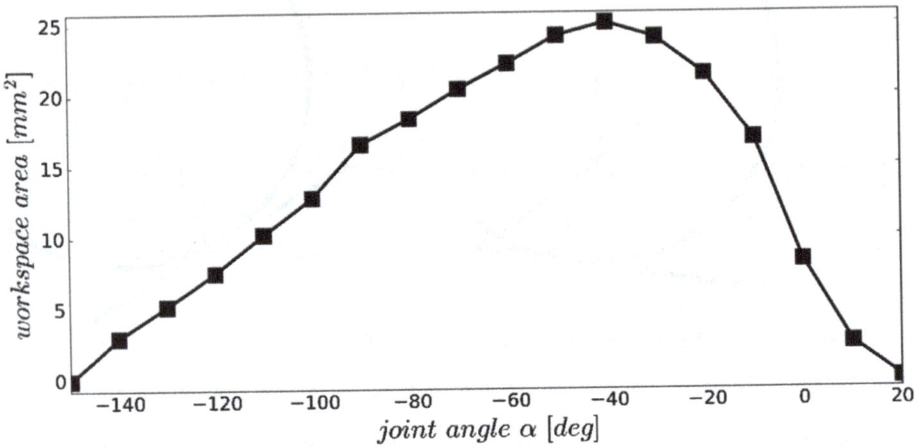

Fig. 5. The workspace area [mm^2] versus the joint angle α [deg].

The complete workspace, for all the considered values of the angle α, was moon-shaped - see Fig. 6. Its femur-side was a perfect circle due to the assumed contact pairs, while the tibia/fibula side was of varying thickness. The smallest thickness was observed in hyperextension and deep flexion.

The results obtained using the assumed planar model of the knee should be further discussed. Probably the most important simplification in the model was it not covering the change in the shape of the condyles in the deep flexion phase - this could potentially increase the somewhat limited workspace in this part. Including other soft-tissue elements, such as the meniscus, in the model would modify the contact geometry of the cartilage and allow it to transfer loads of higher magnitudes. Furthermore, the actual knee is a complex three-dimensional structure. If the assumed planar model was replaced by a spatial one, the workspaces of its elements would be spatial as well (for ligaments - spheres). This would allow us to study the workspace out of the sagittal plane of the knee. Finally, substituting the ligaments with quadratic cables is simple and provides good results. Nevertheless, to obtain more accurate results, the ligaments would have to as having multiple attachments to the bones. We suspect that their resulting workspaces in this case would resemble ellipsoidal shapes.

Despite the simplifications assumed in the study, the proposed methodology and the obtained results presented new insights regarding the behavior of the knee and could be applied in planning novel surgical procedures for the joint.

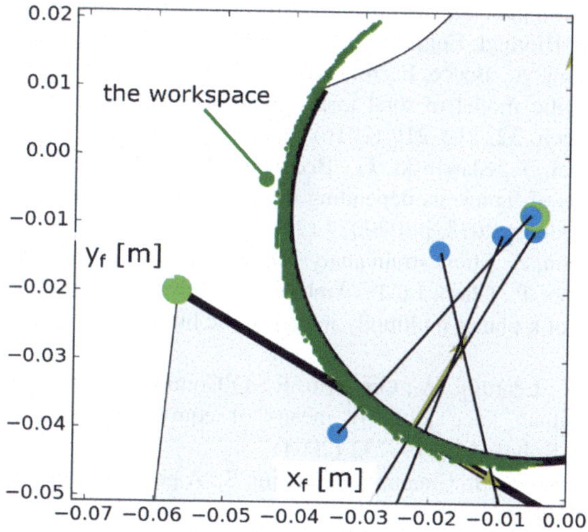

Fig. 6. The workspace for the TF model at all the considered angles, units: [m].

4 Conclusion

In the paper, a method for computing the workspace of the TF joint was presented and analyzed. The proposed algorithm was implemented in Python and tested using an established, planar TF joint model with custom geometrical data. The main conclusions obtained from the results were as follows:

- the obtained workspace of the tibiofemoral joint was moon-shaped with varying thickness,
- the femur-side of the workspace was a perfect circle determined by the assumed Hertzian contact pairs,
- the thickness of the tibia/fibula side reflected the relative geometry of the nonlinear cables representing the ligaments,
- the largest workspace area was observed for a partially bent knee with $\alpha = 40.00°$,
- significantly reduced workspace was noted for hyperextension and deep flexion.

Further studies should include a more accurate description of the bone contact in the deep flexion phase. The meniscus should also be incorporated into the model.

References

1. Ciszkiewicz, A., Milewski, G.: Path planning for minimally-invasive knee surgery using a hybrid optimization procedure. Comput. Methods Biomech. Biomed. Eng. **21**, 47–54 (2018)
2. Klekiel, T., Będziński, R.: Finite element analysis of large deformation of articular cartilage in upper ankle joint of occupant in military vehicles during explosion. Arch. Metall. Mater. **60**, 2115–2121 (2015)
3. Tak-Man Cheung, J., Zhang, M., An, K.N.: Effects of plantar fascia stiffness on the biomechanical responses of the ankle-foot complex. Clin. Biomech. **19**, 839–846 (2004)
4. Szkoda-Poliszuk, K., Żak, M., Pezowicz, C.: Finite element analysis of the influence of three-joint spinal complex on the change of the intervertebral disc bulge and height. Int. J. Numer. Method Biomed. Eng. **34** (2018)
5. Latypova, A., Arami, A., Becce, F., Jolles-Haeberli, B., Aminian, K., Pioletti, D.P., Terrier, A.: A patient-specific model of total knee arthroplasty to estimate patellar strain: a case study. Clin. Biomech. **32**, 212–219 (2016)
6. Arkusz, K., Klekiel, T., Sławiński, G., Będziński, R.: Pelvic vertical shear fractures: the damping properties of ligaments depending on the velocity of vertical impact load. In: AIP Conference Proceedings 2078, p. 020077 (2019)
7. Kubicek, M., Florian, Z.: Stress strain analysis of knee joint. Eng. Mech. **16**, 315–322 (2009)
8. Machado, M., Flores, P., Claro, J.C.P., Ambrósio, J., Silva, M., Completo, A., Lankarani, H. M.: Development of a planar multibody model of the human knee joint. Nonlinear Dyn. **60**, 459–478 (2009)
9. Parenti-Castelli, V., Leardini, A., Gregorio, R., O'Connor, J.: On the modeling of passive motion of the human knee joint by means of equivalent planar and spatial parallel mechanisms. Aut. Robot. **16**, 219–232 (2004)
10. Leardini, A., O'Connor, J.J., Catani, F., Giannini, S.: A geometric model of the human ankle joint. J. Biomech. **32**, 585–591 (1999)
11. Ciszkiewicz, A., Milewski, G.: Ligament-based spine-segment mechanisms. Bull. Polish Acad. Sci. Tech. Sci. **66**, 705–712 (2018)
12. Ciszkiewicz, A., Knapczyk, J.: Load analysis of a patellofemoral joint by a quadriceps muscle. Acta Bioeng. Biomech. **18**, 111–119 (2016)
13. Ciszkiewicz, A., Milewski, G.: A novel kinematic model for a functional spinal unit and a lumbar spine. Acta Bioeng. Biomech. **18**, 87–95 (2016)
14. Gudavalli, M.R., Triano, J.J.: An analytical model of lumbar motion segment in flexion. J. Manipulative Physiol. Ther. **22**, 201–208 (1999)
15. Delp, S.L., Anderson, F.C., Arnold, A.S., Loan, P., Habib, A., John, C.T., Guendelman, E., Thelen, D.G.: OpenSim: open-source software to create and analyze dynamic simulations of movement. IEEE Trans. Biomed. Eng. **54**, 1940–1950 (2007)
16. Montefiori, E., Modenese, L., Di Marco, R., Magni-Manzoni, S., Malattia, C., Petrarca, M., Ronchetti, A., de Horatio, L.T., van Dijkhuizen, P., Wang, A., Wesarg, S., Viceconti, M., Mazzà, C.: An image-based kinematic model of the tibiotalar and subtalar joints and its application to gait analysis in children with Juvenile Idiopathic Arthritis. J. Biomech. **85**, 27–36 (2019)
17. Apkarian, J., Naumann, S., Cairns, B.: A three-dimensional kinematic and dynamic model of the lower limb. J. Biomech. **22**, 143–155 (1989)
18. Zuk, M., Syczewska, M., Pezowicz, C.: Influence of uncertainty in selected musculoskeletal model parameters on muscle forces estimated in inverse dynamics-based static optimization and hybrid approach. J. Biomech. Eng. **140**, 121001 (2018)

19. Sancisi, N., Parenti-Castelli, V.: A 1-Dof parallel spherical wrist for the modelling of the knee passive motion. Mech. Mach. Theory **45**, 658–665 (2010)
20. Moeinzadeh, M.H., Engin, A.E., Akkas, N.: Two- dimensional dynamic modeling of human knee joint. J. Biomech. **316**, 253–264 (1983)
21. Abdel-Rahman, E., Hefzy, M.S.: A two-dimensional dynamic anatomical model of the human knee joint. J. Biomech. Eng. **115**, 357–365 (1993)
22. Caruntu, D.I., Hefzy, M.S.: 3-D anatomically based dynamic modeling of the human knee to include tibio-femoral and patello-femoral joints. J. Biomech. Eng. **126**, 44 (2004)
23. Blankevoort, L., Huiskes, R.: Ligament-bone interaction in a three-dimensional model of the knee. J. Biomech. Eng. **113**, 263–269 (1991)
24. Machado, M., Flores, P., Ambrosio, J., Completo, A.: Influence of the contact model on the dynamic response of the human knee joint. Proc. Inst. Mech. Eng. Part K J. Multibody Dyn. **225**, 344–358 (2011)
25. Crowninshield, R., Pope, M.H., Johnson, R.J.: An analytical model of the knee. J. Biomech. **9**, 397–405 (1979)
26. Zheng, N., Fleisig, G.S., Escamilla, R.F., Barrentine, S.W.: An analytical model of the knee for estimation of internal forces during exercise. J. Biomech. **31**, 963–967 (1998)
27. Ciszkiewicz, A., Lorkowski, J., Milewski, G.: A novel planning solution for semi-autonomous aspiration of Baker's cysts. Int. J. Med. Robot. e1882 (2018)
28. Hertz, H.: On the contact of solids—on the contact of rigid elastic solids and on hardness. In: Miscellaneous Papers, pp. 146–183. Macmillan, London (1896)
29. Galvin, C.R., Perriman, D.M., Newman, P.M., Lynch, J.T., Smith, P.N., Scarvell, J.M.: Squatting, lunging and kneeling provided similar kinematic profiles in healthy knees—a systematic review and meta-analysis of the literature on deep knee flexion kinematics. Knee **25**, 514–530 (2018)
30. van der Walt, S., Colbert, S.C., Varoquaux, G.: The NumPy array: a structure for efficient numerical computation. Comput. Sci. Eng. **13**, 22–30 (2011)
31. Skorupa, A., Skorupa, M.: Wytrzymałość materiałów: wybrane zagadanienia dla mechaników. Wydawnictwo AGH, Kraków (2002)

Investigation of Helmet-Head Interaction in the Aspect of Craniocerebral Tissue Protection

Monika Ratajczak[1(✉)], Tomasz Klekiel[1], Grzegorz Sławiński[2], and Romuald Będziński[1]

[1] Faculty of Mechanical Engineering, University of Zielona Gora, Szafrana 4 Street, 65-516 Zielona Gora, Poland
m.ratajczak@iizp.uz.zgora.pl

[2] Faculty of Mechanical Engineering, Department of Mechanics and Applied Computer Science, Military University of Technology, Gen. Witolda Urbanowicza 2 Street, 00-908 Warsaw, Poland
http://www.zib.uz.zgora.pl

Abstract. The paper is focused on the problem of brain protection for the soldiers exposed to the effects of improvised explosive devices explosion under military vehicles. The soldier uniform predict the helmet whose main function is ballistic protection. In this article, the paddings mechanical properties were analyzed used in the helmet in context to brain protection against impact head inside the vehicle. As the paddings material the polyurethane foams with different density and mechanical properties were used. For finite element method analysis, a head with helmet model was created. The results suggest strong correlation between paddings characteristic and protection efficiency.

Keywords: Biomechanics of brain tissues · Finite element method · Head protection systems

1 Introduction

Brain injuries are a serious problem among soldiers participating in armed conflicts. The main cause of brain tissue damage is an effect of violent forces during the explosion of explosives, as well as gunshots. It should be noted that in recent years there has been a significant increase in the incidence of head injuries among soldiers. In the years 2001–2005, during the war in Iraq and Afghanistan, 21% of all injuries among American soldiers were focused on a head, face and a neck. In addition, until 2005, 35% of the victims in Iraq and Afghanistan died as a result of traumatic brain injury [1]. Nevertheless, it should be emphasized that the size of soldiers injuries is of a benign type and appears in the form of neurological symptoms some time after the injury. This is due to the fact that structures are deformed, as a result of mechanical overloads, which activates the cascade of neurochemical reactions at the cellular level. These changes may imply brain disorders. The result of these injuries may include cerebrovascular damage, neuronal deformation, hypoxia, cerebral edema, or elevated intracranial pressure.

J. Korbicz et al. (Eds.): PCBEE 2019, AISC 1033, pp. 308–315, 2020.
https://doi.org/10.1007/978-3-030-29885-2_28

Destructive changes of a brain tissue, resulted from mechanical overload, are the cause of serious neurological and neurobehavioral disorders.

Mechanical degradation of the brain soft tissues follows the overload, which is depend on properties of the individual components of the brain microstructures. They represent a variety of materials and interact between components. One of the ways to protect these biological structures is the helmet. The main task of the helmet is to separate the soldier's head from elements that can cause injury, and take away the energy of the impact. The hard shell of the helmet is made of modern composites, such as carbon, and protects against abrasions, cuts of the skin, and small objects. Nevertheless, it should be emphasized that current solutions of head protection systems do not allow full protection of craniocerebral tissues, in particular, soft elements. Helmets are usually a good protection against skull fractures. The main parts of the helmet are an outer shell and an inner lining. The outer part is usually made of Kevlar. The inner layer is usually made of polyurethanes. To reduce the risk of brain injuries, helmets are equipped with a crumple zone, made of mostly foam, extend the duration of the deformation, thereby lowering the energy of load impulse [2]. The role of the outer layer of the helmet is protection against the head penetration, before it is wounded and broken. Another task of the shell is to distribute the energy on a larger helmet area, which is more beneficial than the action of the concentrated loads [3]. However, the rigid outer layer is not able to provide adequate protection for the brain itself. The inertia force, generated when the brain is moved, can cause injuries to the delicate brain tissues. The difference between brain and skull velocities can lead to their multiply collisions [4]. The basic principle of head protection is to limit the force value of the force that could damage it by absorbing some of the kinetic energy due to deformation or destruction of the helmet shell and padding [5]. Full protection of the soldiers, especially their heads, requires the use of appropriate absorbers or energy rejection systems in the opposite direction to the overload impact, the task of which will be to suppress the shock wave acting on the head.

Biomechanical investigations of the brain tissues brought the knowledge on the mechanism of injury and a description of structures damaged as a result of mechanical loads. The contemporary research methods in which physical models are used (head impactors), as well as preparations of human and animal tissues, do not provide a full insight into the response of brain tissues resulting from excessive loads. Therefore, currently, one of the best ways to identify and describe the effort of these structures is computational analysis. One of the convincing approaches to the modelling of tissue destruction with nonlinear characteristics is the finite element method (FEM). This method can be used for the models with irregular geometry, composite materials and models taking into account complex loads and boundary conditions, and, hence, it is the preferred method of research effort of the tissue [6]. Therefore, it seems necessary to develop numerical models that will serve, among other things, as a more effective alternative to the currently used physical models. It should be noted, however, that in literature, there are not many numerical models of human head used for dynamic studies in combat conditions, although the head numerical models are still developed [7]. More difficult is the modelling the head in the context of improving the construction of its protection, in particular Polish soldiers participating in armed conflicts outside the country. The statistical data suggested that not using the helmet correlates with a higher number of fractures and admissions in the ICU, TBI severity and with a higher GCS value [8]. Many brain injury mechanisms are still unknown, therefore, a further research is needed that can improve the safety of combat helmet users and further validate models.

The aim of the paper is analyze an influence of the mechanical material properties of the helmet padding on the biomechanical response of the craniocerebral tissues. The protection system was developed to better dissipation energy created from the impact as result of a head impact on the vehicle structure during IED explosion under the military vehicle. The initial conditions were prepared based on the data from other investigations [9,10].

2 Materials and Methods

Head tissue geometry was obtained on the basis of DICOM images obtained from micro computed tomography CT. Using the segmentation software, a simplified 3D geometry of the skull and the brain was created. The model was imported into CAx class programs, where additional geometry of structures, such as cerebrospinal fluid, was created. Based on the helmet construction descriptions and its digital geometry, a helmet model was created. The helmet consists of two layers: the outer layer and the inner layer. In next step, the helmet' construction was put on the head.

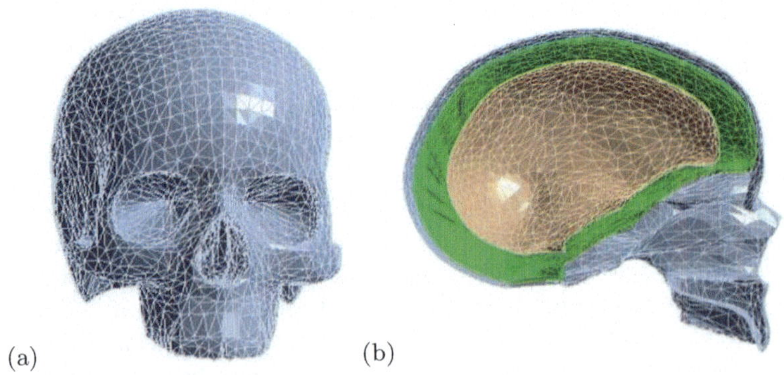

Fig. 1. Head numerical model: (a) skull (b) cross-section view: brain, cerebrospinal fluid, bridging veins.

The entire model was imported into the Ansys software, in which individual sections in the model were separated and material properties as well as boundary conditions were assigned and bridging veins were defined as a cable element.

Bridging veins design was based on anatomical descriptions Kedzia [11], Oka [12] and Kleiven [13]. In the numerical model, the bridging veins were distinguished between the frontal, parietal and occipital parts. The bridging veins in the frontal region are 16–17 mm long and go into the superior sagittal sinus at an angle of about 110°. The first bridging veins in the frontal region is approximately 15 mm from the length of the superior sagittal sinus. Bridging veins in the occipital region are the longest (over 20 mm) and go into the superior sagittal sinus at an acute angle. The geometry of the model is shown in Fig. 1. The schematic of creating a human head numerical model is shown in Fig. 2. The material properties for the head structures, such as the skull, cerebrospinal fluid, brain and bridging veins, used in the model, are summarized in Table 1. Bridging veins were modeled according to Monea descriptions [14].

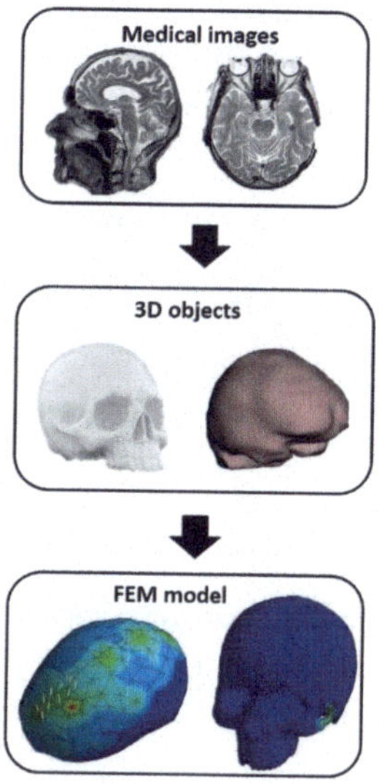

Fig. 2. Diagram of creating a human head numerical model

Table 1. Mechanical properties for head tissues used in the numerical model [7].

Element	Young's modulus [MPa]	Density [kg/m^3]	Poisson ratio
Skull	15000	2000	0.22
Cerebrospinal fluid	$K^* = 2200$	1000	0.49
Brain	Mooney-Rivlin hyperelastic model. $G = 2C_{10} + C_{01}$, where $C_{10} = 62$ Pa and $C_{01} = 69$ Pa		

2.1 Mechanical Properties of Materials Used for the Interior Lining of the Helmet

The work analyzed energy absorption properties for materials, with different stiffness, used for the lining of military helmets. The material used in this investigation is three different forms of polyurethane (PU) foams. The study verified the effect of various mechanical properties of polyurethane foams and assessed the damage to the brain tissues. This padding creates an inner layer of the helmet and can significantly affect the energy absorption coming from the impact. The study analyzed the foams modeled as a linear elastic material.

The PU foams are widely used as core materials in the sandwich constructions as isolating materials in cooling and freezing systems etc. The rigid foams is characterized by a low heat conduction coefficient, low density, low water absorption, and also by a relatively good mechanical strength [15]. The PU foams are applied as a core material of sandwich structures with a steel plates, and its density influences the mechanical stability. These PU foams features indicate good properties needed in the impact energy dissipation.

Table 2. Mechanical properties of the tested polyurethane foams.

Material symbol	Young's modulus [MPa]	Density [kg/m^3]	Poisson ratio	Ref.
mat1	20	800	0.45	[16]
mat2	8.4	63	0.3	[17,18]
mat3	0.8	61	0.3	[17,18]

Table 2 includes the mechanical properties of three selected types of PU foams. For the foams called mat2 and mat3, the Poisson ratio was determined based on the work [18]. The material with similar density and Poisson ratio measurements was selected.

3 Research Results

The paper investigates the effect of different stiffness of PU materials on the brain biomechanical response. Two types of simulations at the speed of 0.7 m/s and 6 m/s were carried out in the study. The results of studies on the influence of foam different stiffness at the speed of 0.7 m/s on the response of the intracranial pressure are presented in Fig. 3. The results for the speed of 6 m/s are shown in Fig. 4. The results presented on both the graphs suggest that the material called mat3 better dissipated energy than other for the high load velocity. The pressure for this material (Fig. 4) increased slowly and reached a lower maximal value. The maximal pressure for materials mat1 and mat2 was reached after 3 ms, however, for material mat3 after 11 ms. As a result, the brain tissues on the outside space are deformed in the shorter time so the energy including to these structures are less and the safety of the brain is increased.

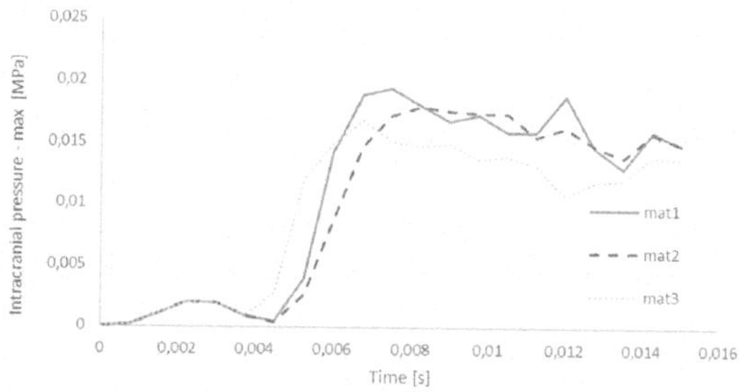

Fig. 3. Intracranial pressure - velocity 0.7 m/s

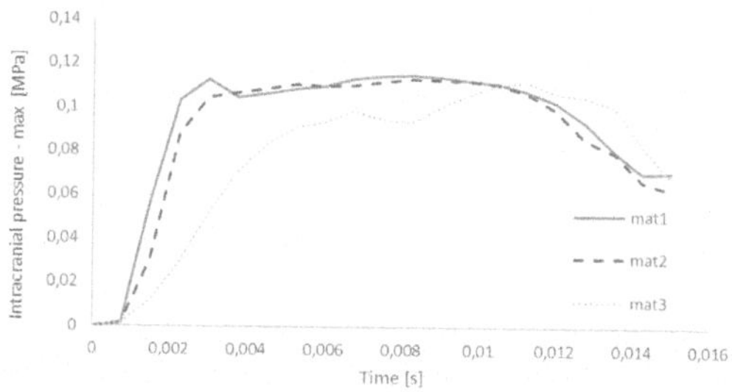

Fig. 4. Intracranial pressure - velocity 6 m/s

4 Summary and Discussion

The helmet lining significantly influences the energy absorption resulting from the impact. The paper analyzes various types of padding with PU foams. The studies have shown that the foam with the lowest stiffness (0.8 MPa) best protects the brain tissue. This effect is partly related to the fact that this foam is subjected to the greatest displacement during the impact. Table 3 lists brain damage thresholds for intracranial pressure. The research shows that at the higher velocity with the placement of materials mat1 and mat2 (with higher stiffness), brain damage could have occurred.

Table 3. Thresholds of brain damage for intracranial pressure.

Damage threshold	Ref.
<235 kPa - injury	[19]
<173 kPa - very mild injury/no injury	[19]
≥300 kPa - mild brain injury	[20]
66–114 kPa - injury	[8]
44–78 kPa - no injury	[8]

Nevertheless, it should be noted that the values given in the literature as acceptable are approximate values calculated on the basis of statistics and examinations during autopsy. Brain injuries can occur below, sometimes above these values. The tests showed that the material used for the inner layer of the helmet should be designed in such a way that it deforms stably under the impact. Foam with too low density will not absorb the right amount of energy, while the foam with too high density will not start to deform under impact, therefore, too much energy will reach head. However, it should be emphasized that while the currently used materials do not protect the soldier against brain injuries, the risk of serious injury is much greater for the unprotected head.

References

1. Friedman, R., Epstein, Y., Gefen, A.: Traumatic brain injury in the military: biomechanics and finite element modelling. In: Gefen, A., Epstein, Y. (eds.) The Mechanobiology and Mechanophysiology of Military-Related Injuries, pp. 209–233. Springer, Cham (2016)
2. Post, A., Hoshizaki, B., Gilchrist, M.D.: Finite element analysis of the effect of loading curve shape on brain injury predictors. J. Biomech. **45**, 679–683 (2012). https://doi.org/10.1016/j.jbiomech.2011.12.005
3. Di Landro, L., Sala, G., Olivieri, D.: Deformation mechanisms and energy absorption of polystyrene foams for protective helmets. Polym. Test. **21**, 217–228 (2002). https://doi.org/10.1016/S0142-9418(01)00073-3
4. Flanagan, S.P.: Biomechanics: A Case-Based Approach. Jones & Bartlett Learning, Burlington (2014)
5. Newman, J.A.: Biomechanics of head trauma: head protection. In: Nahum, A.M., Melvin, J.W. (eds.) Accidental Injury, pp. 303–323. Springer, New York (2002)
6. Miller, K.: Biomechanics of the Brain. Springer Science & Business Media, New York (2011)
7. Ratajczak, M., Ptak, M., Chybowski, L., Gawdzińska, K., Bedzinski, R.: Material and structural modeling aspects of brain tissue deformation under dynamic loads. Materials (Basel) **12**, 271 (2019). https://doi.org/10.3390/ma12020271
8. Zhang, L., Yang, K.H., King, A.I.: A proposed injury threshold for mild traumatic brain injury. J. Biomech. Eng. **126**, 226–236 (2004)
9. Klekiel, T.: Biomechanical analysis of lower limb of soldiers in vehicle under high dynamic load from blast event. Ser. Biomech. **29**, 3–14 (2015)
10. Sławiński, G., Niezgoda, T., Barnat, W., Wojtkowski, M.: Numerical analysis of the influence of blast wave on human body. J. KONES Powertrain Transp. **20**, 381–386 (2013)
11. Kędzia, A., Kedzia, E., Kedzia, W.: Vascular catastrophes and the venous system of the human brain. Adv. Clin. Exp. Med. **19**, 163–176 (2010)
12. Oka, K., Rhoton, A.L., Barry, M., Rodriguez, R.: Microsurgical anatomy of the superficial veins of the cerebrum. Neurosurgery **17**, 711–748 (1985)
13. Kleiven, S., von Holst, H.: Consequences of head size following trauma to the human head. J. Biomech **35**, 153–160 (2002)
14. Monea, A.G., Baeck, K., Verbeken, E., et al.: The biomechanical behaviour of the bridging vein-superior sagittal sinus complex with implications for the mechanopathology of acute subdural haematoma. J. Mech. Behav. Biomed. Mater. **32**, 155–165 (2014). https://doi.org/10.1016/j.jmbbm.2013.12.007

15. Randall, D., Lee, S.: The Polyurethanes Book. Huntsman Polyurethanes, Everberg, Belgium (2002)
16. Przekwas, A., Tan, X.G., Chen, Z.J., et al.: Computational modeling of helmet structural dynamics during blunt impacts. ASME. In: ASME International Mechanical Engineering Congress and Exposition. Biomedical and Biotechnology Engineering, vol. 2, pp. 447–449 (2009). https://doi.org/10.1115/IMECE2009-12958
17. Zhang, L., Makwana, R., Sharma, S.: Brain response to primary blast wave using validated finite element models of human head and advanced combat helmet. Front. Neurol. **4** (2013). https://doi.org/10.3389/fneur.2013.00088
18. Witkiewicz, W., Zieliński, A.: Properties of the polyurethane (PU) light foams. Adv. Mater. Sci. **6**(2), 35–51 (2006)
19. Ward, C., Chan, M., Nahum, A.: Intracranial Pressure-A Brain Injury Criterion. Published in HS-030 682 (SAE-P-88), "Stapp Car Crash Conference (24th) Proceedings," Warrendale, Pennsylvania, 1980, pp. 161–185. Conference held in Troy, Michigan, 15-17 October 1980. Research sponsored by National Highway Traffic Safety Administration (1980)
20. Newman, J., Barr, C., Beusenberg, M.C., et al.: A new biomechanical assessment of mild traumatic brain injury. Part 2 - Results and conclusions. SAE Int 2000-13-00:2–12 (2000)

European Research on Magnetic Nanoparticles for Biomedical Applications: Standardisation Aspects

Peter Schier[1(✉)], Craig Barton[2], Simo Spassov[3], Christer Johansson[4], Daniel Baumgarten[1], Olga Kazakova[2], Paul Southern[5], Quentin Pankhurst[5], Marco Coisson[6], Cordula Grüttner[7], Alex Price[8], Roman Rüttinger[9], Frank Wiekhorst[10], James Wells[10], and Uwe Steinhoff[10]

[1] UMIT - Private University for Health Sciences, Medical Informatics and Technology, 6060 Hall in Tirol, Austria
peter.schier@umit.at
[2] National Physics Laboratory, Teddington TW11 0LW, UK
[3] Institut Royal Meteorologique De Belgique, 1180 Brussels, Belgium
[4] RISE Research Institutes of Sweden AB, 400 14 Gothenburg, Sweden
[5] Healthcare Biomagnetics Laboratory, University College London, London W1S 4BS, UK
[6] Istituto Nazionale di Ricerca Metrologica, 10135 Turin, Italy
[7] micromod Partikeltechnologie GmbH, 18119 Rostock, Germany
[8] BSI Standards Limited, London W4 4AL, UK
[9] DIN Deutsches Institut für Normung e. V., 10787 Berlin, Germany
[10] Physikalisch-Technische Bundesanstalt, 10587 Berlin, Germany

Abstract. Magnetic nanoparticles have many applications in biomedicine and other technical areas. Despite their huge economic impact, there are no standardised procedures available to measure their basic magnetic properties. The International Organization for Standardization is working on a series of documents on the definition of characteristics of magnetic nanomaterials. We review previous and ongoing European research projects on characteristics of magnetic nanoparticles and present results of an online survey among European researchers.

Keywords: Magnetic nanoparticles · Standardisation · European research

1 Introduction

Liquid suspensions of magnetic nanoparticles (MNPs) are used in many technical areas like loudspeakers, mobile phones, vacuum sealings, metal separation and water remediation. In biomedical applications, MNPs play a very important role in in-vitro diagnostics for the separation of cells, bacteria, viruses, protein, nucleic acids and other compartments from blood and body liquids. MNPs are used as contrast agents in Magnetic Resonance Imaging, as well as tracers in Magnetic Particle Imaging and sentinel lymph node detection, as well as in MNP based therapies, where they act as heating agents in magnetic field hyperthermia or as drug carriers in magnetic drug

J. Korbicz et al. (Eds.): PCBEE 2019, AISC 1033, pp. 316–326, 2020.
https://doi.org/10.1007/978-3-030-29885-2_29

targeting and magnetic gene therapies [1]. The economic impact of MNP based biomedical products of European companies alone amounts so far to more than 2 billion € per year [2]. The largest part of this economic impact is generated by in-vitro diagnostics applications. Another large application field is the use of nanostructured iron oxide in pigments for cosmetics, structural engineering, and many other purposes. The annual production of those pigments alone in the European Union is more than 100,000 tons per year [3]. Obviously, this creates a demand for international standards on the main characteristics of MNPs and the respective measurement procedures.

The International Organization for Standardization (ISO) currently prepares document standards on magnetic nanomaterials to provide harmonised definitions for commercial trade, application development, regulation and science in the MNP sector. Current activities focus on magnetic nanosuspensions (ISO 19807-1) [4] and magnetic beads for DNA extraction (ISO 19807-2) [5].

In a new field such as nanotechnology, the foremost interest is the definition of terms and characteristics. These terms can then be used by MNP manufacturers for labelling their products and designing technical data sheets, or by scientists exchanging information on MNPs. While the definitions for chemical composition and mechanical fluid properties could be taken from existing standard documents, many of the magnetic properties of MNP suspensions needed a new definition.

The selection of important MNP characteristics should be based on reviews of the scientific work on MNPs and of economically relevant MNP based applications. Outstanding in this field have been two pre-normative European projects: the EU FP7 project NanoMag, devoted to metrological groundwork on MNP standardisation and the EU COST Action RADIOMAG, that dealt with the standardisation of the measurement of MNP hyperthermia performance. Both projects will be briefly summarised in this paper. Furthermore, this paper is a result of the co-normative EURAMET project MagNaStand, which is supporting the ISO standardisation of MNPs. In the framework of MagNaStand, we have evaluated the European CORDIS database for previous and running EU-projects concerned with MNPs. In addition to the pure database work, we have also performed a survey of researchers involved with MNPs.

2 Pre- and Co-normative EU Projects on Magnetic Nanoparticles

2.1 The Pre-normative EU NanoMag Project

The NanoMag project was funded by the EU FP7 research program in the years 2013–2017 and had a budget of about 11 M€. The NanoMag project objectives were to standardise and harmonise ways to measure and analyse the data for MNP systems. NanoMag brought together leading experts in: synthesis of magnetic single- and multicore nanoparticles, characterisation of magnetic nanoparticles, and national metrology institutes. The project defined standard measurements and techniques which are necessary for defining a magnetic nanostructure and quality control [6]. NanoMag was focused on biomedical applications, for instance bio-sensing (detection of different biomarkers), contrast substance in tomography methods (Magnetic Resonance Imaging and Magnetic Particle Imaging) and magnetic hyperthermia (for cancer therapy).

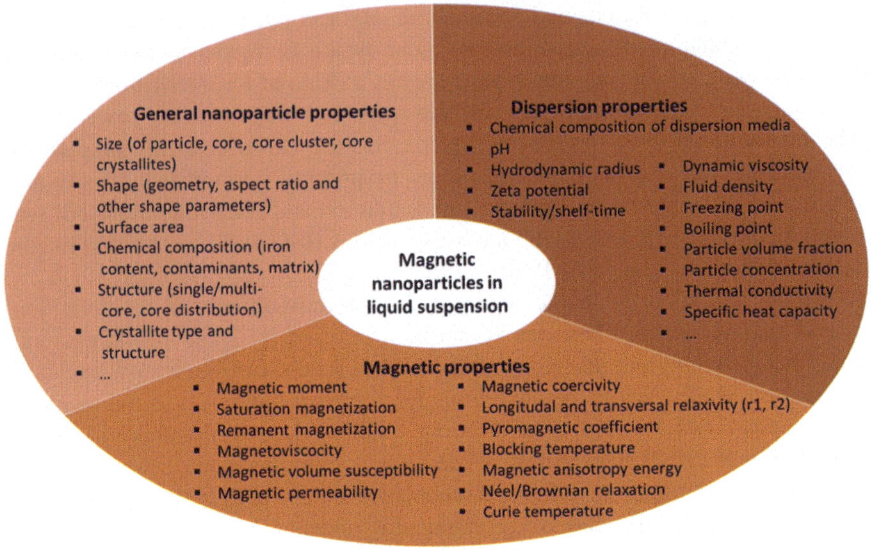

Fig. 1. Important parameters of liquid MNP suspensions that can be standardised.

Initially, NanoMag was designed as a pre-normative project for the preparation of measurement standards. When the International Organization for Standardization started a normative activity for material properties of MNP suspensions, NanoMag members entered the ISO committee TC 229 “Nanotechnologies” as technical experts and made substantial contributions to ongoing ISO standardisation work in the field of MNP systems.

The knowledge of the NanoMag experts has been summarised in four electronic-learning modules that can be accessed via the internet [7]. To support MNP standardisation the NanoMag project has:

- developed a clearer nomenclature to describe the structure and magnetic properties of MNPs and MNP ensembles (see Fig. 1);
- performed surveys of measurement methods for MNPs and their pros and cons, classification of these methods including also classification of different types of MNPs (single-core, multicore, nanoflowers);
- used Monte-Carlo simulations to explain the experimental results, which have led to improved modelling of MNP magnetic properties, especially the dynamic magnetic behaviour; and
- performed initial round robin measurements (using the same analysis methods on the same MNP samples but in different laboratories utilising the developed standard operating procedures (SOPs)) using different types of MNP systems.

A standardisation roadmap developed during the NanoMag project forecasted the development of measurement standards for MNPs to start in the year 2021. Many results of the NanoMag project were summarised in a review paper in [8].

2.2 EU Transdomain COST Action RADIOMAG

The EU COST Action RADIOMAG was a network of over 140 scientists in the field of tumour therapy and it was dedicated to research on multifunctional nanoparticles for Magnetic Field Hyperthermia and Indirect Radiation Therapy. The RADIOMAG project duration was from 2014–2018. An important task of RADIOMAG was the harmonisation and standardisation of the assessment of heat generation by MNPs in an alternating magnetic field, since this is the basic principle in Magnetic Field Hyperthermia [9].

Magnetic Field Hyperthermia, especially in combination with radiotherapy, has been demonstrated as an effective tool to slow down or stop tumour growth and to support anti-cancer therapy in difficult tumour cases like glioblastoma. Pilot studies in humans are underway in a number of clinics in Europe. In April 2019 a new magnetic field hyperthermia treatment centre for brain tumours has been opened at the Independent Public Clinical Hospital No. 4 (SPSK 4) in Lublin, Poland [10].

These continuous developments have also created a niche market for several small and medium-sized enterprises for manufacturing magnetic field hyperthermia test devices. These devices are mainly used by academics testing in-vitro and ex-vivo the efficiency of MNP suspensions to deliver heat. The relevant physical quantity is the Specific Loss Power (SLP). For calorimetric determinations, the SLP is deduced from temperature versus time curves T(t), measured with the SLP test device.

The RADIOMAG activities concerning the standardised characterisation of MNPs for magnetic field hyperthermia focused on SLP measurements:

1. Survey amongst RADIOMAG members on available SLP test devices/setups and their field/frequency combinations;
2. Development of a standard operating procedure for the calorimetric SLP determination, i.e. from T(t) measurements;
3. A comparative SLP determination on water-based ferrofluids between 21 participating laboratories (SLP ring test) and evaluation of SLP calculation methods;
4. Study of the field dependence, i.e. SLP(H); and
5. Design of a possible calibration sample for SLP test devices.

The results showed that a large majority of groups determine the SLP from calorimetric measurements with non-adiabatic setups, also commercially available on the market. In contrast, only a few producers exist for non-calorimetric devices using AC hysteresis and a single laboratory used a “home-made” nearly adiabatic setup.

The RADIOMAG work demonstrated that there are no common procedures available for carrying out T(t) measurements in magnetic field hyperthermia setups. Typically, different laboratories use their own individual best-practice protocols, or follow instructions given in the SLP test device manual, in case of commercial setups. RADIOMAG has therefore developed a SOP for measurements, including a questionnaire for instrument specific parameters.

Furthermore, RADIOMAG performed a ring test on the determination of SLP values, where the same MNP formulation was investigated by 21 different laboratories. The results showed a significant variation of quantitative SLP values, even for an identical MNP suspension. Further analysis of these results is still ongoing, and the publication of detailed results is expected in 2019/2020.

2.3 The EURAMET Project MagNaStand

The EURAMET organisation has established the co-normative project MagNaStand, running from 2017–2020. The objective of MagNaStand is to collect the available knowledge on standardised measurements of MNPs, to create it where it is not readily available, to make this knowledge available for the standardisation of MNPs at ISO level, and to involve stakeholders from industry and academia.

Specifically, this includes the preparation of standard operating procedures for static and dynamic magnetisation measurements and for specific loss power assessment in magnetic hyperthermia. The MagNaStand project works on the definition of long-term stability of MNP suspensions and of SLP in magnetic hyperthermia in a metrologically sound and traceable way.

Another objective of the MagNaStand project is the preparation of future measurement standards for MNPs by summarising the existing metrological knowledge in a structured form for static magnetic susceptibility, dynamic magnetic susceptibility and specific loss power in magnetic hyperthermia.

The MagNaStand projects enables the continuous participation of European experts in the standardisation work at ISO/TC229. Thus, the scientific results of the pre-normative EU projects “NanoMag” and “RADIOMAG” can be introduced into the international standardisation process. This includes the definition of terms for magnetic quantities and the compartments of magnetic nanoparticles, actual versions of SOPs for magnetic measurements, and surveys of industrial requirements on standardisation of magnetic nanoparticles. MagNaStand experts are co-leaders of the development of ISO/TS 19807-2 “Magnetic beads for DNA extraction”, together with experts from China.

3 Search of the EU CORDIS Database for Research Projects that are relevant for MNP Standardisation

In this section we present the evaluation of the Community Research and Development Information Service (CORDIS) database on research projects funded by the EU. The aim of this activity was to provide a short summary of the most important EU FP7 and H2020 projects on MNPs. The search terms are explained in Table 1.

3.1 Results of the Search of the CORDIS Database

The search of the CORDIS website was conducted using all search terms of Table 1 in quotation marks and connected by Boolean OR operators (i.e. “*magnetic *particle” OR “*magnetic *bead*” OR etc.).

Afterwards the result was filtered by the homepage's "Refine by:" option. Only results of the "Content Type: Project" which were part of "Programme: Horizon 2020" or "Programme: FP7" were downloaded in tabular form as CSV-files and - for the full description of the project objectives – as PDF-booklets. The search resulted in a total of 108 EU research projects with summarised project budgets of roughly € 267 million. In comparison, a search in the Research Portfolio Online Reporting Tools (RePORT) of the U.S. Department of Health & Human Services using the same search terms from Table 1 resulted in 214 projects with a total budget volume of $ 67 million. A close inspection of the search results revealed that despite the wide range of the search terms, a number of relevant EU projects concerned with MNPs that were already known to the authors were not captured. They were later added manually to the final list of relevant EU projects concerned with MNPs. The final total number of EU projects concerned with MNPs was 118 with a total budget of € 348 million.

Table 1. Search terms in the CORDIS database for identification of relevant EU projects for MNP standardisation.

Search term	Description
*magnetic *particle*	Since the standardisation of magnetic nanoparticle characteristics is our main goal, it is only natural to find all projects containing any variant form of "*magnetic *particle*" (e.g. "superparamagnetic nanoparticles")
*magnetic *bead*	Some research groups use the notation "bead" instead of "particle". Otherwise, the reason to use this search term is the same as "*magnetic *particle*"
iron oxide nanoparticle	Iron oxide nanoparticles are by far the most commonly used magnetic nanoparticles
superparamagnetic*	A salient feature of MNPs is their superparamagnetic behaviour

4 An Online Survey of Researchers Involved in MNP Projects

We compiled a list of contact addresses of the leaders of the identified EU MNP projects, enhanced by contacts from the NanoMag and RADIOMAG networks. Altogether, over 100 European researchers were asked to participate in an online survey on MNP standardisation. We have received 32 responses, of which the most relevant results are summarised in this chapter in a question-answer scheme.

4.1 Survey Results

Q. 01: Which of the following were your application areas of MNPs? **A. 01**: 32 biomedical applications, 14 MNP synthesis, 6 environmental applications, 9 pharmaceutical applications, 6 other.

Q. 02: What was the Technology Readiness Level (TRL) of your MNP project? **A. 02**: 22% TRL1 – basic principles observed and reported, 19% TRL2 – technology concept and/or application formulated, 30% TRL3 – characteristic proof-of-concept, 23% TRL4 – component validation in lab environment, 2% TRL5 – component validation in relevant environment, 2% TRL6 – prototype demonstration in relevant environment, 1% TRL7 – prototype demonstration in application environment, and 0 in TRL8 – complete system in test and demonstration and TRL 9 – complete system in successful operation.

Q. 03: What kinds of materials for MNPs were used during the project? Please specify core- and coating material. A free-text answer was possible. **A. 03**: Magnetite (13) and maghemite (9) were the most common core materials, other ferrites (Barium-Ferrite, Nickel-Ferrite, Nickel-Cobalt-Ferrite, etc.) (10) played also a role. The coating material came from these material groups:

- Polymer/Organic which included plastics, polysaccharides and organic acids (53)
- Metal/Alloy which included metals, metalloids and alloys (10)
- Biofunctionalized which included bacteria and proteins (7)
- No coating for uncoated MNPs (1)

Q. 04: In which environment(s) were the MNPs used? **A. 04**: 42% laboratory, 33% in-vitro (cell cultures), 25% in-vivo.

Q. 05: Properties of the applied magnetic field: specify field strength and frequency during your MNP application. **A 05**.: see Fig. 2.

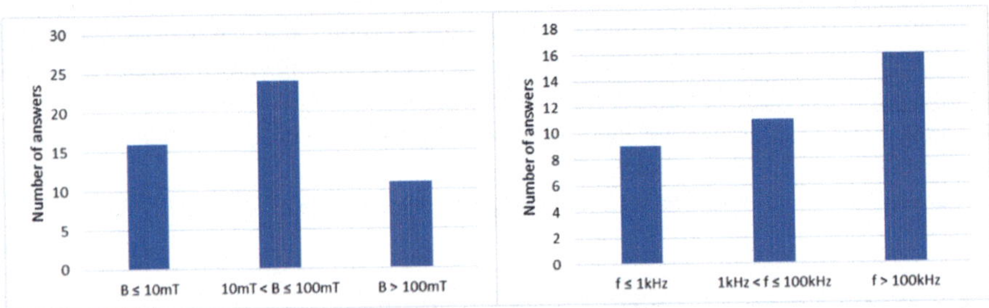

Fig. 2. Magnetic field strength and frequency during the MNP application.

Q. 06: Did you use a static magnetic field or an alternating magnetic field? **A.06**: 25 alternating magnetic field, 18 static magnetic field.

Q. 07: What are the sources of the MNP characteristics that you used, please rank. **A. 07**: Summary rank: 1. own measurements, 2. literature values 3. technical data sheet, 4. custom measurement (see Fig. 3).

Q. 08: Please rate the importance of proposed characteristics for your project (1 = low, 5 = high). **A. 08**: see Fig. 3.

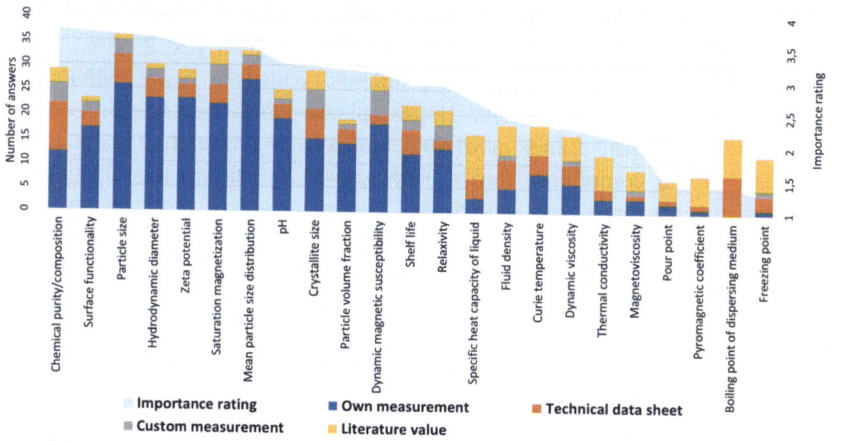

Fig. 3. Sources of different MNP parameters sorted by importance rating.

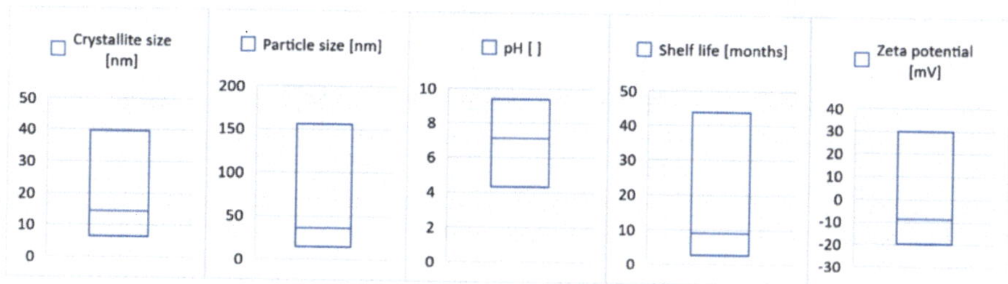

Fig. 4. Mean value ranges for several MNP parameters (minimum, typical and maximum).

Q. 09: Please enter the approximate value range of the characteristics. **A. 09**: see Fig. 4.

Q. 10: Did you encounter any serious problems during the measurement process of important characteristic properties? **A. 10**: 64% no, 21% unclear measurement procedure, 15% measurements were not reproducible.

Q. 11: Are you aware of reference laboratories, where you could check or cross-validate your own measurement results? **A. 11**: 50% yes, 37% no, 13% no answer.

Q. 12: Please rank the most important measurement technique to characterise MNPs. **A. 12**: 1. transmission electron microscopy (TEM), 2. magnetorelaxometry (MRX), 3. Hysteresis loops (DC magnetometry), 4. ZFC/FC curves (temperature dependent DC magnetometry) and 5. X-ray powder diffraction (XRD).

Q. 13: What are the most important characteristics? **A. 13**: 36 particle size, 27 saturation magnetisation, 13 hydrodynamic size, 9 specific absorption rate, 8 stability of suspension, 8 chemical composition, 6 biological properties, 5 dispersity, all other <5.

Q. 14: Did you use a standardised measurement protocol or were there any other standardisation aspects in your project? **A. 14**: 50% no, 25% yes, 25% no answer.

Q. 15: Did you encounter any problems with the used MNPs due to erroneous/unspecified characteristics? If yes, please elaborate. **A. 15**: People were dissatisfied with numerous things involving erroneous or unspecified characteristics of MNPs. Most of the complaints can be summarised by stating that the information provided by manufacturers was incomplete. Complaints involved missing expiration dates, wrong or missing concentration values, wrong or missing magnetite/maghemite indications, unknown surface compositions, stability change over time and recommended storage conditions. Several participants also described problems reproducing previously obtained or published results.

Q. 16: How would you rate the current state of standardisation of MNPs? (1 = low, 4 = high). **A. 16**: see Fig. 5.

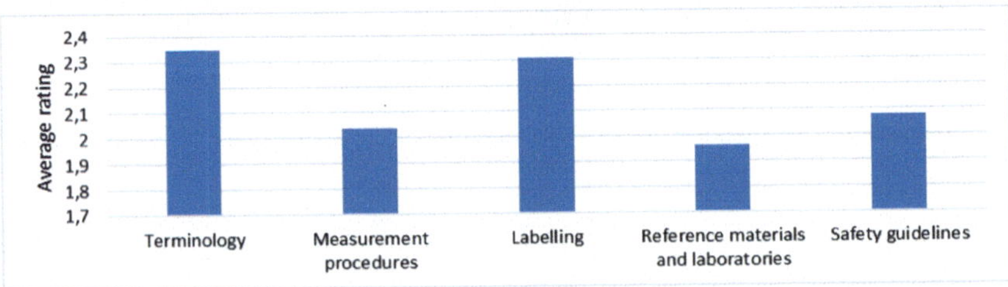

Fig. 5. Rating of current standardisation of different aspects of MNP handling.

4.2 Discussion of the Survey Results

The survey, although it is still based on a limited data base, shows clearly the wide range of applications and parameters that have to be considered when the standardisation of MNPs is intended. It reveals that EU funded MNP research so far is restricted to very early or early Technology Readiness Levels. This might be one reason that the need for standardisation has been underestimated so far. On the other hand, the survey demonstrates that even in early research in scientific labs people often encounter problems with non-standardised parameters and procedures and non-reproducible measurement results. The need for further standardisation is clearly demonstrated. The survey gives valuable hints, which measurement methods and parameters are most important and would most benefit from standardised procedures.

5 Summary

Despite the huge economic importance of magnetic nanoparticles, there exists no standardised description and measurement of their basic magnetic properties. This is especially surprising with respect to the 118 research projects and the budget of € 348 million that the European Union has invested into MNP related projects since the FP7 program. The pre- and co-normative EU projects NanoMag, RADIOMAG and MagNaStand have made decisive steps to change this situation. Currently, Europe is an important contributor to the standardisation of magnetic nanomaterials that is performed in the ISO/TC229 "Nanotechnologies" committee. The efforts for harmonisation are based on surveys and other interactions with academic and industrial stakeholders.

We have presented the detailed content of such a survey in order to stimulate a discussion on the most urgent needs in MNP metrology. In addition to the basic description that is currently developed at ISO, standard operating procedures for magnetic measurements, calibrated measurement devices and certified reference materials for magnetic MNP properties are desirable outcomes of future research.

Acknowledgments. This work was supported by the EMPIR program co-financed by the Participating States and from the European Union's Horizon 2020 research and innovation program, grant no. 16NRM04 "MagNaStand".

References

1. Mohammed, L., Gomaa, H.G., Ragab, D., Zhu, J.: Magnetic nanoparticles for environmental and biomedical applications: a review. Particuology **30**, 1–14 (2017)
2. Roche Diagnostics GmbH: Socio-Economic analysis. Use of diglyme as a process chemical in the manufacture of one specific type of Dynabeads® for use in immunodiagnostic assays (IVD), p. 19 (2016). Authorisation application for use of diglyme at the European Chemical Agency (ECHA). https://echa.europa.eu/documents/10162/8e0b0272-73d1-4e9a-8616-97ad9ee813d9. Accessed 30 Mar 2019
3. Hynes, J., Novotný, T., Nič, M., Danihelka, P., Kocurkova, L., Přichystalová, R., Brzicová, T., Bernatikova, S.: Literature study on the uses and risks of nanomaterials as pigments in the European Union. European Chemicals Agency, p. 23 (2018). ISBN 978-92-9020-623-1
4. ISO/TS 19807-1 Nanotechnologies – Magnetic nanomaterials – Part 1: Specification of characteristics and measurements for magnetic nanosuspensions. International Organization for Standardization (2019)
5. ISO/TS 19807-2 Nanotechnologies – Magnetic nanomaterials – Part 2: Specification of characteristics and measurements for nanostructured superparamagnetic beads for nucleic acid extraction. International Organization for Standardization, under development
6. Bogren, S., Fornara, A., Ludwig, F., del Puerto Morales, M., Steinhoff, U., Hansen, M., Kazakova, O., Johansson, C.: Classification of magnetic nanoparticle systems—synthesis, standardization and analysis methods in the NanoMag project. Int. J. Mol. Sci. **16**(9), 20308–20325 (2015)
7. Magnetic Nanoparticles: Standardisation and Biomedical Applications. 1.5 days training course. National Physics Laboratory. https://training.npl.co.uk/course/magnetic-nanoparticles-standardisation-and-biomedical-applications/. Accessed 30 Mar 2019
8. Wells, J., Kazakova, O., Posth, O., Steinhoff, U., Petronis, S., Bogart, L.K., Southern, P., Pankhurst, Q., Johansson, C.: Standardisation of magnetic nanoparticles in liquid suspension. J. Phys. D Appl. Phys. **50**(38), 383003 (2017)

9. Simo Spassov, D., Ortega, S., Dutz, U., Steinhoff, E., Natividad, J., Wells & RADIOMAG team: RADIOMAG – A COST networking project in experimental cancer treatment research, combining magnetic fluid hyperthermia and radiotherapy. In: Schütt, W., Zborowski, M., Häfeli, U. (eds.) Scientific and Clinical Applications of Magnetic Carriers, p. 42 (2016)
10. Nowa metoda leczenia nowotworów głowy. Otworzyli ośrodek w Lublinie. In: Gazeta Wyborcza w Lublinie, 4 April 2019

A Factorial Design for Assessment of the Effect of Selected Process Variables on the Impulse Electrostatic Droplet Formation

Katarzyna Kramek-Romanowska(✉), Marcin Grzeczkowicz, Paulina Korycka, and Dorota Lewińska

Nalecz Institute of Biocybernetics and Biomedical Engineering Polish Academy of Sciences, Warsaw, Poland
kkramek@ibib.waw.pl

Abstract. The aim of this study was to investigate the effect of changing processing variables on the impulse electrostatic droplet formation, being the first and crucial step during manufacture of novel particles and capsules intended for drug delivery. A factorial design was performed to assess the consequences of altering selected parameters: liquid dynamic viscosity, nozzle inner diameter, liquid flow rate and electric voltage. Pure glycerol and glycerol aqueous solution (c = 85 %wt) were used as model substances and the process of droplet formation was analyzed. The results of this study can be applied to controlled formulation of droplets during impulse electrostatic atomization.

Keywords: Electrostatic droplet formation · Factorial design · Biomedical product engineering

1 Introduction

A detailed understanding of the elementary processes that occur during electrostatic atomization is crucial to control the size of obtained droplets. The capability to predict the characteristics of pharmaceutical products allows for the optimal design of medical formulations, the key issue in the discussion about cost-effective and time-saving manufacture of therapeutic materials (Hamad et al. 2010). A growing need for such research was the reason for dedicating this study to the investigation of the influence of several processing variables on the size of droplets which are generated by electrostatic impulse method and which are the first step product during manufacture of particles and capsules, intended for various biomedical applications.

It should be underlined, that among many application fields for bioencapsulation (e.g. biotechnology, food industry, pharmacy), the stringent requirements concerning capsule size should be met in particular in microencapsulated cell transplantation (Hunkeler et al. 2001). Obtaining desired size and the uniformity of capsules are particularly important for example to avoid the effects of incorporation of encapsulated cells into the capsule membrane or protrusion out of the beads. These effects cause the direct activation of the immunological system of the host and microencapsulated graft rejection (Chang and Prakash 1999). That is way, the droplet formation process, used in

J. Korbicz et al. (Eds.): PCBEE 2019, AISC 1033, pp. 327–336, 2020.
https://doi.org/10.1007/978-3-030-29885-2_30

this application, must ensure the manufacture of beads with adequate and fully controllable size distribution parameters.

One of the techniques, successfully employed for the microencapsulation of living cells, is the electrostatic method with impulse mode (Lewińska et al. 2004, 2012). Droplet formation in this method takes advantage of electrohydrodynamic phenomena occurring in flowing conductive liquids after introduction of an electric field to the system. During the process a flowing liquid is subjected to a high electrical potential difference (at least several kV) in impulse mode, normally applied between the source of the flow (i.e. the nozzle or needle) and an earthed or oppositely charged substrate. Provided the liquid is electrically conductive, this results in the dripping or micro-dripping mode of the formation of droplets, which have decreased sizes as compared to free flow without the electrostatic field.

There are numerous evidences that altering process conditions influences significantly the physical characteristics of the final product, i.e. droplet size. The key parameters are supposed to be the chemical composition and physical properties of the processed solution, its flow rate, the nozzle diameter, the distance over which the potential difference is applied and the magnitude of potential difference (Ahmad et al. 2008; Kim et al. 2014). It should be stressed, that the whole process is of a complex character and the great number of possibly influencing factors complicates considerably the effective control of the product properties.

Because of that, although the effect of physical parameters during the electrostatic droplet formation has an undeniable role in manipulation of produced droplet size, there is still lack of extensive research on this topic, covering possibly broad range of the process aspects. Several studies concentrate mostly on individual variables, as for example the liquid viscosity, being one of prominent physicochemical parameters, was studied by Ku and Kim (2002). The authors compared diameters of produced droplets with existing scaling laws (Ganan-Calvo 1997) and demonstrated an evident discrepancy in the droplet size due to high viscosity of dispersed liquid. In fact there are still too few works describing in details the phenomenology of the process of electrostatic droplet formation in the dripping or micro-dripping modes, as the majority of existing theoretical studies are focused on the cone-jet mode in which the liquid emerges from the nozzle in the form of a regular, symmetric cone with a thin jest at its apex (Roine et al. 2017). An interesting experimental study, related to micro-dripping mode, is presented in work by Lee et al. (2013), but it discusses only liquids with relatively low viscosities (in the range 0.31–1.54 mPas). As a result it cannot be directly applied to the atomization of polymers or encapsulation of biomaterials.

Consequently, in this study, the factorial design, being a well-established technique to identify the most significant information concerning the influence of factors on a specific problem, was employed, as it allows for analyzing complex systems (Costa et al. 2007). Experimental design allows to obtain product or process with desirable characteristics in an efficient way. This means it gives the possibility to: (i) understand the effects of the analyzed factors and/or (ii) model the relationship between the output and input variables, with a minimum of experiments. In the discussed case, the results of factorial experiments enable to outline the conditions for the production of droplets with desired sizes.

2 Materials and Methods

2.1 Materials

Pure glycerol was purchased from Sigma Aldrich. Water used for dilution purposes was purified by reverse osmosis (Puricom, USA).

2.2 Experimental System

A detailed description of the experimental setup (the electrostatic droplet generator with impulse voltage of own construction) can be found in works by Lewińska et al. (2004, 2012). In this study tested liquid was forced by a syringe pump Alaris Asena (GH) to flow through a stainless steel nozzle with a given diameter. Droplets were formed at the nozzle tip and detached to the droplet collector bellow. The nozzle was connected to the positive end of a high voltage supply, coupled to a frequency modulator, whereas the collector was grounded. The setup enabled to apply pulsed electric voltages to the nozzle and, thus to charge flowing liquid in a controlled manner. Regulated electric parameters included: the applied voltage value U (in the range 0–21 kV, stepwise), duration time of the voltage impulse τ (in the range 1–9 ms, stepwise) and frequency of impulse application f (in the range 1–100 Hz, stepwise). The distance between the tip of the nozzle and the droplet generator was set at 5 cm in all cases. The droplet generator was accompanied by a camera mtv-1361CA (Mintron Enterprise Co., Taiwan) and DVD recorder, coupled to a video monitor, in order to record the process of droplet formation at the tip of the nozzle. The ambient temperature was equal to 25 °C and relative humidity was ca. 60% during all the experiments. The droplet diameter was derived from measured time of a single droplet formation. The calculations were done according to the assumption that the droplet is spherical at the moment of detachment from the nozzle tip. All the measurements were done in triplicate. Standard deviation (*SD*) and coefficient of variation (*CV*) were determined.

2.3 Application of the Factorial Design Method in Discussed Case

The aim of this work was to assess the influence of selected process parameters of electrostatic droplet formation on the droplet diameter. In order to reach this goal it was necessary both to determine the direct influence of the experimental variables as well as their possible interaction effects. For this, two-level factorial design is the appropriate tool. The method is widely used in the analysis of various technological problems, also related to biomedicine and bioengineering (Costa et al. 2007; Kramek-Romanowska et al. 2011). In this work, the impact of four process variables on the electrostatic droplet formation was investigated, i.e. the effect of solution dynamic viscosity, nozzle inner diameter, solution flow rate and electric voltage. In order to conduct a full factorial design, high and low levels of mentioned parameters were selected. The respective values are presented in Table 1.

High level of the solution viscosity is the viscosity of pure glycerol at $T = 25$ °C and low level of the parameter, which is one order of magnitude smaller than the value for pure glycerol, is the viscosity of glycerol solution in water with concentration

Table 1. Process variables and experimental domain used in the factorial study.

Parameter [Unit]	Low level (−1)	High level (+1)
Liquid dynamic viscosity (M) [mPas]	81.5	945
Nozzle inner diameter (D) [mm]	0.445	0.680
Liquid flow rate (Q) [ml/s]	0.00139	0.00417
Electric voltage (U) [kV]	0	11

$c = 85$ %wt (Hodgman 1959). The values of D are in the range of nozzles inner diameters already applied in earlier authors' works (for example Lewińska et al. 2004, 2012). High level of the solution flow rate was the biggest attainable value with the use of applied syringe pump for the investigated liquid. Low level of the parameter was chosen so as the time of droplet formation would not exceed ca. $t = 10$ s. The selected value of duration time of the voltage impulse, τ, represents the centre of the range available in the experimental system. In addition, it was observed in earlier works (Lewińska et al. 2004) that application of this value ensured no satellite droplets formation. The chosen value of frequency of impulse application, f, was the biggest value, which allowed to obtain electrostatic dripping regime for given U and τ values and selected substances. Finally, the high level of electric voltage was selected as the centre of the range available in the experimental system. All the values are in the range applied in abovementioned studies by Lewińska et al.

3 Results and Discussion

3.1 Experimental Results

All the experimental variants are summarized in Table 2. In Table 3 the results of conducted measurements of mean time of droplet formation, as well as mean frequency of droplet formation with corresponding standard deviation and variation coefficient values are listed.

As one may notice analyzing Table 3, the range of obtained values of mean time of droplet formation covers almost two orders of magnitude. This means that the process is indeed highly dependent on the variables selected in the discussed investigation. Besides, during conducted experiments, stabile droplet formation was achieved for majority of investigated variants. However, in cases of high liquid viscosity and with applied electric voltage some additional effects were sometimes observed, such as bedewed nozzle or droplet oscillations before detaching from the nozzle tip. Therefore, the time of droplet formation for these variants elongated (see exp. no. 13 and 15 in Table 3).

Table 2. Experimental variants for a 2^4 factorial design.

No.	Glycerol concentration [%wt]	Liquid dynamic viscosity [mPas]	Type of nozzle	Nozzle inner diameter [mm]	Liquid flow [ml/s]	U [kV]	τ [ms]	f [Hz]
1	85	81.5	7	0.445	0.00139	11	4	5
2	85	81.5	7	0.445	0.00139	0	0	0
3	85	81.5	7	0.445	0.00417	11	4	5
4	85	81.5	7	0.445	0.00417	0	0	0
5	85	81.5	10	0.680	0.00139	11	4	5
6	85	81.5	10	0.680	0.00139	0	0	0
7	85	81.5	10	0.680	0.00417	11	4	5
8	85	81.5	10	0.680	0.00417	0	0	0
9	100	945	7	0.445	0.00139	11	4	5
10	100	945	7	0.445	0.00139	0	0	0
11	100	945	7	0.445	0.00417	11	4	5
12	100	945	7	0.445	0.00417	0	0	0
13	100	945	10	0.680	0.00139	11	4	5
14	100	945	10	0.680	0.00139	0	0	0
15	100	945	10	0.680	0.00417	11	4	5
16	100	945	10	0.680	0.00417	0	0	0

Table 3. Mean values of time of droplet formation and frequency of droplet formation.

No.	Mean time of droplet formation [s]	Standard deviation [s]	Variation coefficient [%]	Mean frequency of droplet formation [1/s]	Standard deviation [1/s]	Variation coefficient [%]
1	85	81.5	7	0.445	0.00139	11
2	85	81.5	7	0.445	0.00139	0
3	85	81.5	7	0.445	0.00417	11
4	85	81.5	7	0.445	0.00417	0
5	85	81.5	10	0.680	0.00139	11
6	85	81.5	10	0.680	0.00139	0
7	85	81.5	10	0.680	0.00417	11
8	85	81.5	10	0.680	0.00417	0
9	100	945	7	0.445	0.00139	11
10	100	945	7	0.445	0.00139	0
11	100	945	7	0.445	0.00417	11
12	100	945	7	0.445	0.00417	0
13	100	945	10	0.680	0.00139	11
14	100	945	10	0.680	0.00139	0
15	100	945	10	0.680	0.00417	11
16	100	945	10	0.680	0.00417	0

3.2 Response Surface Model

Presented above experimental results enabled to obtain equations describing relationships between the droplet diameter and four investigated process variables (*M, D, Q, U*) These dependences, being response surface models, are in the form of Eq. (1):

$$y = b_0 + \sum_{i=1}^{4} b_i \cdot x_i + \sum_{i=1}^{4}\sum_{j=1}^{4} b_{ij} \cdot x_i \cdot x_j + \sum_{i=1}^{4}\sum_{j=1}^{4}\sum_{k=1}^{4} b_{ijk} \cdot x_i \cdot x_j \cdot x_k + b_{1234} \cdot x_1 \cdot x_2 \cdot x_3 \cdot x_4 \quad (1)$$

where: b_0-b_{1234} – the model coefficients; x_1-x_4 – the process parameters (*M, D, Q, U*, respectively); y – the investigated response (droplet diameter). After solving Eq. (1) by least square method the model coefficients for the droplet diameter were obtained. The values are listed in Table 4.

According to Table 4, apart from the constant term, b_0, the most significant influence on the analyzed parameters was the electric voltage (*U*). The relationship was inversely proportional as the corresponding coefficient, b_U, was negative. The greater the electric voltage, the more energy was supplied for breaking up the liquid into smaller droplets. Regarding the influence of *M*, *D* and *Q* on the droplet diameter, *d*, the impact of these process variables was considerably weaker as compared to the electric voltage. For all the parameters the effect was positive and of a similar magnitude. Whereas this was commonly observed (Jain et al. 2011; Lewińska et al. 2012; Faraji et al. 2017), the importance of the processing variables' interactions is not that obvious. Analyzing Table 4, it can be observed that out of all interaction terms for droplet diameter, *d*, there were only two – b_{MU} and b_{MDU} considerably stronger than the rest interaction terms and comparable to single parameter ones by value. This indicates that the significance of these interactions cannot be omitted in the discussion.

Table 4. Coefficients of response surface models for droplet diameter, *d* [mm] (M – liquid dynamic viscosity, D – nozzle inner diameter, Q – liquid flow rate, U – electric voltage).

Model coefficient	Value	Model coefficient	Value
b_0	2.064	b_{DQ}	−0.051
b_M	0.206	b_{DU}	0.047
b_D	0.119	b_{QU}	0.054
b_Q	0.122	b_{MDQ}	−0.048
b_U	−0.7	b_{MDU}	0.123
b_{MD}	0.042	b_{MQU}	−0.027
b_{MQ}	0.027	b_{DQU}	0.016
b_{MU}	0.186	b_{MDQU}	0.014

Obtained response surface model can be used to plot three-dimensional graphs. Figures 1, 2, 3 and 4 show a comparison of 3D plots for different combinations of variables. Colours used in the surface plots emphasize the way the values of droplet diameter changed with processing parameters – the lighter the colour, the bigger the diameter value.

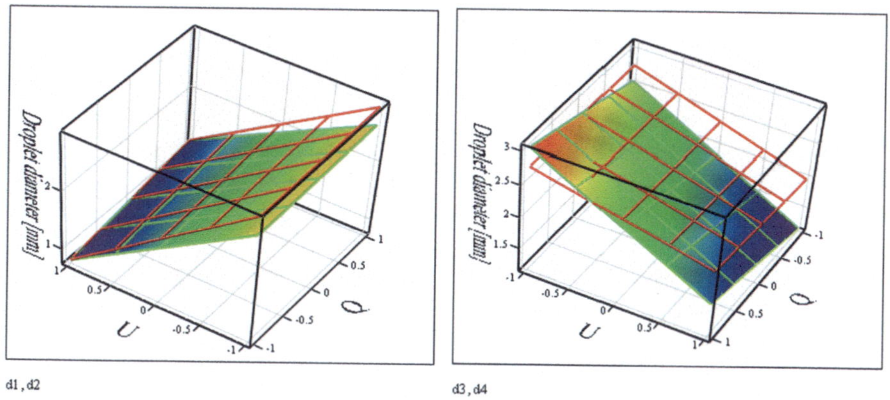

Fig. 1. Response surface plots showing the effects of processing parameters on the droplet diameter. d1, d3 – colour maps; d2, d4 – grids. Liquid dynamic viscosity (M) and nozzle inner diameter (D) were kept at the following levels: d1 (M = −1, D = −1), d2 (M = −1, D = +1), d3 (M = +1, D = −1), d4 M = +1, D = +1). Q, liquid flow rate; U, electric voltage; +1, high level of the parameter; −1, low level of the parameter.

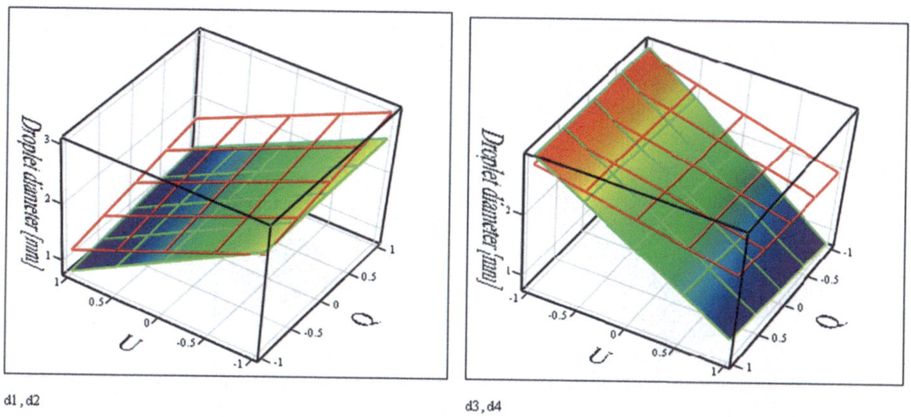

Fig. 2. Response surface plots showing the effects of processing parameters on the droplet diameter. d1, d3 – colour maps; d2, d4 – grids. Liquid dynamic viscosity (M) and nozzle inner diameter (D) were kept at the following levels: d1 (M = −1, D = −1), d2 (M = +1, D = −1), d3 (M = −1, D = +1), d4 M = +1, D = +1). Q, liquid flow rate; U, electric voltage; +1, high level of the parameter; −1, low level of the parameter.

Grid plots were also graphical representations of the surface model but presented in a different way to make the three-dimensional graphs more legible. Analyzing Fig. 1, it can be noticed, that generally the impact of changing inner nozzle diameter, while *M* remained constant, on the shape of the graphs was rather weak and increased only a bit for the higher value of the liquid dynamic viscosity. Similar observation can be

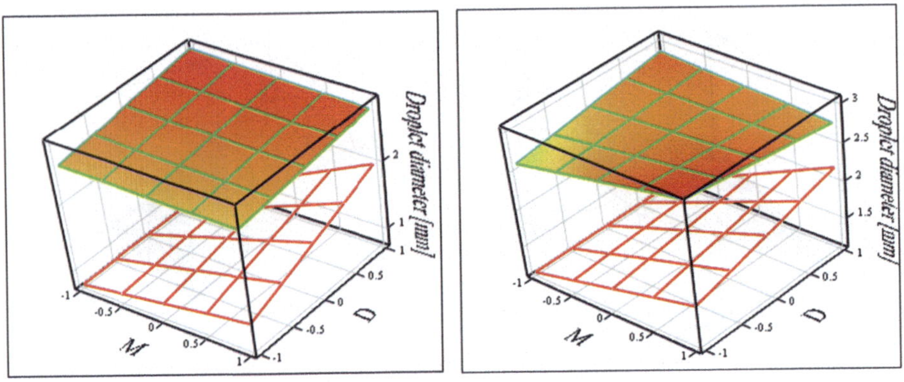

d1, d2 d3, d4

Fig. 3. Response surface plots showing the effects of processing parameters on the droplet diameter. d1, d3 – colour maps; d2, d4 – grids. Liquid flow rate (Q) and electric voltage (U) were kept at the following levels: d1 (Q = −1, U = −1), d2 (Q = −1, U = +1), d3 (Q = +1, U = −1), d4 Q = +1, U = +1). M, liquid dynamic viscosity; D, nozzle inner diameter; +1, high level of the parameter; −1, low level of the parameter.

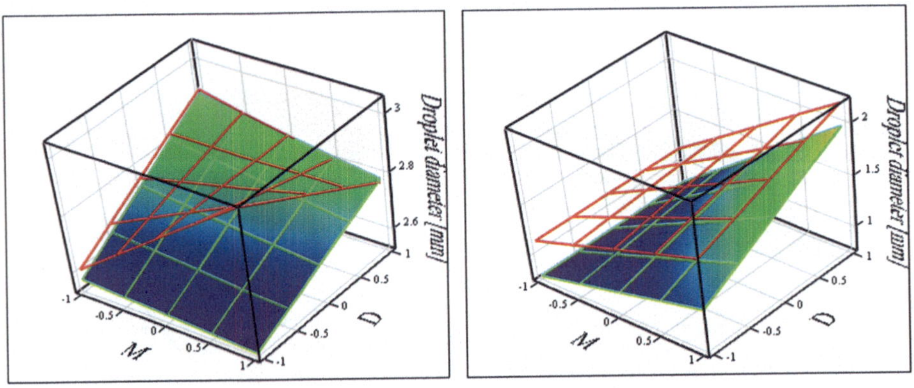

d1, d2 d3, d4

Fig. 4. Response surface plots showing the effects of processing parameters on the droplet diameter. d1, d3 – colour maps; d2, d4 – grids. Liquid flow rate (Q) and electric voltage (U) were kept at the following levels: d1 (Q = −1, U = −1), d2 (Q = +1, U = −1), d3 (Q = −1, U = +1), d4 Q = +1, U = +1). M, liquid dynamic viscosity; D, nozzle inner diameter; +1, high level of the parameter; −1, low level of the parameter.

made in case of the effect of changing liquid viscosity, while *D* remained constant (Fig. 2).

For all parameter combinations minimum droplet diameter was obtained for the high value of *U* and low value of *Q*, regardless the actual values of *M* and *D*. As it could be expected, the electric voltage had the most significant effect out of all discussed process parameters. The confirmation of this statement can be seen in graphs presented in Figs. 3 and 4, where the influence of high and low levels of Q and U is depicted. Plots shown in Fig. 3 indicate, that enhancement of the electric voltage, U, made the $d(M, D)$ dependency stronger. For all the parameter combinations, minimum value of the droplet diameter was obtained for the high level of U and low levels of M and D. The impact of changes in the liquid flow rate, Q, was hardly noticeable and variations in the electric voltage indeed predominated all other parameters.

However, still some additional valuable information can be achieved from Fig. 4, presenting the influence of Q while U is constant. In case of the low level of electric voltage, the liquid flow rate influenced the shape of the plot to great extent, though the combination of parameters resulting in the minimum d value remained unchanged. Similarly to Fig. 3, the enhancement of Q made the $d(M, D)$ dependency more significant.

4 Conclusions

The effect of four variables (liquid dynamic viscosity, nozzle inner diameter, liquid flow rate and electric voltage) on the droplet diameter produced by impulse electrostatic droplet generator was evaluated by 2^4 factorial design. The greatest influence on the discussed parameter was observed for electric voltage. The greater the electric voltage, the more energy was supplied for breaking up the liquid into smaller droplets, resulting also in shorter time of droplet formation. Regarding the influence of other process variables on the droplet diameter, the impact of M, D and Q was considerably weaker, as compared to the electric voltage. For all the parameters the effect was positive and of a similar magnitude. In addition, the significance of interactions between process variables were noticed and consequently they cannot be a priori omitted in the process analysis. The ideas presented in this paper certainly require further development in terms of solutions with different physicochemical properties and other process parameters, such as nozzle length or distance between the nozzle tip and the droplet collector. This work demonstrates that, when electrostatic droplet formation is involved in the manufacture and development of capsules for biomedical purposes, the process conditions should be carefully studied as they impact significantly product properties, which are crucial for the effectiveness of medical applications.

References

Ahmad, Z., Zhang, H.B., Farook, U., Edirisinghe, M., Stride, E., Colombo, P.: Generation of multilayered structures for biomedical applications using a novel tri-needle coaxial device and electrohydrodynamic flow. J. R. Soc. Interface **5**, 1255–1261 (2008)

Chang, T.M.S., Prakash, S.: Removal of urea in uremia and ammonia in liver failure with emphasis on the use of artificial cells for encapsulation of genetically engineered cells. In: Kuehtreiber, W.M., Lanza, R.P., Chick, W.L. (eds.) Cell Encapsulation. Technology and Therapeutics, pp. 379–391. Birkhauser, Boston-Basel-Berlin (1999)

Costa, C.B.B., Rivera, E.A.C., Ferreira Rezende, M.C.A., Wolf Maciel, M.R., Maciel Filho, R.: Prior detection of genetic algorithm significant parameters: coupling factorial design technique to genetic algorithm. Chem. Eng. Sci. **62**, 4780–4801 (2007)

Faraji, S., Sadri, B., Vajdi Hokmabad, B., Jadidoleslam, N., Esmaeilzadeh, E.: Experimental study on the role of electrical conductivity in pulsating modes of electrospraying. Exp. Therm. Flu. Sci. **81**, 327–335 (2017)

Ganan-Calvo, A.M.: On the theory of electrohydrodynamically driven capillary jets. J. Fluid Mech. **335**, 165–188 (1997)

Hamad, M.L., Bowman, K., Smith, N., Sheng, X., Morris, K.R.: Multi-scale pharmaceutical process understanding: From particle to powder to dosage form. Chem. Eng. Sci. **65**, 5625–5638 (2010)

Hodgman, C.D.: Handbook of Chemistry and Physics, 40th edn. Chemical Rubber Publishing Co., Cleveland (1959)

Hunkeler, D., Rehor, A., Ceausoglu, I., Schuldt, U., Canaple, L., Bernard, P., Renken, A., Rindisbacher, L., Angelova, N.: Objectively assessing bioartificial organs. In: Hunkeler D., Cherrington A., Prokop A., Rajotte R. (eds.) Bioartificial Organs III. Tissue Sourcing, Immunoisolation and Clinical Trials, vol. 944, pp. 456–471. Annals of the New York Academy of Sciences (2001)

Jain, V., Prasad, D., Jain, D., Mishra, A.K., Singh, R.: Factorial design-based development of measlamine microspheres for colonic delivery. Biomatter **1**(2), 182–188 (2011)

Kim, H.H., Teramoto, Y., Negishi, N., Ogata, A., Kim, J.H., Pongrác, B., Machala, Z., Gañán-Calvo, A.M.: Polarity effect on the electrohydrodynamic (EHD) spray of water. J. Aerosol Sci. **76**, 98–114 (2014)

Kramek-Romanowska, K., Odziomek, M., Sosnowski, T.R., Gradoń, L.: Effects of process variables on the properties of spray-dried mannitol and mannitol/disodium cromoglycate powders suitable for drug delivery by inhalation. Ind. Eng. Chem. Res. **50**(24), 13922–13931 (2011)

Ku, B.K., Kim, S.S.: Electrospray characteristics of highly viscous liquids. J. Aerosol Sci. **33**, 1361–1378 (2002)

Lee, M.W., Kim, N.Y., Yoon, S.S.: On pinchoff behavior of electrified droplets. J. Aerosol Sci. **57**, 114–124 (2013)

Lewińska, D., Rosiński, S., Weryński, A.: Influence of process conditions during impulsed electrostatic droplet formation on size distribution of hydrogel beads. Artif. Cells Blood Substit. Biotechnol. **32**(1), 41–53 (2004)

Lewińska, D., Chwojnowski, A., Wojciechowski, C., Kupikowska-Stobba, B., Grzeczkowicz, M., Weryński, A.: Electrostatic droplet generator with 3-coaxial-nozzle head for microencapsulation of living cells in hydrogel covered by synthetic polymer membranes. Sep. Sci. Technol. **47**(3), 463–469 (2012)

Roine, J., Murtomaa, M., Salonen, J.: Influence of parallel nozzle electroencapsulation parameters on microcapsule properties – a case study using the Taguchi robust design method. J. Electrostatics **90**, 91–105 (2017)

Author Index

J. Korbicz et al. (Eds.): PCBEE 2019, AISC 1033, pp. 337–338, 2020.
https://doi.org/10.1007/978-3-030-29885-2

The manufacturer's authorised representative in the EU is Springer Nature Customer Service Centre GmbH, Europaplatz 3, 69115 Heidelberg, Germany. If you have any concerns regarding our products, please contact ProductSafety@springernature.com

Printed and bound by CPI Group (UK) Ltd, Croydon, CR0 4YY
22/09/2026
02221068-0002